z	.00	.01	.02	.03	.04	.05	.06	.07	.08	.09
0.0	.0000	.0040	.0080	.0120	.0160	.0199	.0239	.0279	.0319	.0359
0.1	.0398	.0438	.0478	.0517	.0557	.0596	.0636	.0675	.0714	.0753
0.2	.0793	.0832	.0871	.0910	.0948	.0987	.1026	.1064	.1103	.1141
0.3	.1179	.1217	.1255	.1293	.1331	.1368	.1406	.1443	.1480	.1517
0.4	.1554	.1591	.1628	.1664	.1700	.1736	.1772	.1808	.1844	.1879
0.5	.1915	.1950	.1985	.2019	.2054	.2088	.2123	.2157	.2190	.2224
0.6	.2257	.2291	.2324	.2357	.2389	.2422	.2454	.2486	.2517	.2549
0.7	.2580	.2611	.2642	.2673	.2704	.2734	.2764	.2794	.2823	.2852
0.8	.2881	.2910	.2939	.2967	.2995	.3023	.3051	.3078	.3106	.3133
0.9	.3159	.3186	.3212	.3238	.3264	.3289	.3315	.3340	.3365	.3389
1.0	.3413	.3438	.3461	.3485	.3508	.3531	.3554	.3577	.3599	.3621
1.1	.3643	.3665	.3686	.3708	.3729	.3749	.3770	.3790	.3810	.3830
1.2	.3849	.3869	.3888	.3907	.3925	.3944	.3962	.3980	.3997	.4015
1.3	.4032	.4049	.4066	.4082	.4099	.4115	.4131	.4147	.4162	.4177
1.4	.4192	.4207	.4222	.4236	.4251	.4265	.4279	.4292	.4306	.4319
1.5	.4332	.4345	.4357	.4370	.4382	.4394	.4406	.4418	.4429	.4441
1.6	.4452	.4463	.4474	.4484	.4495	.4505	.4515	.4525	.4535	.4545
1.7	.4554	.4564	.4573	.4582	.4591	.4599	.4608	.4616	.4625	.4633
1.8	.4641	.4649	.4656	.4664	.4671	.4678	.4686	.4693	.4699	.4706
1.9	.4713	.4719	.4726	.4732	.4738	.4744	.4750	.4756	.4761	.4767
2.0	.4772	.4778	.4783	.4788	.4793	.4798	.4803	.4808	.4812	.4817
2.1	.4821	.4826	.4830	.4834	.4838	.4842	.4846	.4850	.4854	.4857
2.2	.4861	.4864	.4868	.4871	.4875	.4878	.4881	.4884	.4887	.4890
2.3	.4893	.4896	.4898	.4901	.4904	.4906	.4909	.4911	.4913	.4916
2.4	.4918	.4920	.4922	.4925	.4927	.4929	.4931	.4932	.4934	.4936
2.5	.4938	.4940	.4941	.4943	.4945	.4946	.4948	.4949	.4951	.4952
2.6	.4953	.4955	.4956	.4957	.4959	.4960	.4961	.4962	.4963	.4964
2.7	.4965	.4966	.4967	.4968	.4969	.4970	.4971	.4972	.4973	.4974
2.8	.4974	.4975	.4976	.4977	.4977	.4978	.4979	.4979	.4980	.4981
2.9	.4981	.4982	.4982	.4983	.4984	.4984	.4985	.4985	.4986	.4986
3.0	.4987	.4987	.4987	.4988	.4988	.4989	.4989	.4989	.4990	.4990

ELEMENTARY STATISTICS

Addison-Wesley Publishing Company

READING, MASSACHUSETTS • MENLO PARK, CALIFORNIA
DON MILLS, ONTARIO • WOKINGHAM, ENGLAND • AMSTERDAM • SYDNEY
SINGAPORE • TOKYO • MEXICO CITY • BOGOTA • SANTIAGO • SAN JUAN

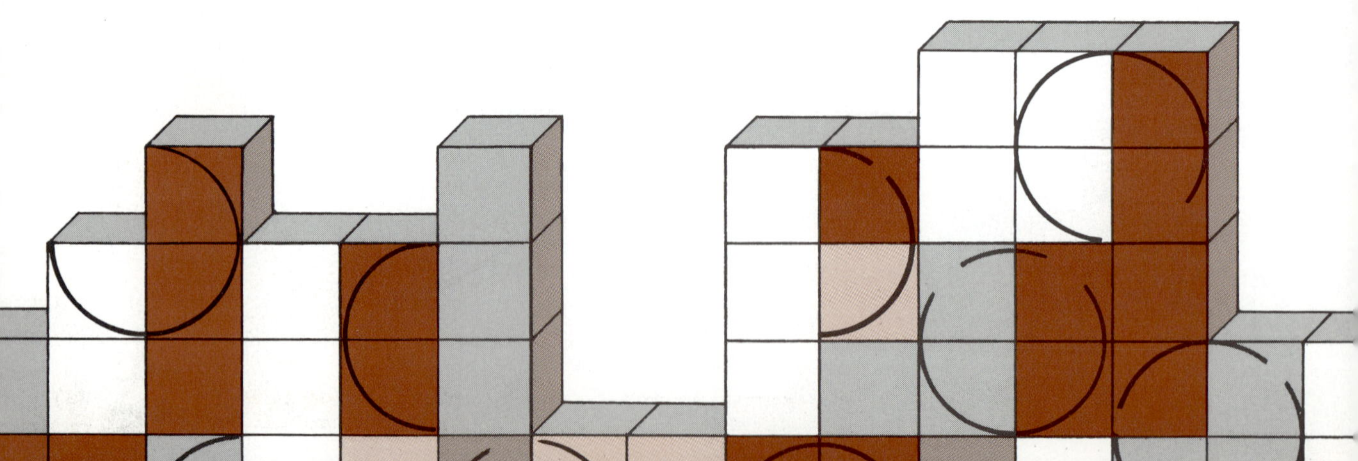

ELEMENTARY STATISTICS

Donna H. Skane

Catonsville Community College

To
My Best Friends

Jeffrey Pepper Sponsoring Editor

Ann E. DeLacey Manufacturing Supervisor
Robert C. Forget Art Coordinator
Maureen Langer Cover and Text Designer
Stephanie Argeros Magean/Peter Petraitis Production Editors
Jerrold Moore Copy Editor
Martha K. Morong Production Manager
Martha H. Stearns Managing Editor
Textbook Art Associates Illustrator

Library of Congress Cataloging in Publication Data

Skane, Donna, 1947–
Elementary statistics.
Includes index.
1. Statistics. I. Title.
QA276.12.S57 1985 519.5 84-9335
ISBN 0-201-06751-X

ABCDEFGHIJ-HA-8987654

Preface

All students are usually inexperienced with statistics when they enroll in a first course in the subject. Any latent feelings of math anxiety come to the fore and are reinforced by a new vocabulary and reasoning processes that require assumptions and probabilistic interpretation. Thus I concentrated on maintaining an open, informal style of writing and used a wide variety of examples in order to aid comprehension of the material. A pedagogical use of the second color highlights key words, boxed definitions, and other "helpers" students should understand. A single year of high school algebra should be sufficient mathematical preparation. Mathematical rigor is avoided, but explanations of theory through examples or at an intuitive level are presented.

Attention Grabber Introductory Sections

The first two chapters are attention grabbers for the student—to catch their interest without the detail of technical processes. They consist of many examples of misuses and abuses of statistical methods, which students can easily read and understand. Chapter 2 provides a discussion of sampling procedures and a brief introduction to experimental method, both of which are lacking in most current elementary statistics texts. Students need to develop a healthy skepticism toward the uses of statistics they encounter all the time and to begin to think about the uses of numbers before they get involved in calculations; to ease their minds with verbal descriptions before they become embroiled in equations.

News & Views Give Real-life Uses of Statistics

Each chapter includes sections entitled *News & Views.* These are presented as boxed supplements to illustrate or amplify the topic under discussion. They consist of current research, past work, or historical development; they provide variety for the student and continue the approach used in Chapters 1 and 2.

Exercises/Worked Examples/Study Notes

Starting in Chapter 3, each section is followed by exercises. Sections contain worked examples, both as a means of introducing subject matter and as additional examples of the techniques introduced. Each chapter concludes with several different study aids called *Study Notes:* First is a list of key terms, then an outline that stresses the main points, and finally a set of review problems. The section exercises are easily identified with the equations and techniques of that section, whereas chapter problems require more analysis by the student regarding methods of solution.

Stem-and-Leaf Diagrams Provide a Transition into Graphing

Chapter 3 presents frequency distributions, with an additional section on stem-and-leaf diagrams that provides a transition into graphing. There are different viewpoints on the types and amount of graphing that should be introduced in a statistics course. I have included all types: the minimum necessary to aid in the solution of text examples, exercises, and problems (in Chapter 3) and the other types of graphs in Appendix A for teachers and students who wish to use them. This appendix provides the instructor with the flexibility to expand instruction in graphing to include bar graphs, line graphs, and circle graphs—complete with exercises.

Throughout the text an attempt is made to refer to topics already covered and to the later use of techniques under discussion. Many ideas are intuitively broached early; for example, when the ogive is developed in Chapter 3, it is used to find the median merely as an interpretation of the graph. Then when the median and measures of position are introduced in Chapter 4, the student refers back to the ogive and reinforces, now with formal vocabulary, what he or she learned before. Exercises in Chapter 7 point toward techniques of estimation in Chapter 8, and so on.

Chapter 5 introduces notions about probability that are necessary in order to develop probability distributions. I made a conscious effort to avoid having students bog down in this chapter by including only those ideas that are necessary for later work.

In Chapter 6, general probability distributions derived from the discrete examples in Chapter 5 introduce the characteristics of probability distributions; this chapter also ties in work from

Chapter 3 regarding frequency distributions and graphing. The student can then begin to put together the abstract ideas of Chapter 5 and the concrete graphing work of Chapter 3.

Early Introduction of "t" Distribution

Note that this text introduces the t distribution in Chapter 7, following the discussion of the central limit theorem. This is done in order to allow students to do problems concerning means without breaking stride to learn another table. That will occur soon enough with work on σ^2. Students need to learn early to decide when to use the normal distribution and when to use the t distribution. My intent in putting the t distribution in Chapter 7 is to require students always to think about the distinction; they will not be lulled into always using z. Pattern and format are emphasized, both for the overall structure of a hypothesis test and for confidence interval estimates. Students can proceed from Chapter 8 through half of Chapter 10 using only variations of the

$$z(\text{or } t) = \frac{\text{Sample results} - \text{Parameter from } H_0}{\text{Appropriate standard error}}.$$

Chapter 8, the introductory chapter to the standard techniques of hypothesis testing and estimation, sets the tone for the second half of the book.

Chapter 9 extends the work in Chapter 8 to two populations, which reaches its logical conclusion for 3 or more populations in Chapter 11 with ANOVA. Chapter 10, with its work on proportions, standard deviations, and variances, completes the discussion of univariate data. Even though Chapter 11 logically follows from Chapter 9, it is necessary to use the F distribution introduced in Chapter 10 to perform the ANOVA procedure in Chapter 11. Chapter 12, on correlation and regression, and Chapter 13, on chi-square analysis, provide work with bivariate and multivariate data and teach the student the methods of establishing relationships. Chapter 14 completes the text with a quick look at some techniques of non-parametric statistics.

Appendixes A, B, and C supplement the student's knowledge with instructional material and exercises on graphing, summation notation, and factorial notation; and all tables necessary for work using this text are contained in Appendix D.

Acknowledgements

The staff at Addison-Wesley have been very supportive and helpful. They deserve much credit for the conception and development of this book, especially Wayne Yuhasz, who convinced me that I could write it.

I wish to thank the following people for their suggestions and detailed reviews: Ronald Barnes, University of Houston; Keith Craswell, Western Washington University; Howard Dachschlager, Saddleback Community College; Richard Fritz, Moraine Valley Community College; Michael Karelius, American River College; L. Michael Perry, Appalachian State University; Maxine Reed, State Technical Institute at Memphis; Larry J. Ringer, Texas A & M University; John Rushton, Metropolitan State College; Gerald L. Sievers, Western Michigan University; William Stines, North Carolina State University.

Catonsville, Maryland D.H.S.
December 1984

Supplementary material

Solutions Manual contains answers to even-numbered problems (for instructors only).

Student Study Guide by Ron Barnes provides review material for each section in the text as well as sample questions.

Introductory Statistics Software Package by William Frankenberger, Thomas Blakemore, and Kenneth Heilman is a software package for use on the Apple II Plus, Apple IIe, or Apple IIC microcomputer (48K minimum, one disk drive, DOS 3.3). Programs include: Grouped frequency distribution, histogram and frequency curves, mean and standard deviation, correlation, regression and scatter diagram, sampling distribution, independent *T*-test, correlated *T*-test, one-way analysis of variance, two-way analysis of variance, and chi-square.

Contents

PART II *DESCRIPTIVE STATISTICS* 27

3 Frequency distributions and graphs 28

4 Descriptive measures—summarizing data 49

5 Probability 86

PART III INFERENTIAL STATISTICS 165

8 Inference concerning the means of one population 166

9 Statistical inference concerning the means of two populations 197

PART I

INFORMATION: THE BASIS FOR STATISTICS

Avoiding the pitfalls

1

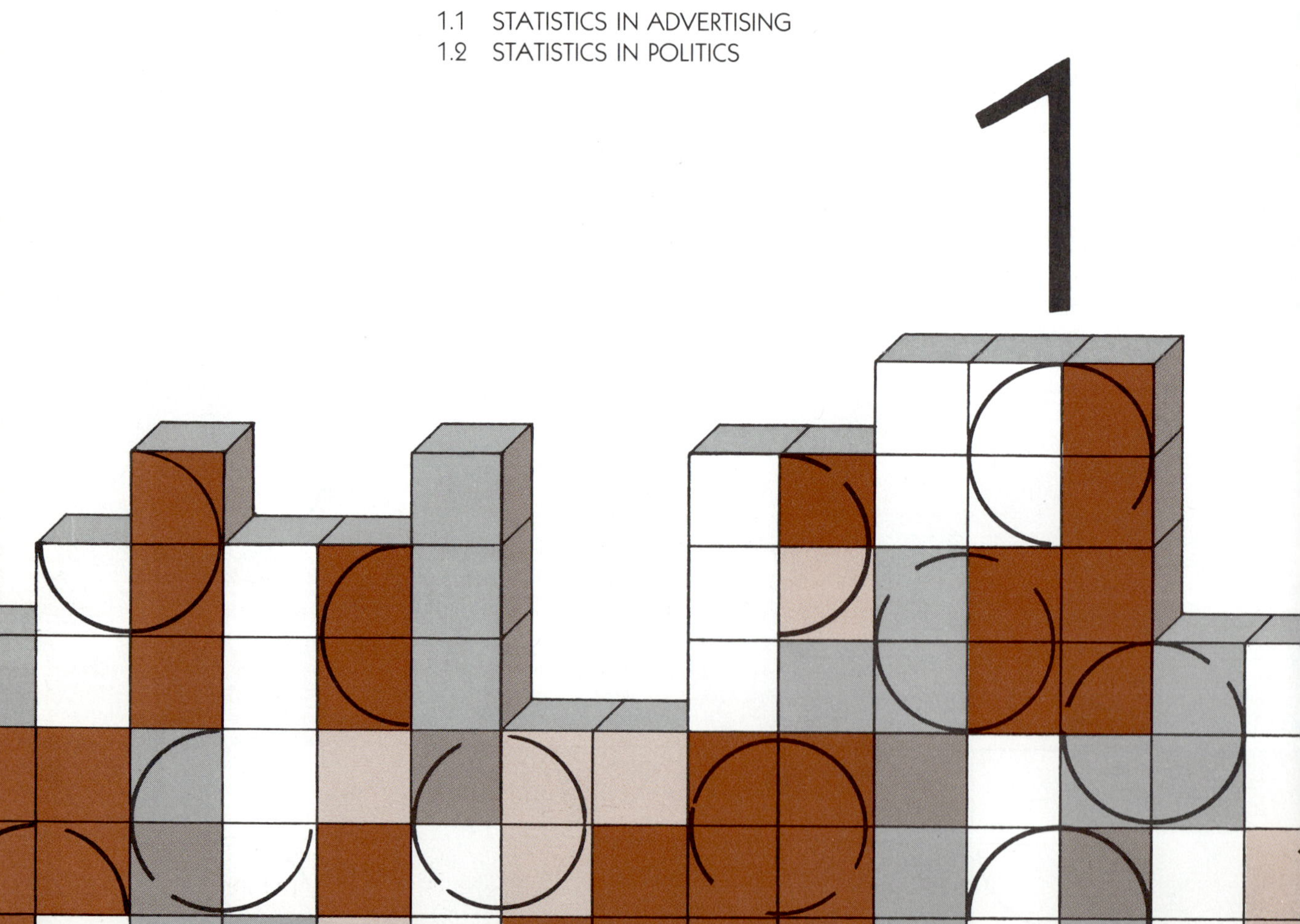

Everyone knows what statistics *are:* the facts and figures quoted and bandied about to make points or impress listeners that we use and are exposed to daily. We are fed statistics in newspapers, on television, and in advertisements. Many statistics are used deceptively, and many are used legitimately, often with no clue about which is which. So we need to find out what statistics *is*, which is what this book is all about. It will help us begin to understand how information is gathered; what can be done to organize and present information after it is collected; and, most importantly, how to interpret and make judgments about the information we have at hand.

Gathering data began a long time ago: The sizes of flocks and armies, ages at death, and numbers of children in families are found in the Bible. However, it took a sixteenth-century Englishman, John Graunt, to begin to *interpret* data about births and deaths published weekly on Bills of Mortality for each parish. He drew some conclusions about such vital statistics as the chances of dying a particular death, the "fitness of the Country for long life," the threats of occupation upon longevity, and the surplus of males leading to a "natural bar to Polygamy." As any good researcher, he even left some unanswered questions: How many married versus single people were there? How many were in various age brackets? What were the sizes of populations of various areas? Why did burials in London exceed christenings (whereas the opposite occurred in the country)?

The idea of interpreting information for social and political purposes began a field of study that eventually incorporated ideas from the gambling tables of France (probability), quality control in an Irish brewery (the *t* distribution), agricultural studies in England (analysis of variance), the heights of sons compared to those of their fathers (correlation and regression analysis), and many others—all giving us some very powerful tools to use in studying the world in which we live. We will begin to use some of these tools in the coming chapters.

Why do you as an individual need to learn about statistical methods? Because so much of your life, whether you know it or not, is influenced by the results of polls and polling. Polls tell us about the popularity of political candidates we are asked to support, the television shows we watch or do not watch, and the products we are likely to buy (and even the colors their packages should be). As consumers and voters, we need to know something about statistical methods in order to judge critically the daily barrage of information about products, politics, and other aspects of our society. H. G. Wells once said, "Statistical thinking will one day be as necessary for efficient citizenship as the ability to read and write." Anyone who reads a newspaper or watches television knows that that day is here.

Let us examine some uses and abuses of statistics in two areas: advertising and politics. This will help us to develop a healthy skepticism and an understanding of the uses and limitations of statistics. No tool is really useful unless we know what it *cannot* do and how it should *not* be used. We need to examine, as Disraeli said, some examples of the "lies, damned lies, and statistics" of our world.

1.1 Statistics in Advertising

Our society is very impressed with numbers. They seem to have mystifying and mystical properties in and of themselves that lend authority to any statement containing them. Few people seem to question that, at best, numbers may be useless and, at worst, deceptive.

Let us examine some free-floating, and consequently useless, statistics from recent advertisements. They leave the viewer with an impression of scientific precision, but that impression is created without any basis for it.

A tire manufacturer advertised that its tires stopped "35% faster." Faster than what? Faster than the competitors' tires? Faster than its own tires did last year? Where did that number come from? A percent must always refer to something: 35% of what?

A large auto manufacturer advertised on the radio that over a five-year period its average gasoline mileage had increased 50%. This certainly sounds impressive, doesn't it? But 50% of what? If the old mileage were two miles per gallon and now the average is three miles per gallon, we have a 50% increase. So what? We have no basis on which to judge that free-floating percent.

For years we have been told that Ivory soap is 99 and 44/100% pure. Pure what? Pure soap? Is this good or bad? What is the other 56/100%? How does this compare with other soaps?

Statistics do not have to be entirely free-floating to be useless. A foreign car manufacturer advertised that 95% of all its autos sold during the past 11 years were still on the road. This is not an entirely baseless statistic. We are told that 95% refers to all sales within a specific period of time. What we are to assume is that those cars were sold in approximately equal numbers each year. This is definitely not the case. Most of the cars were sold within the preceding three or four years.

Very large numbers also need to be put in perspective. When we hear about the millions or billions of dollars that industry is spending to clean up the environment or to explore for oil and gas, we are impressed. But, how do these figures compare with the total amounts that need to be spent? With what other companies in the same field are spending? With the total budget of a company and its profits? Don't let the large numbers create an atmosphere of precision and credibility!

The aura of scientific study pervades the advertising and marketing practices of pharmaceutical companies. We hear that an over-the-counter medication "contains twice as much pain reliever." Does this mean that it was more effective in relieving pain? The manufacturer of one popular pain reliever offering this claim attempted to prove product effectiveness by conducting tests on women who had just given birth. They were given aspirin (in itself not a very good idea because of aspirin's possible interference with blood clotting) and the manufacturer's product. The women were then asked to report the *percent* of relief obtained every hour. (Can you report relief in percentage terms?) The results of the study were so confusing that the commercials still refer to "containing twice as much pain reliever" rather than saying anything about the amount of relief obtained. And, lest you be fooled by the arthritic woman who can now open jars after taking the pain reliever, do you recall ever seeing a negative response on a commercial?

Have you ever been impressed by the "four out of five doctors (dentists) surveyed . . ."? These surveys are not always very scientific. How many dentists or doctors were surveyed? (These surveys may not have used a large group.) How were they chosen? (This is important, as we will see in the next chapter.) Many times when doctors are asked to recommend drugs, they are given a list of ingredients from which to choose. Since many over-the-counter drugs contain many of the same ingredients, such as aspirin and caffeine, the doctors' recommendations are restated as brand-name preferences if the ingredients they selected from the list are contained in the brand-name medication. An answer of *yes* to the question, "Do you ever prescribe an aspirin substitute?" gets counted as an endorsement of a particular product.

The Federal Trade Commission has examined advertisements for years. It has found, for example, that claims of product preference among repairpeople have been derived by asking only technicians who work for the promoting company's authorized service agencies to state their preference. How objective are these people? How widely experienced are they in repairing many different brands? How qualified are they to compare brands?

Be careful of such claims as, "No diet pill works harder to make you lose weight," or "No hair dryer will dry your hair more quickly." You are *supposed to think* that the diet pill is the most effective and that the hair dryer dries faster than other brands. What the statements *actually say* is that the diet pill works as hard as other brands and that the hair dryer dries just as quickly as others.

1.2 Statistics in Politics

Political polls play an increasingly active role in our lives. These polls are used to predict the electability of major candidates in an election year. Politicians accord the polls great significance when the results favor their viewpoints and choose to ignore them when they don't. Let us look at some polling results that serve to buttress the opinion of one-time Prime Minister of Great Britain, Harold Wilson: "Accord to the polls interest, but not idolatry."

News & Views 1.1 is the first of a series in this book that illustrate definitions or concepts presented in the text and provide examples of the uses of statistics in the newspapers or magazines. This one on polling should help put political polls in perspective for you.

The most disastrous example of political polling occurred during the presidential election of 1936. In an attempt to forecast election results, the *Literary Digest* decided to take a straw vote by sending ten million ballot postcards to owners of registered automobiles and people listed in telephone books throughout the country. About two million returned their postcards; the results indicated that Roosevelt would lose—with only about 41% of the vote. Since we have never had a President Landon, Roosevelt's opponent, something obviously went wrong. Roosevelt won—with almost 61% of the vote. Just exactly what happened? Several things conspired to affect the results of the poll. First of all, who owned cars or had telephones during the Depression? Were these people more likely to vote Democrat or Republican? Secondly, who usually takes the trouble to respond to a questionnaire, even one as simple as a postcard ballot? Those who are satisfied with the status quo or those who are dissatisfied? Roosevelt, a Democrat, was the incumbent. Who were most likely to want their voices heard in this situation, fellow Democrats or opposing Republicans? Even though a seemingly large sample of two million people expressed their preference, this was not sufficient to overcome the bias inherent in the group.

2 polls were 'poles apart' on attitudes toward tax bill

In the past few weeks, two major polls gave substantially different readings of Americans' views on the recently passed tax-increase bill—a disparity that highlights some of the potential problems of public opinion surveys.

An Associated Press-NBC News survey, taken August 9–10, said the public was narrowly split on the tax bill, with many people still unfamiliar with the legislation.

But a *Washington Post*-ABC News poll taken a week later—the night after President Reagan appealed to the public to support the tax bill—said a majority of Americans opposed the measure.

Do the results mean that Mr. Reagan's nationally televised speech had the paradoxical effect of turning the public against the bill he was supporting?

Not necessarily.

The two polls used completely different questions that emphasized different aspects of the complex tax bill, which was approved by Congress Thursday and sent to President Reagan for his certain signature.

The AP-NBC News poll mentioned the bill's role in cutting the federal deficit and outlined its major provisions. A series of questions in the *Post*-ABC poll emphasized Mr. Reagan's support for the bill, raised the question of its fairness and asked whether it would cut the deficit.

Often, a difference in wording makes substantial differences in poll results. Research has found that people do listen to the poll questions and try to respond to the precise question asked. The order of questions can also have an impact on the results.

The question in the AP-NBC poll, asked of 1,594 adults nationwide, was: "Have you heard or read about the new tax law being debated by Congress—a proposal designed to lower federal deficits by increasing some taxes on business and also increasing taxes on things like cigarettes, telephone calls and airline tickets?

If the answer was yes, the next question was: "Do you generally approve or disapprove of this new tax proposal?"

The responses showed:

- 25 percent had not heard or read of the bill.
- 37 percent approved.
- 32 percent disapproved.
- 6 percent were not sure.

The *Post*-ABC poll, based on telephone interviews with 913 persons August 17, used a series of questions on the tax bill, starting with: "As you may know, Reagan called for public support of a tax bill now being considered by Congress. Reagan said it is more a tax-reform bill than a tax-increase bill. Would you say the bill is more tax reform or more tax increase, or don't you have an opinion on that?"

It then asked: "Just your best guess: Do you think the tax-increase bill would make the federal tax system fairer for all Americans or not?"

Next, it asked: "Supporters of the tax bill say it's needed to reduce coming federal budget deficits. If the tax bill is passed by Congress, do you think it will reduce federal deficits—a great deal, a fair amount, very little, or not at all?"

And then the key question: "Would you say you approve or disapprove of the tax bill?"

To the last question, the responses were:

- 35 percent approved.
- 54 percent disapproved.
- 11 percent expressed no opinion.

Jeffrey Alderman, director of polling for ABC News, said he saw two reasons for the differences:

"One, different question-wording; and two, time had passed. During that period, a lot more attention was focused on the bill and we polled right after [the president's speech]. It is difficult to tell which is more important, but I would say the more attention focused on the bill is probably the more important factor."

Source: *Baltimore Sun*, August 21, 1982.

A triumphant Truman was photographed holding up a newspaper that proclaimed Thomas Dewey the winner in 1948. Again, political polling methods fell short. And, as recently as the 1980 election, polls failed to predict the landslide victory of Ronald Reagan over Jimmy Carter by ten percentage points.

Finally, let us end this brief look at the pitfalls of using statistics with another example from the world of politics that affects all social policy, that of budget and taxation.

We are all awed by huge numbers, which are usually incomprehensible in their largeness: "A billion here, a billion there," as Everett Dirksen said. Why does it always seem as though the president and Congress can't agree on the nation's budgetary needs, whether or not a tax cut is needed, what effect such a cut would have on the economy, and so on? Consider a hypothetical tax cut of $20 billion that is intended to increase savings and thereby provide more capital for plant modernization and more efficient production. Twenty billion dollars is impressive; it should have an impact on the economy. But, would people save or spend this money? The $20-billion tax cut may translate to $17.67 more in take-home pay every two weeks. Will anyone actually save the $17.67 extra that shows up in the paycheck every other week? If not, are the claims made about the effect of the tax cut really valid?

Both the administration and Congress throw around huge numbers—$28 billion here, $41 billion there—giving us vastly different estimates of what is needed, what the budget deficit will be, or how big a tax cut will be. Who is right and who is wrong? Are they lying to us? Probably not, and neither is right or wrong. Both groups work from different economic assumptions, different definitions of the problems, different priorities in funding, and different prescriptions for fixing whatever needs to be fixed. As voters and taxpayers we need to listen for those definitions, priorities, and solutions—and attempt to put the numbers associated with them in perspective.

Sampling

2

2.1 Introduction

The basic ideas behind research are relatively simple: We want to gather information about some group of people or objects; then organize, describe, and summarize the data; and finally, use these results to make judgments and predictions. This last step is the most important and, perhaps, the most difficult. It relies heavily on how well we gathered the information in the first place.

Research in the social sciences, or in any other field that gathers data, requires that the information be as unbiased as possible, be unprejudicially obtained, and represent what is going on in a population at large. That **population** may consist of all voters in the United States, everyone killed in World War II, all school-age children, speeders on the nation's highways, the types of television shows watched by children under six—in short, any group of individuals or objects about which we are interested. We need to define precisely what our population is: who or what is in it, as well as who or what is not. This tells us where to begin to collect data and, in the end, to whom or what we apply our results.

population ■ The complete collection of people or objects about which we are interested.

In News & Views 2.1, the population is defined as "every one of the 33,637,548 death certificates recorded during the 20-year period in the United States." Such large studies are usually unnecessary, but it is necessary to define a population this precisely.

A classic example of wasted effort because of poor definition of the population is that of the pain-reliever commercial referred to in Chapter 1. We would assume that the target population in this case would contain anyone who might take an aspirin or a similar over-the-counter pain reliever. After all, that is the group at which the commercial is aimed. From whom were the data gathered? Women who had just given birth. Is this group unbiased? Does it represent the population as a whole? It obviously leaves out a large segment of the population. Do post-partum pains represent the type of pain for which one might take an analgesic? Are after-birth pains similar to headache pains? Because the population was not properly represented, any results of this study are open to question and, in fact, turned out to be worthless (except, perhaps, to women who have just given birth).

NEWS & VIEWS 2.1

Killer diseases acting mysteriously

Death rates from heart disease, stroke and seven other leading killers of Americans all declined during the last two decades, but no one seems to know why.

Experts say the figures show that some important changes must have occurred in national life style, behavior or environment, but no one knows just what the crucial changes were.

"We all find it mysterious and energizing," said Dr. Harold Margulies, deputy administrator of the Health Resources Administration in the Department of Health, Education and Welfare.

He said many of the trends are surprising and called for new research to seek out explanations. It was in that sense that they used the term "energizing." Something has been happening to produce the new patterns of death. If scientists understood the causes of these changes the implications for public health could be profound.

The data covered the period of the 1950's. They show declines in 9 of the 15 leading causes of death among Americans and rises in the six others. Some, notably suicides, homicides and accidents were all declining throughout much of the 1950's but then, inexplicably, turned upward again and have continued to rise.

Changes such as the 15 percent drop in heart disease death rate are so surprising that they seem to fly in the face of current health dogma. But there is little room for doubt as to their validity.

"The trends are real and well established inasmuch as most have been going on for at least 5 years and some for 10," said A. Joan Klebba, chief author of the study. She said, furthermore, that the original figures show that the trends are continuing into the decade of the 1970's. The study, by the national center for health statistics, is not based on a population sample as are most such studies. Instead, it involved analysis of every one of the 33,637,548 death certificates recorded during the 20-year period in the United States.

All of the data are expressed in what are called age-adjusted death rates; that is, deaths per 100,000 population adjusted to compensate for the changing age profile of the American population. Without adjusting for national age patterns, death rates would give an unrealistic picture because the average age of the American population is increasing. Without compensating for this, statistics would emphasize unduly the diseases of middle and old age. This may be one reason why the decline in heart disease has gone largely unnoticed.

Harold M. Schmeck Jr., "Killer Diseases Acting Mysteriously," *New York Times*, May 4, 1974.

2.2 Sampling Techniques

How do we make sure that a population is fairly represented in any study we conduct? This is probably one of the most important questions to be asked in statistics. No computer can fix up the data from a bad sample or from the use of a poorly written, ambiguous questionnaire. The computer science people have a phrase for this situation: "Garbage in, garbage out." Why not include everyone in the population or measure everything we want to know? With populations of any size, this can be an extremely time-consuming and expensive process. It may also be impossible. Not everyone is willing to cooperate in answering questions or giving out information. Even the census, which is supposed to include every citizen of the United States, misses people who are born or die during the census taking, who refuse to return their census forms, or who refuse to cooperate out of distrust or fear. It may also be impossible to include an entire population because the study is in some way destructive. Crash testing the bumpers of *all* cars to be driven in the United States is not very feasible.

Should we just be sure to include a large number of people in our survey? The *Literary Digest* example in Chapter 1 shows us that this approach alone is inadequate. Ten million ballots were sent out, and two million were returned. These are certainly large numbers but, because of the methods used to obtain those ten million names, the results were biased and gave a faulty prediction.

So how do we go about meeting the goal of representing fairly the population of our study? Keeping in mind the necessity of representativeness, we turn to samples and *techniques* of sampling. If populations were homogeneous like blood, that is, not changing from one part to another, we could sample a small group and obtain a miniature picture of the whole. Unfortunately, homogeneity often is not present, particularly when we study groups of people, so we must carefully consider various approaches to choosing a sample.

A **sample** is some subset of a population. We want it to represent the entire population in ways that will help us to understand what we want to know about the population. We want a sample to be unbiased so that our conclusions can reflect accurately the characteristics of the population.

> ***sample*** ■ A subgroup or subset of a population.

The characteristic to be studied is called a **variable.**

> ***variable*** ■ The characteristic of a population or sample to be studied.

In algebra, variables are usually labelled with letters such as X and Y, as is done in this text. Variables take on values from the specific responses of sample numbers: If the variable is height, a value of the variable is 5′8″.

Let us now discuss the various techniques of choosing a sample. We do so by presenting a particular bit of research to be done.

In an effort to assess constituents' opinions with regard to the defense budget (Should it be increased, decreased, or remain the same?), the senator from a large industrial state asks us to conduct a poll, the results of which will help the senator decide how to vote. We select a sample size of 1500 people from the senator's state, based on the amount of time and money available for the work; we are free to choose the way in which we will contact these people.

First, we must decide exactly who is in the population about which the senator is concerned. Should we talk with teenagers? Should we question people on welfare who may not pay taxes? There are several ways to answer such questions, but the important thing is to answer them: to define our population well so that we know where to go for our sample. The way we define our population determines the way we can draw conclusions in the end.

We decide to make our population the registered voters in the senator's state. This is a political matter, they are the people who hold the senator accountable, and so they are the ones to be questioned.

Now, how should we choose the 1500 registered voters to question? We want to choose a group that is a miniature reflection (microcosm) of the population, not letting our biases with regard to amount of taxes paid (you do not have to pay taxes to vote), level of education, age, race, sex, and other factors, affect our selection. We cannot allow the individual who actually chooses our sample to introduce a **selection bias** and thereby influence the selection.

selection bias ■ A property of a sampling technique that introduces prejudice into the sample.

Random Sample

Ideally we would like to choose what is known as a **random sample.** The word *random* is used in two different ways: (1) in everyday talk, meaning unplanned or haphazard; and (2) in the technical sense in statistics, meaning that an individual data value or a sample has an equal chance with every other possible one of being selected. Thus when choosing a sample of 1500 from our population, we will get a random sample if we can guarantee that every possible sample of 1500 had an equal chance with every other one of being chosen as our sample. If all samples have an equal chance of being chosen, whichever one we choose will be random. We will then be able to determine probabilities associated with the sample. This means that, in theory, we should list every possible different *group* of 1500 registered voters and then choose one of the groups in lottery fashion. Obviously, this is an overwhelming task and ultimately boils down to giving every member of the population (every registered voter) an equal chance of being chosen. We can focus on *individuals* and choose by lot. We need a complete list of all registered voters in the senator's state. If there are 40 million inhabitants of the state and 60% are registered voters, we will have a list of 24 million names. We do not have to write these names on pieces of paper and put them in a container before choosing the 1500 for our study. A computer can easily perform this part of the selection, but we do need the complete list in order to guarantee every registered voter an equal chance of being chosen—and thus giving us a random sample. Then we have to locate the 1500 people whose names the computer has given us.

Already you can see some problems. Do you think that it is possible to get a *complete* list? If we leave any registered voter or group of voters out, they have no chance of being in our sample and the sample is no longer random. However, it may be too expensive to verify that the list is complete and then to find the individuals selected. We cannot just use the telephone books from the senator's state to give us the lists because not everyone has an equal chance of being chosen. Some people do not have phones (from what economic brackets?); some have unlisted numbers; and some, such as doctors and lawyers, may be listed more than once (for both home and office). We cannot even use a telephone book to pick a quick and dirty sample because we would probably just open to a page and point to a name somewhere in the middle, giving people at the extreme top or bottom very little chance of being chosen.

Let us digress briefly from our study for the senator to consider one of the last times an actual attempt was made to choose a sample randomly from a very large population: the draft lottery in the 1960s. The population consisted of all men registered for the draft. A truly random sample would have guaranteed that all such men would have had an equal chance of being chosen. The lottery that was held consisted of a drawing of birthdays, 366 dates, from a drum. This method would have given every young man an equal chance of being chosen if the number of men born on any particular day were the same as the number born on any other day. This is not the case. Some months seem to be more popular than others for births. (Check this out for yourself the next time you are with a group of 20 or more people.) The draft lottery did not give the man born in August the same chance as the man born in April. The drawing was further complicated by the fact that the capsules containing the dates were not well mixed. Thus this attempt was just that, an attempt that did not quite fulfill all statistical requirements.

To obtain a random sample can be a very expensive and time-consuming process. Let us consider some alternatives to the totally random method, keeping in mind that we still want an unbiased, representative sample.

Stratified Sample

The first alternative is known as a **stratified sample.** The population of the senator's state is divided into *strata* or *layers based on some characteristic related to the study at hand.* For example, we might divide the state into geographic districts, such as voting precincts. If we then choose our sample from the various precincts, we would be sure to include voters from rural, urban, and suburban areas; from wealthy and poor areas; and from various ethnic groups. We could stratify the population on the basis of age: 18–25, 26–35, 36–45, 46–55, 56–65, and 66 and older. We could group registered voters on the basis of income: those making \$0–10,000, \$10,001–20,000, \$20,001–\$30,000, and \$30,001 and above.

A broader type of stratification similar to that of income is one of **socioeconomic status** (SES).

> ***socioeconomic status*** ■ A form of stratification of a population that takes into account income, education, and type of job.

It is entirely possible for a plumber to be making as much money as a pediatrician, but their educational levels and probably their outlooks are different, so they would not be grouped together. The groupings for SES are:

1. Professional and managerial positions.
2. Clerks and retail sales people.
3. Skilled craftsmen and kindred workers (plumbers, carpenters, electricians, etc.).
4. Laborers and unskilled service workers.
5. Farmers and farm laborers.

One important aspect of stratification is that we should know the proportion of the population that falls into each group of the characteristic we have chosen. We should be able to find out from the Census Bureau or state agencies the proportion of the population that lives in each voting precinct, that is 18–25 years of age, or makes $20,001–30,000 per year. The U.S. Bureau of Labor Statistics or the corresponding state office can help us with SES data.

For our study let us choose to stratify the number of registered voters on the basis of geography. We group all rural precincts in one category, put all suburban precincts in another, and place all urban precincts in a third, to complete our listing of the population. We can choose 500 registered voters from each of these three groups to get the 1500 members of our sample—a stratified sample. We can refine this basic technique by taking into account the proportion of our population in each of the groups. There are probably more registered voters living in the urban precincts than in the suburban precincts and more in the suburban precincts than in the rural ones. For example, if 40 percent of the registered voters of the state live in cities, then 40 percent of our sample

$$0.40 \cdot 1500 = 600$$

should be chosen from the cities. If we find that 35 percent live in suburban areas, then 35 percent of our sample

$$0.35 \cdot 1500 = 525$$

should be suburbanites. That leaves 25 percent to come from rural areas, or

$$0.25 \cdot 1500 = 375.$$

We then have a microcosm of the population to use as a sample, reflecting where our population lives. When we choose a stratified sample on this basis, we are choosing a **proportional** or **quota sample.**

We could have chosen age as the basis of our stratification, assuming that different age groups have different opinions about the size of the defense budget. If we had done so, we would have a proportional sample like the following one.

Age	Percent of Population
18 to 25	20%
26 to 35	30
36 to 45	20
46 to 55	15
56 to 65	10
66 plus	5

And our sample would contain the following numbers of people in each age group.

Age	Number in Sample
18 to 25	0.20 · 1500 = 300
26 to 35	0.30 · 1500 = 450
36 to 45	0.20 · 1500 = 300
46 to 55	0.15 · 1500 = 225
56 to 65	0.10 · 1500 = 150
66 plus	0.05 · 1500 = 75

The important thing to remember when choosing a proportional sample is that *the characteristic you choose for stratifying your population must be one about which you can find some data from external sources and one that is related to what you are studying.*

Convenience Sample

All this seems rather involved. Why don't we just go to a local movie theater on five Saturday nights and question 300 people each night as they exit? Or perhaps go to a shopping center and ask the opinions of 1500 people who walk by? The reason we don't is that all the people questioned in this manner would represent only one part of a stratification. The sample at the movie would be predominantly from the younger age brackets. At the shopping center, we would obtain a sample primarily from one geographic area. This is a relatively easy way to collect data, but chances are that it will not meet the goal of representing fairly the entire population. Using a college class or classes to gather data or questioning laborers leaving a factory are two more examples of **convenience samples.** They are convenient but may not be conclusive if the population includes more than sophomores or steel workers. Be careful of convenience samples. They may result from conscious or unconscious biases on the part of the data gatherer, as well as deciding "to do it the easy way."

Systematic Sample

Finally, we examine a **systematic sample.** Choosing a sample in a systematic way requires that members of the population be presented in some kind of order. We decide in advance the pattern we want to use selecting a sample from the population as it is presented to us. This can be as simple as choosing every tenth name on a list or picking every fiftieth person who passes a certain street corner. We could choose every fifth house on a block or the first name on every page of a telephone directory. The important thing is to decide on the system in advance of actual sampling, so that you do not introduce any known bias.

Although this sampling technique avoids known biases, it is not truly random. It is impossible for two people sitting next to each other both to be chosen for the same sample. Thus they do not have an equal chance of being chosen. There will be no samples containing both, so not all samples have an equal chance of being chosen. Systematic sampling assumes that all kinds of people are evenly distributed throughout the population. Studies of sampling methods have found that names of ethnic groups tend to cluster in an alphabetic list of the population; systematic sampling can miss these groups altogether or pick up too many of them in proportion to their numbers in the population.

The Gallup organization, perhaps the oldest active polling organization, samples nationwide to determine voter opinion about candidates and many national issues. It uses samples of 1500 people, considering only adults 18 years of age or older. A multistage sampling procedure is used. Random selection of individuals is ideal but unattainable, since any list of people (even voter registration lists) can never be complete; 300 voter precincts are randomly chosen instead. This list of precincts remains constant over time and is easy to use. Once the precincts are chosen the interviews are conducted at systematically chosen houses: every fifth, eighth, or some other previously selected number. The voting age occupant to be interviewed is chosen randomly in advance, so as to eliminate choice and thus bias on the part of the interviewer.

The Nielsen television rating service chooses its sample proportionately using demographic characteristics, such as income and age, that are important to commercial television broadcasting. Boxes are attached to television sets to record the stations the television is tuned to and the times the set is on. The boxes do not record who is watching, if anyone. Some rating services make an attempt through diaries to have members of the sample fill out a diary of viewing: times, shows, and the characteristics of those who watched.

2.3 Data Gathering

Methods of Questioning Sample Members

Now that we have decided who should be in our sample and how to go about choosing them, we must determine how to question them. There are three widely used methods: the **telephone survey, mailed questionnaires,** and **personal interviews.** The telephone survey is convenient but does not allow us to consider those who do not have telephones or those who have an unlisted number.

Questionnaires mailed to individual homes with provision for returning them (stamped, addressed envelope enclosed) have a notoriously poor rate of return. A 20-percent return is considered good. Most are simply misplaced or thrown out. Now we must ask ourselves: "Who took the effort to return the questionnaire?" After carefully choosing what we thought was an unbiased sample, did we end up with biased results because only certain types of people returned the questionnaires? (Perhaps those who are dissatisfied in some way or wish to complain.) A very vocal group, even though a small minority, can prejudice results. Twenty percent of our 1500-member sample for the senator's study is 300. If only 151 reply strongly in one direction, we have a majority (of the respondents) who favor one position. But consider that 151 is only about 10 percent of our carefully chosen sample.

The *Literary Digest* example presented in Chapter 1 is just such a case of giving a minority opinion too much emphasis. More recently, syndicated columnist Ann Landers asked her readers, "If you had it to do over again, would you have children?" Seventy percent of the parents who replied said, "No." These results were based on the replies of those parents who were interested enough to reply; they were not confirmed when the same question was posed in professionally conducted polls.

Personal interviews or questionnaires that are delivered and picked up by the data gatherer are more likely to elicit opinions that are representative of the sample. Thus the data are more useful in forming accurate judgments about the sample and, if the sample is representative of the population, about the population (our real goal). These methods are time-consuming and expensive, but they have proved to be reliable in gathering information that leads to accurate predictions. Thus they are used by reputable polling organizations, which recognize that credibility is the most important asset of any statistician.

For the senator, we will use registration lists to randomly sample our population of voters, with volunteers delivering and retrieving short questionnaires. We will examine methods to analyze this data in succeeding chapters.

Are there other aspects of data gathering that can taint the results? The answer is yes—and sometimes they are hidden. Take, for example, the survey that was done to determine the most popular spectator sport in America. Pro football won. It turns out that the survey was taken right before a Super Bowl game. Would the results have been the same in July?

In an attempt to determine which soft drink was more popular, Pepsi or Coke, interviewers gave participants glasses of the soft drinks labeled with letters to try and then asked them to state their preference. We are told on Pepsi commercials that it won. It turned out later (when an independent study was conducted) that the letter labeling the Pepsi cup was a more popular letter. How do we know that the letter did not win, rather than the Pepsi? (People also seem to prefer even numbers over odd numbers when asked to choose a number, and certain colors are more popular than others.)

How reliable are any product endorsements that are backed by so-called scientific evidence? Have you ever seen negative endorsements on TV? In "man-in-the-street" type interviews, do you see unedited versions? And, of course, all such interviews ignore all the principles of sound sampling methods.

Overcoming Obstacles in Data Gathering

Testing of new drugs and vaccines is a complicated process. Perhaps none was more complicated than the 1954 experiment to test the Salk vaccine for polio. Polio is a disease that mainly struck children aged 5–9 in seemingly randomly occurring epidemics. These epidemics occurred in the summer, but it was quite possible for a community to escape entirely any incidence of the disease during a particular summer (the actual incidence of polio was relatively rare). Polio is caused by a virus and can exhibit symptoms similar to many other flu-type diseases: fever, weakness, and respiratory difficulty.

Problems that faced the experimenters included the following:

1. The experiment had to be carried out on a very large scale. Since only about 5 people in 10,000 ever got the disease, the sample had to be large enough to include, potentially, a number of cases. More than one million children actually took part in the experiment.
2. Two experimental groups were needed because not everyone was to be given the vaccine. If everyone were given the vaccine, the experimenters could not establish whether the vaccine had prevented the disease or whether there had been no epidemic that particular summer. There had to be a group which did not get the vaccine, so that the experimenters could compare what happened to the two groups.
3. Parental consent was necessary for ethical reasons.
4. Doctors evaluating the effects of the vaccine might bias the experiment if they assumed, even subconsciously, that the children getting the vaccine would do better than those who did not.

These problems were solved using methods that are an important part of statistical experimentation. The two experimental groups were chosen by randomly assigning children to the group to receive the vaccine or to the group that was not to receive it, called the **control group.**

> ***control group*** ■ In an experiment, a group that does not receive any experimental treatment in order to provide baseline data.

This control group provided baseline data against which to compare the vaccinated group. If the vaccinated group had a significantly lower incidence of polio than the control group, then the vaccine would be (and was) judged a success.

Treating the control group differently from the vaccinated group could have led to the introduction of biases. For example, parents and children might have reacted differently to the disease, depending on whether or not the children had been vaccinated; it is well known in medical circles that mental attitude affects the severity of a disease. And, of course, doctors might diagnose the disease differently, attributing minor symptoms to the flu rather than to polio in a vaccinated child. To solve this problem, the control group was given a **placebo,** an innoculation that appeared to be identical to the vaccination but was really an inactive salt solution. No one—not parents, children, or the diagnosing doctors—knew until after the experiment which code identified the active innoculation and which code the inactive.

> ***placebo*** ■ A substance that matches identically an experimental substance, except for an active ingredient. It is administered to control for psychological effects in interpreting results.

When everyone directly involved in making observations is ignorant of the item being tested, the experiment is called a *double-blind* experiment. Observer bias is minimized.

News & Views 2.2 and 2.3 present two amusing examples of what is known as the placebo effect. They demonstrate the importance of controlling for it.

The issue of the effect of smoking on health is one that has caused much discussion. Again scientists were faced with many problems, such as:

1. Smoking is not like polio in that smoking is voluntary; getting polio is not. For the sake of an experiment people cannot be *randomly* assigned to a smoking group or a nonsmoking

control group and given no choice about whether they must smoke.

2. The effects of smoking seem to take a long time to accumulate in the body. The experiment cannot be conducted over one summer.
3. Polio is caused by a virus. Heart disease and cancer are caused by many things acting together: heredity, diet, blood pressure, occupation, and others, as well as whether or not a person smokes.

All of these factors can influence the disease. They are known as **confounded variables.** R. A. Fisher, an Englishman who was instrumental in experimental design, suggested use of the randomized study with two groups to control for confounded variables. If each group were made up randomly, these factors would tend to balance each other out.

> ***confounded variables*** ■ Factors for which effects cannot be separated in order to learn the individual effects of each.

NEWS & VIEWS 2.2

Brand names important

How important are brand names in the treatment of illness? In many cases, they are as important as the medicines they identify.

In a recent issue of "British Medical Journal," researchers at Keele University wrote of performing the following experiment:

They recruited 835 women who regularly took tablets for their headaches. Half said they used one popular brand of aspirin. The medical researchers thereupon prepared four sets of identical-looking tablets. They consisted of (1) aspirin in the manufacturer's package with the brand name; (2) aspirin in a plain packet, (3) dummy tablets (placebos) in the manufacturer's package and (4) dummy tablets in a plain packet.

The inert placebos in the plain packet provided some relief for 73% of the headaches. The placebos in the manufacturer's packaging performed 10% better. The genuine aspirin in the plain packet rated another 6% higher; and the genuine aspirin in the manufacturer's packaging proved best of all.

The conclusion is that many people who believe (from advertising, word-of-mouth recommendations and hearsay) that one brand of aspirin is more effective than another will find that true—largely because they have been so programmed. The power of suggestion, the basic ingredient of advertising, cannot be discarded by science.

Source: Lloyd Shearer in *Parade,* October 25, 1981.

NEWS & VIEWS 2.3

Thinking makes it so

People who think they are drinking alcohol but really aren't act in certain stereotyped ways as though they were becoming intoxicated, according to two experts on the psychological effects of alcohol.

"The think-drink effect is as dramatic as a placebo's seemingly miraculous curative power," said Dr. G. Alan Marlatt of the University of Washington and Dr. Damaris J. Rohsenow of the University of Wisconsin, writing in the December issue of *Psychology Today*.

They gave plain tonic water to people who thought the drink was vodka and tonic, and the subjects became less anxious, more aggressive and more sexually aroused, they said.

Various studies of existing data giving cause of death were made, but they did not approach the randomized procedure. They could only provide counts of smokers among lung cancer victims. A more convincing study was done through the American Cancer Society. The experimenters enlisted 22,000 women volunteers, each to keep track of ten healthy men aged 50–69. During a four-year period, 12,000 of these men died. The experimenters could then count the number of smokers who died of lung cancer and thus get a death rate among smokers. They also noted the presence of other disorders, such as heart disease, that seemed to be more prevalent among smokers. They could focus on smokers, follow their progress rather than focus on a cause of death (lung cancer), and try to work backwards, as analysts in earlier studies had done. The results showed a lung cancer rate almost 24 times higher for heavy smokers than for nonsmokers. The general results of this study were also obtained in a British study. Such *replication of results* is a desirable feature in any statistical work.

Because of the ethical problems involved with random studies of human beings, such studies must be and are being done with animals. The proverbial white mice are used in experiments because they are all genetically similar, and thus many factors that might confound results are kept constant. News & Views 2.4 and 2.5 contain articles that point out the concerns and problems associated with animal research and confounding variables.

Most of the procedures discussed from this point on presume simple random sampling. As you continue, please critique problems and exercises that describe data; try to determine whether the samples used were randomly chosen. Many times we must work with data as given, but it is important to keep in mind that randomness in sampling is a basic assumption.

NEWS & VIEWS 2.4

The dowager mouse of all the labs

To the uninitiated, all laboratory mice look alike. That's no coincidence, according to a report in a recent issue of Nature magazine: in fact, the nine inbred strains used in today's experiments are all descended from a single female.

Three California scientists arrived at this result by analyzing DNA in the tiny energy-producing organelles in the cells of the mice. These organelles, or mitochondria, contain molecules of DNA inherited solely from the mother (the only thing that sperm contribute to the next generation is half the genes in the fertilized egg's nucleus). When researchers Stephen Ferris, Richard Sage and Allan Wilson of the University of California at Berkeley examined the mitochondrial DNA of old inbred strains, they got a surprise. All nine of the strains contained exactly the same DNA in their mitochondria, which means they were descendants of the same maternal ancestor. That ancestor and her daughters mated with males of different wild strains to produce the inbred mice scientists use today. The dowager mouse could have lived anytime between 3,200 years ago, when mice were domesticated in China, and the early 1900s, when the first inbred strains were created for laboratory use.

Source: *Newsweek,* February 8, 1982, p. 93.

NEWS & VIEWS 2.5

The mice they reared

A pedigree is of more importance in the world of lab animals than it is in any royal family. When medical researchers try to unravel the secrets of the cell, essential to understanding cancer, they must be absolutely certain of the genetic "purity" of their test subjects. Thus the biological community was rocked last week by the news that a strain of albino lab mice used by cancer investigators everywhere was genetically contaminated. The tainted mice were discovered by University of Wisconsin Biologist Brenda Kahan and her colleagues while they were growing a primitive type of tumor called a teratocarcinoma. A puzzling enzyme uncharacteristic of the breed kept showing up in the host animals, a common strain known as BALB/C. Careful genetic tests soon confirmed suspicions: the mice were not the purebreds promised by their supplier, the Massachusetts-based Charles River Breeding Laboratories, Inc. Since the company is the world's largest producer of lab animals, similar mongrels are presumably frolicking in the cages of many other research centers. And the long, tedious work of determining where they are and what experiments they compromised must now begin.

Source: *Time,* July 26, 1982, p. 41.

EXERCISES

1. In the library find *Statistics: A Guide to the Unknown* by Tanur, et al., which presents applications of statistics in many fields. Choose an article and critique it for sampling methods.
2. You are asked to determine the major field of study of students at your school. How would you choose your sample using each of the methods discussed in this chapter? Be specific.
3. You are to determine the average price of a share of stock by using the stock reports in the daily newspaper. This is a very long list. How would you sample it using each of the methods discussed?
4. You are asked to determine the effect of television on preschoolers. How would you choose your sample using the methods discussed?
5. You want to consider sales records and age of salespersons in a firm. How would you choose a sample using each of the methods discussed?
6. Find two newspaper or magazine articles using survey or polling results and bring them to class for discussion.
7. Find two advertisements that claim scientific evidence for their products and bring them to class for discussion.
8. A large auto manufacturer uses drivers of a competing car to test drive its own car. Are comments such as "It drives as smoothly as mine," or "It's OK" convincing? Do you think that the drivers know who is conducting the test and paying the bill? Are they likely to give negative responses or polite ones?
9. Data seemed to indicate that wartime activities were more hazardous to civilians of a particular unoccupied country than to military personnel. During the period of the war, amputations were performed on 120,000 civilians but on only 18,000 military personnel. Please comment on the relative sizes of these numbers.

Study Notes

KEY TERMS

confounded variables ■ Factors for which effects cannot be separated in order to learn the individual effects of each.

control group ■ In an experiment, a group that does not receive any experimental treatment in order to provide baseline data.

placebo ■ A substance that matches identically an experimental substance, except for the absence of an active ingredient. It is administered to control for psychological effects in interpreting results.

population ■ The complete collection of people or objects about which we are interested.

sample ■ A subgroup or subset of a population.

selection bias ■ A property of a sampling technique that introduces prejudice into the sample.

socioeconomic status (SES) ■ A form of stratification of a population that takes into account income, education, and type of job.

variable ■ The characteristic of a population or sample to be studied. As in algebra variables are usually labeled with letters such as X and Y. Variables take on values from the specific responses of sample members: If the variable is height, a value of the variable is 5′8″.

OUTLINE

I. Goal of sampling—representativeness of the population.

II. Sampling techniques
   - A. Random sample—a sample chosen in such a way as to guarantee that every member of the population has an equal chance of being chosen. This requires a complete list of all members of the population and a sample chosen in a way similar to a lottery.
   - B. Stratified sample—After a population is divided into groups based on some characteristic related to the study, a sample is obtained by selecting members from each of the groups.
   - C. Proportional or quota sample—a form of stratified sample in which members of the sample are chosen from each of the strata in proportion to the number of members in the population in each group.
   - D. Convenience sample—a sample completely chosen from only one group in a stratification.
   - E. Systematic sample—a sample chosen by selecting every *n*th member of the population as the population is presented to the chooser.

III. Methods of questioning sample members
   - A. Telephone survey.
   - B. Mailed questionnaires.
   - C. Personal interviews.

IV. Overcoming obstacles in data gathering
   - A. Salk vaccine test.
   - B. Smoking and health studies.

PART II

DESCRIPTIVE STATISTICS

Frequency distributions and graphs

3

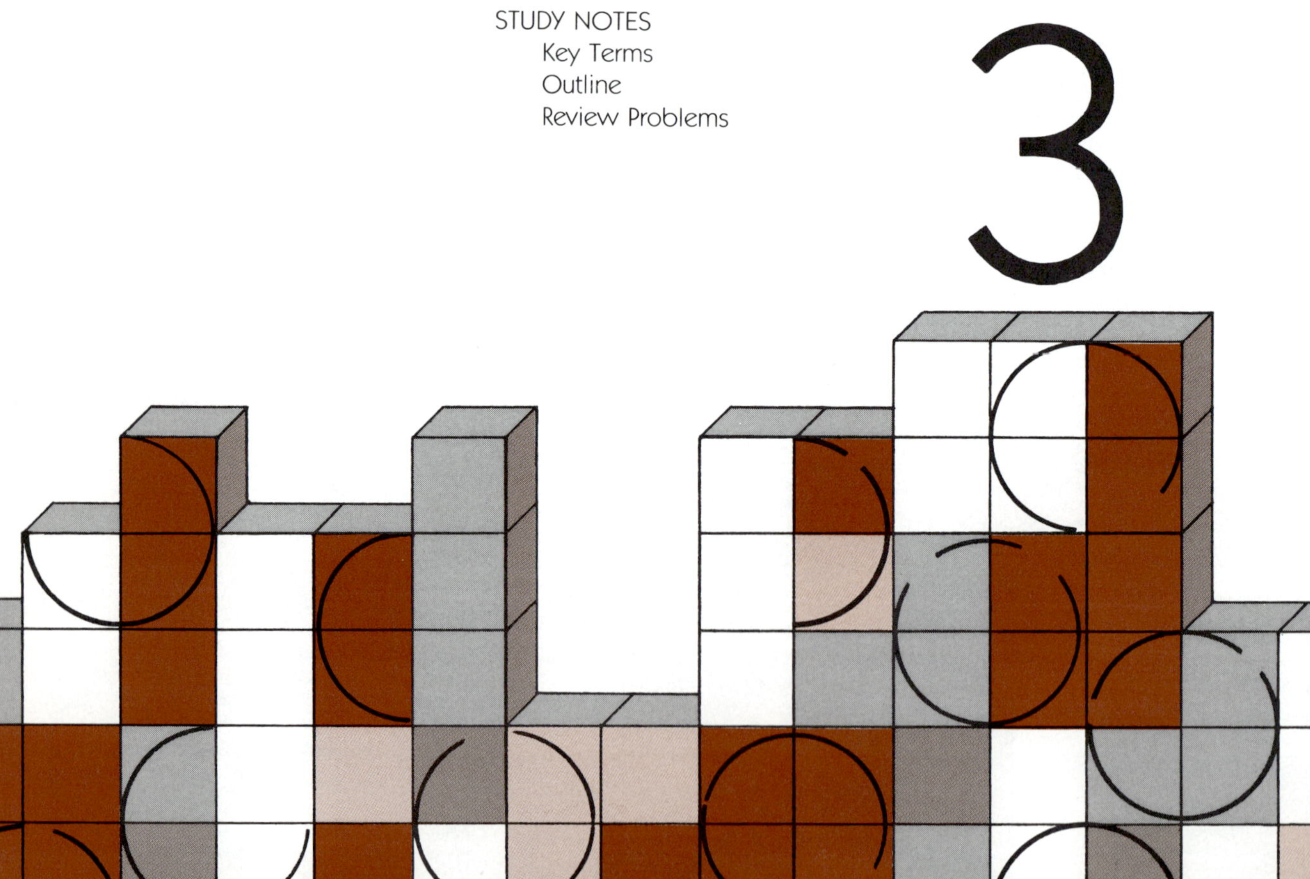

Data collection often results in large amounts of data, which are difficult to absorb quickly. For the information to be useful, we need to organize, summarize, and display it. One of the best ways to do this is to condense the data into a table and then present it graphically, as discussed in this chapter. In Chapter 4 we will develop some numerical measures to summarize the data.

3.1 Frequency Distributions

A **frequency distribution** is a list of possible values of the characteristic or variable we are considering, along with a tally or count (frequency) of how many times each value shows up in the data.

> ***frequency distribution*** ■ A list of possible values of the variable under consideration and the frequency with which these values occur in the data.

When only a few values of the variable are possible, we can construct an *ungrouped frequency distribution.* Consider the following example.

EXAMPLE 3.1 In a survey of the American family in the 1980s, 25 married adults were contacted and asked about the number of children in their families. The following responses were obtained.

0, 1, 4, 2, 0, 3, 2, 3, 1, 2, 4, 2, 0, 1, 0, 3, 1, 5, 6, 1, 2, 3, 7, 3, 2

Construct an ungrouped frequency distribution for this information.

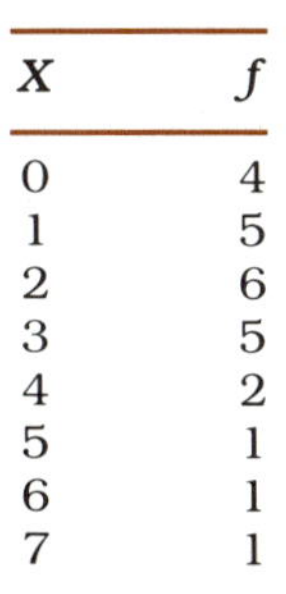

X	f
0	4
1	5
2	6
3	5
4	2
5	1
6	1
7	1

Solution We start by making a list of the possible values for X, letting X stand for the number of children in a family. Then we count the number of times each of the values occurs and label this column frequency, or f.

Note that 2 is the number of children per family most frequently reported: It occurs 6 times. If we add up the f column, we should get 25. This serves as a check to make sure that we have counted the entire sample. □

Most of the time there are many possible values for a characteristic; too many, in fact, to list them all easily. When this situation occurs, we need to form some preliminary groupings before we do the counting. The procedure for grouping is fairly easy and allows for much individual choice in selecting groups, so long as three fundamental rules are followed:

RULES

1. There should be between 5 and 12 groups or classes.
2. The classes should be of equal width.
3. Every piece of data should fit into one and only one class.

The following example illustrates this procedure.

EXAMPLE 3.2 Table 3.1 lists the weights of 50 adult males. Construct a *grouped frequency distribution* from these data.

Solution The data values range from 108 to 247, or through 247 − 108 = 139 lbs. It is usually a good idea to extend the list in both directions, say from 105 to 255, to be sure that the two endpoints of 108 and 247 are included. Now we are considering a range of 255 − 105 = 150 lbs. If we let each class contain 15 lbs.,

TABLE 3.1 Weights of adult males (lbs)

108	144	155	167	181
111	144	156	168	182
121	145	158	169	184
123	145	159	169	185
125	146	159	170	185
130	147	160	171	186
130	148	160	171	187
131	148	161	172	188
134	149	161	174	190
138	149	162	175	192
138	150	162	175	193
140	151	162	175	195
140	152	163	176	196
141	153	164	178	198
142	154	164	178	200
142	154	166	179	212
144	154	166	180	225
144	155	167	181	247

we will have ten classes:

105–120

120–135

135–150

.

.

.

240–255

You probably have noticed that we satisfy the first two rules but not the third. Where should we put the data values of 120 and 135? In the first class or the second class in which each appears? We cannot have overlapping intervals and must avoid this ambiguity. This can be fixed by changing the second column of numbers to:

105–119

120–134

135–149

.

.

.

240–254 □

We now have the smallest and largest numbers from the data that go into each class. These are called the **class limits.**

lower class limit. ■ The smallest possible data value in a class.

upper class limit. ■ The largest possible data value in a class.

One other set of numbers needs to be discussed: the **class boundaries.** For the types of graphs we will draw in Section 3.3, it is important that there be no gaps, if consecutive classes actually contain data. However, if we limit our discussion to class limits, it looks as though we have a gap between the upper class limit of one class and the lower class limit of the next class (such as between 119 and 120). This is where class boundaries help.

The numbers in Table 3.1 represent measurements of weights in pounds. Measurements are often rounded because of the measuring instrument's lack of precision. If someone actually weighed 119.3 lbs, what would you record for their weight? Probably, 119. Similarly, if someone actually weighed 119.8 lbs, you would probably record their weight as 120. So the apparent gap between 119 and 120 does not really exist. At what point would you stop putting numbers in the 105 to 119 class and start putting them in the 120–134 class? At 119.5. Thus 119.5 is the upper class boundary of the 105–119 class and the lower class boundary of the 120–134 class. What about the overlap and ambiguity? Since 119.5 does not actually occur in the data, this problem is avoided. Class boundaries are numbers that occur because of rounding a measurement. They are numbers created by the process and are not actually found as data values.

> ***class boundaries*** ■ Numbers halfway between consecutive upper and lower class limits. They involve rounding in the measurement process and serve to complete or fill the gaps in a set of classes.

A listing of class boundaries would look like this:

104.5–119.5
119.5–134.5
134.5–149.5
.
.
.
139.5–254.5

Now we are ready for a precise definition of **class width.**

> ***class width*** ■ The distance between consecutive lower class limits *or* the distance between class boundaries.

By the first definition we find that the first class has width 120 − 105 = 15. By the second definition the first class has width 119.5 − 104.5 = 15. Similarly, the class width for all classes of our sample data must be the same.

NOTE

> Be careful *not* to subtract class limits: 119 − 105 = 14. This will give the wrong width.

Either a system of class limits or a system of class boundaries may be used for a frequency table. The class-limit system is easier to read because it has fewer decimals, so we will continue with it. Our next step is to count the number of data values that go in each class.

Class	f
105–119	2
120–134	7
135–149	19
150–164	23
165–179	19
180–194	13
195–209	4
210–224	1
225–239	1
240–254	1
	90

Note again that if we add the frequencies, we get the sample size; we have included all the data. The most frequent class is 150–164 because it contains more values (23) than any other.

Before leaving frequency tables, we need to consider another set of numbers that is useful: the **class marks.**

class mark ▪ The midpoint of a class, found halfway between class limits or class boundaries

$$\frac{\text{Lower limit [boundary]} + \text{Upper limit [boundary]}}{2}$$

In the first class the class mark is 112, using class limits. This is found by adding the limits and dividing by 2:

$$\frac{105 + 119}{2} = \frac{224}{2} = 112.$$

Or, using class boundaries:

$$\frac{104.5 + 119.5}{2} = \frac{224}{2} = 112.$$

The class mark is the average of the limits, a point in the middle of the interval, and as such can be used to represent the class. Once a frequency distribution is made, individual numbers within a class are lost. If we need to represent those numbers, we can do it with the class mark. The completed table looks like this.

Class	f	Class Mark
105–119	2	112
120–134	7	127
135–149	19	142
150–164	23	157
165–179	19	172
180–194	13	187
195–209	4	202
210–224	1	217
225–239	1	232
240–254	1	247

Now you can see that the class marks are separated by 15, the class width; this is a good check on your arithmetic. We will add other columns to this table as needed when we get to graphing.

EXERCISES/Section 3.1

1. Toss a die 50 times and record the results. Construct an ungrouped frequency distribution of these results. Let X = number on die.

X	f	X	f
1		4	
2		5	
3		6	

2. Survey eye color in a group of 30 people and record the results. Construct an ungrouped frequency distribution of these results. Let X = eye color.

X	f
Brown	
Blue	
Green	
Hazel	
Other	

3. The following data give the number of years of college completed for newly hired employees of Design, Inc. Construct an ungrouped frequency distribution of this information.

1	0	0	2	4	0
1	0	2	2	4	5
4	2	3	2	2	6
1	3	2	1	1	5
0	5	2	5	0	2

4. Fuel economy, in miles per gallon, was measured for 40 vehicles, with the following results.

22.5	25.0	27.0	21.1	24.1	23.1	23.1	14.5
32.5	24.3	21.7	22.9	24.6	18.6	18.1	23.8
25.3	17.8	28.7	22.6	19.1	18.0	20.1	13.8
30.6	28.0	23.7	16.2	23.4	19.9	25.7	13.2
22.5	27.4	27.9	26.4	21.0	13.9	15.2	21.0

a) Construct a grouped frequency distribution beginning at a first lower class limit of 13.0 and ending with a last upper class limit of 32.9, using classes of width 2.
b) Determine the class boundaries.
c) What are the class marks?

5. An insurance company collected the following data on family income (in thousands of dollars).

15	10	12	26	28
12	17	19	9	30
25	21	22	11	11
20	18	30	14	14
11	9	21	18	9
14	7	20	22	10
16	11	12	17	12

a) Construct a frequency distribution using 6–8 as the first class.
b) What are the class boundaries of the first class?
c) What is the class width?
d) What is the class mark of the first class?

6. The following information represents runs batted in for 50 baseball players.

30	27	14	22	14
25	31	24	31	21
42	43	13	27	32
19	9	41	32	17
18	11	19	51	46
3	21	37	46	4
53	32	42	30	27
21	16	13	25	29
19	15	12	15	7
18	27	8	10	13

a) Construct a frequency distribution beginning at 0 and having a class width of 10.
b) What are the class limits of the second class?
c) What are the class boundaries of the second class?
d) What is the class mark of the first class?

7. The following are starting weekly salaries, in dollars, for 30 jobs in data processing.

150	130	115	175	250	140
130	127	192	200	225	275
225	175	200	180	150	300
150	200	190	150	125	200
120	225	160	275	135	175

a) Construct a frequency distribution using 100–124 as the first class.
b) What is the class width?
c) What are the class boundaries of the second class?
d) What is the class mark of the first class?

8. Forty cities reported the following average temperatures (in °F) in July.

73.9	86.4	72.3	82.3
77.7	88.2	75.1	65.4
82.4	70.1	67.2	79.4
62.4	73.6	62.3	77.2
67.5	81.2	81.9	78.1
64.6	62.4	78.4	76.5
60.1	68.1	73.4	71.2
65.5	74.6	76.8	67.8
80.6	86.4	68.9	83.6
81.7	69.0	70.6	79.4

a) Construct a frequency distribution starting at 60 and use a class width of 3.0.
b) What are the class boundaries of the first class?
c) What is its class mark?

9. The following data give the life of a tire, in thousands of miles, tested on 50 different cars.

10	11.1	9.7	9.3	11.1
12	9.8	10.1	11.3	16.8
13	10	11.1	12.9	14
9	7.4	10.9	16.1	15.1
7	6	9.7	14.2	15.3
8.5	10.2	11.1	13.5	11.2
11.4	17.1	12.1	9.7	12.3
13	13	12.2	10.9	12.4
17	14	10	10.9	11.1
12.2	14.3	14	8.2	9.7

Construct a frequency distribution. Be sure to include class marks.

10. The heights, in inches, of 40 veterans were obtained from their military records.

68	68	66	67	68
68	66	71	68	71
70	74	69	73	79
68	75	69	69	69
67	70	69	67	67
66	69	70	71	68
67	69	71	70	66
69	74	68	68	66

Construct a frequency distribution using classes of width 2. Include class marks.

3.2 Stem-and-Leaf Diagrams

A variation on the frequency distribution is called a **stem-and-leaf diagram.** It allows us to condense data, much as a frequency distribution did, but to retain the individuality of the data.

EXAMPLE 3.3 Records from an insurance company give the following data on the amounts of insurance purchased by 35 people (in units of $1000).

40	21	42	92	55
35	38	35	37	125
100	72	75	34	73
120	55	101	30	50
25	25	57	44	34
50	24	45	45	30
60	36	43	49	35

Draw a stem-and-leaf diagram to display these data.

Solution The "leaves" in the stem-and-leaf diagram are the last digit in each number; in 40, 0 is the leaf. The first digit(s) provide the "stem"; in 40, 4 is the stem. In this example the stems range from 2–12. We list all the stems in order, draw a vertical line beside them, and then record the leaves for each stem in a row to the right of the vertical line.

This diagram shows that we have 4 data values in the 20's in the first row (25, 21, 25, 24), 10 in the 30's in the second row, and so on. We count the number of data values in each row, the leaves, to get frequencies.

Stem	Leaves	f
2	5 1 5 4	4
3	5 8 6 5 7 4 0 4 0 5	10
4	0 2 5 3 4 5 9	7
5	0 5 7 5 0	5
6	0	1
7	2 5 3	3
8		0
9	2	1
10	0 1	2
11		0
12	0 5	2
		35

□

Again, summing the frequencies provides a check on whether we included all the data and only once. Note that all the individual numbers can be recovered by putting the stem with a leaf.

EXAMPLE 3.4 The following data show the annual cost of living for a family of four (in units of $1000) in 20 major metropolitan areas.

13.1	14.6	16.1	13.9
14.4	13.4	13.9	15.1
12.9	14.4	14.1	14.5
15.4	17.0	15.0	13.8
13.8	12.9	14.8	15.0

Draw a stem-and-leaf diagram.

Solution The leaves are the final digits, so in this case the stems range from 12–17.

		f
12	9 9	2
13	1 8 4 9 9 8	6
14	4 6 4 1 8 5	6
15	4 0 1 0	4
16	1	1
17	0	1
		20 □

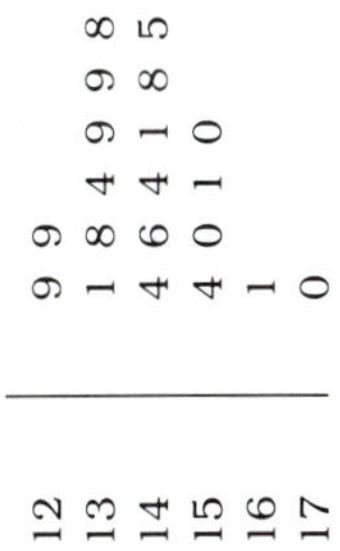

12 | 9 9
13 | 1 8 4 9 9 8
14 | 4 6 4 1 8 5
15 | 4 0 1 0
16 | 1
17 | 0

Figure 3.1

If we turn the stem-and-leaf diagram sideways, we have a likeness of a bar graph (Fig. 3.1). In the next section we will discuss graphs in detail.

EXERCISES/Section 3.2

1. The following are lengths, in inches, of the legs of 40 prehistoric animals as determined by paleontologists using fossil bones.

5	36	40	68	5
43	29	16	3	41
21	22	52	47	51
2	46	51	54	36
31	70	59	19	45
19	2	54	1	1
29	12	16	22	8
71	12	6	16	62

Draw a stem-and-leaf diagram.

2. Average weekly earnings, in dollars, for workers in 20 major metropolitan areas are:

163	220	205	222
191	219	193	219
228	252	179	209
222	163	216	200
199	205	184	199

Construct a stem-and-leaf diagram.

3. Annual salaries (in units of $1000) in 60 engineering jobs are:

27.7	26.0	29.0	28.0	25.8	24.0
27.9	28.9	27.6	32.5	21.2	25.0
31.1	28.2	29.5	26.3	25.1	20.2
25.4	23.1	24.1	21.7	18.0	18.5
26.6	19.6	27.2	22.0	21.5	19.4
28.1	25.3	27.0	24.8	21.8	20.7
27.1	26.7	26.8	26.1	21.4	21.5
23.8	26.7	27.0	25.2	21.0	20.0
32.1	27.0	29.9	25.5	21.6	29.9
27.0	27.3	26.7	27.4	22.3	20.9

Construct a stem-and-leaf diagram.

4. Age at the time of a heart attack for 40 victims was:

31	58	40	63	73	78	31	58	42	69
61	52	49	48	37	52	40	60	61	60
61	52	34	40	73	61	56	60	52	58
44	52	37	51	41	44	52	45	45	59

Construct a stem-and-leaf diagram.

5. The following are the number of points scored in a five-year period by 40 players on a local high school team.

1	18	0	57	0
0	2	59	45	0
0	42	9	3	3
2	60	7	4	1
6	37	35	17	4
48	3	46	8	3
21	2	0	18	2
75	75	2	21	6

Construct a stem-and-leaf diagram.

6. The following are election returns, as a percentage of the vote cast for the Democratic candidate, from 25 precincts during a recent election.

62	28	31	65	17
69	75	86	29	5
39	57	79	29	15
28	22	66	35	39
55	43	69	35	45

Construct a stem-and-leaf diagram.

3.3 Graphs

Graphs are used to present information simply and quickly. They give visual impressions of magnitude, grouping, trends, and patterns in data. The goal in making a graph is to show clearly and accurately the important features of the data that we want to present. We also need to know how to read graphs knowledgeably because first impressions can be wrong or distorted.

Long lists of information make it difficult for us to pick out many characteristics of the data when we just see columns of numbers. However, a well-made graph can present the same data in a form that tells us most of what we need or want to know. All graphs should have the following characteristics.

CHARACTERISTICS OF GRAPHS

1. They should be clear and easy to read.
2. An informative title describing the subject of the graph is basic, and all parts should be carefully labeled. Sometimes a pseudograph is made by superimposing a grid on a line graph, with no indication of what the vertical and horizontal scales mean. This gives it a scientific aura (which is the idea), but it really tells us nothing.

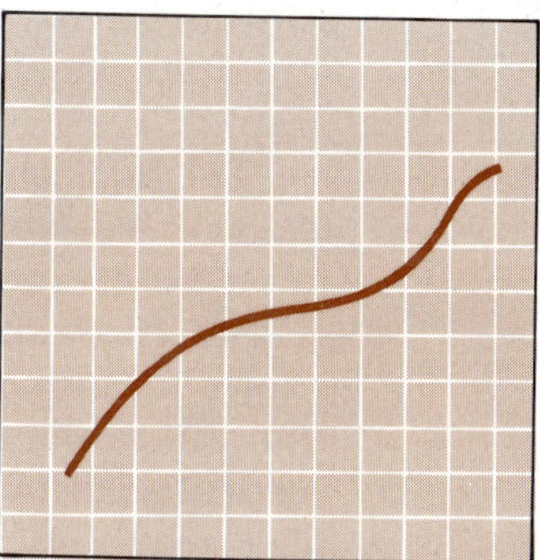

3. Any graph that contains vertical and horizontal scales, or axes, must have equally spaced intervals along those axes. Otherwise distortion occurs, and false impressions can be created if some intervals are larger than others, even though the axes might be clearly labeled; most people just glance at general shapes and do not linger over details. (We will look at examples of this when we examine some specific graphs.) Certain types of technical graphs—probability and logarithmic, for example—deliberately use unequal intervals, but they are specialized in nature and occur infrequently except in technical writing. A slash may be used to indicate the shortening of an axis, as shown in the accompanying diagrams, but this should be clearly done and only on certain types of graphs, namely line graphs or bar graphs.

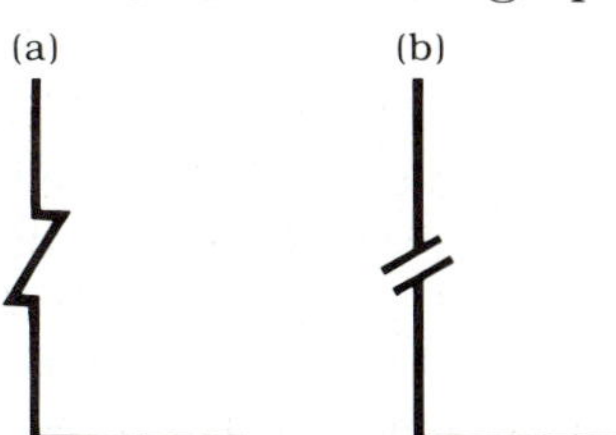

4. Any graph that displays quantities or frequencies on either the vertical or horizontal scale should have a zero point on that scale. Otherwise, again, distortion can easily occur.

Appendix A includes a detailed description of several types of graphs and a discussion of deceptive practices in graphing that can be used to distort data. Specific sets of graphs are often used in statistics. We will now look at two.

Histograms

The histogram is a technical form of a vertical bar graph. It has additional characteristics that allow us to make generalizations and draw conclusions, which we cannot do from an ordinary bar graph.

The horizontal scale represents the *characteristic* of the population or variable that is being graphed. All of the intervals again must be evenly spaced.* Intervals on the horizontal are called classes, just as in a frequency distribution. The vertical scale shows *frequency*, that is, the number of members of the sample that fall into a class. A variation on the use of frequency for the vertical scale is **relative frequency,** or frequency *relative* to sample size. To obtain relative frequency, we merely divide the count (frequency) by the size of the sample. If 5 people in a sample of 50 fall into a class consisting of people who are between 5′8″ and 5′10″ in height, then the frequency is 5 and the relative frequency is 5/50 or 1/10.

> ***relative frequency*** ■ Frequency divided by the size of the sample.

Histograms have the added property that the vertical bars must touch each other if contiguous classes contain some members of the sample. A blank space between bars means that the category is empty; that it has zero frequency. It is not there for aesthetic reasons, as are the blank spaces in ordinary bar graphs. All of the other rules that we discussed at the beginning of the section still apply.

Let us construct a histogram using the frequency distribution we developed from Table 3.1.

Class	*f*
105–119	2
120–134	7
135–149	19
150–164	23
165–179	19
180–194	13
195–209	4
210–224	1
225–239	1
240–254	1

We first list the class limits along the horizontal axis and frequencies along the vertical axis, as shown in Fig. 3.2. Using class limits keeps the graph from looking cluttered (class boundaries or class marks would work just as well). We now draw a vertical bar up to the proper frequency for each class, starting with 2—then 7, then 19, and so on—and ending with 1.

This graph can be easily turned into a relative frequency histogram by dividing each of the numbers on the frequency scale by

* This rule can be modified using densities, but great care must be taken.

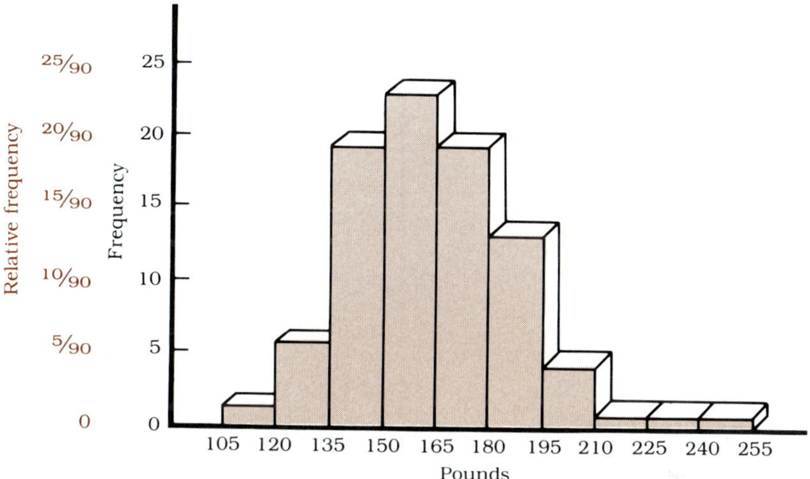

Figure 3.2
Weights of adult males

the sample size of 90. Nothing else needs to be changed. Relative frequencies are shown in color on the vertical scale in Fig. 3.2 for comparative purposes (normally, they would be shown as the single vertical scale on the graph). The relative frequency histogram tells us, for example, that 7/90 of our sample falls between 120 and 135. If we were to choose at random someone from the sample, we would have 7 chances in 90 of obtaining someone weighing between 120 and 135 lbs. This notion will show up again when we discuss probability, and using a histogram in this manner will help us to draw some conclusions about our population.

If the data are ungrouped as in Example 3.1, it is best to mark the values of the classes in the middle of the bars, as in Fig. 3.3. Note that 0 on the horizontal axis was moved to the right from the corner in order to keep the entire first bar on the graph.

Histograms take a variety of shapes, some of which are shown schematically in Fig. 3.4. The symmetric histogram (Fig. 3.4a) indicates that there are as many members of the sample in both the high and low classes; frequencies gradually increase and peak in the middle at the most numerous class. Many naturally occurring phenomena such as height, weight, IQ, and growth patterns in plants have symmetric distributions: few cases at the extremes; most cases bunched in the middle.

The skewed distributions (Fig. 3.4b and 3.4c) indicate that very few cases fall at one end of the scale and that most cases fall at the other end. Wealth or income is skewed to the right. Most of us have salaries at the lower end of the income scale, say below

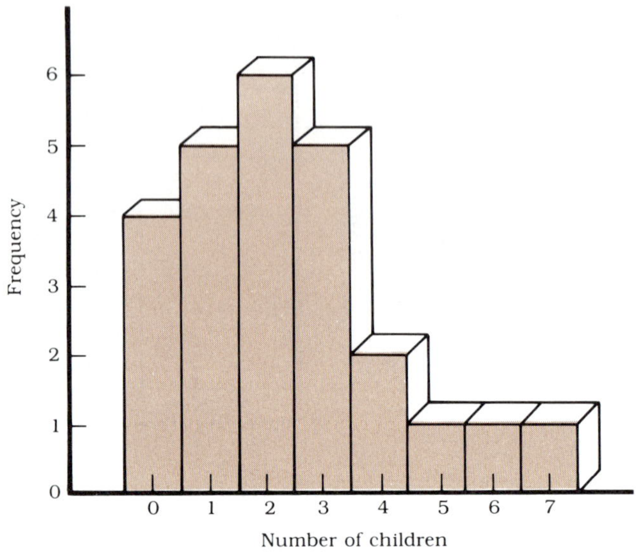

Figure 3.3
Family size

Figure 3.4
(a) Symmetric; (b) skewed to the right, positively skewed; (c) skewed to the left, negatively skewed; (d) uniform; (e) bimodal.

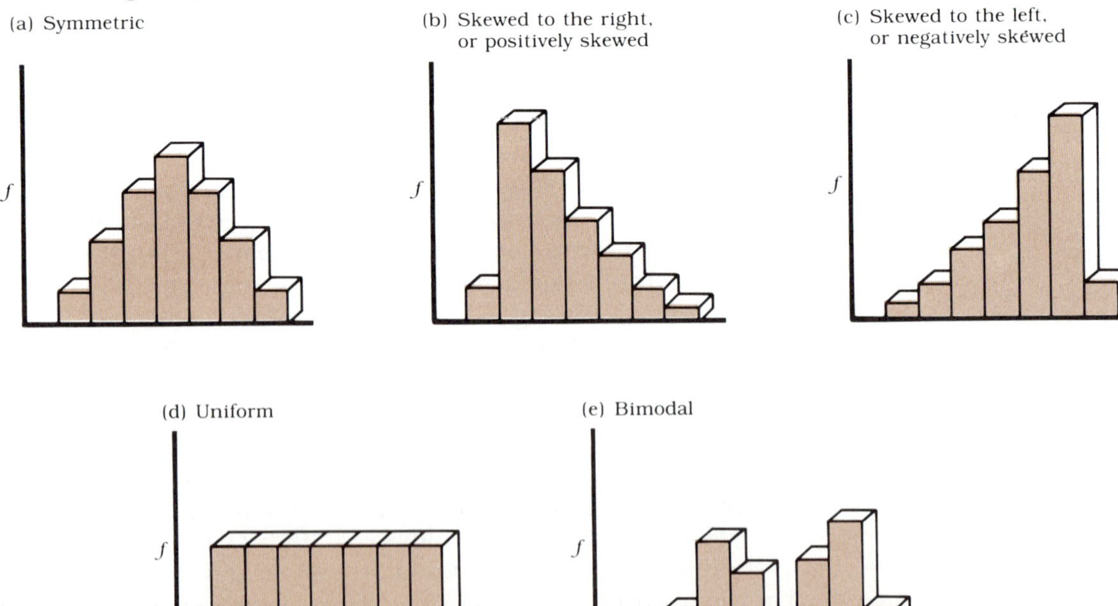

$50,000; there are, relatively speaking, very few millionaires. However, age at death is skewed to the left. There are few cases of death, relative to the size of the population, in the early years; the frequency of death increases as age increases.

A uniform distribution (Fig. 3.4d) is one where all classes contain the same number of members of the sample. If we tossed a fair die sixty times and counted the times that each of the six numbers (dots) showed up, we would expect to see a roughly uniform distribution, that is, each number to occur ten times.

A bimodal histogram (Fig. 3.4e) usually means that we are looking at the combined data for two groups, each of which has its own symmetric distribution. For example, if we used data on weights of both men and women, we would probably have a bimodal histogram showing a separate peak for the most frequent weights for each sex.

Ogives

If we want to accumulate the data through the classes from lowest to highest, we can develop a **cumulative frequency.** For example, using the class and frequency data developed from Table 3.1, we know that there were 2 members in the class having a lower limit of 105 and 7 members in the class having a lower limit of 120, or a total of 9 in the first two classes. If we continue to add the frequency for each class to the total for all preceding classes and write this in chart form, we get:

Class	Frequency	Cumulative Frequency
105–119	2	2
120–134	7	2 + 7 = 9
135–149	19	9 + 19 = 28
150–164	23	28 + 23 = 51
165–179	19	51 + 19 = 70
180–194	13	70 + 13 = 83
195–209	4	83 + 4 = 87
210–224	1	87 + 1 = 88
225–239	1	88 + 1 = 89
240–254	1	89 + 1 = 90

Since the cumulative frequency at the end (90) equals the sample size, we have accounted for the entire sample. Now let us graph this information using a line graph. Remember, we have to pass *through* a class in order to be sure that we have accumulated all the data, before we can count the number of its members. Thus we put a dot at each upper class boundary opposite the corresponding cumulative frequency value and connect the dots, as in Fig. 3.5.

This type of graph is called an **ogive.** It always starts at 0 at the lower class boundary of the first class; it ends at the sample size, its highest point, at the upper class boundary of the last class. It will always climb from left to right—or level off for a class that has no members. It can never come back down the vertical scale because the cumulative frequencies were obtained by adding positive numbers.

> ***ogive.*** ■ A line graph where cumulative frequency or cumulative relative frequency is plotted on the vertical axis and class boundaries are plotted on the horizontal axis.

This graph, like a histogram, can be easily changed to show cumulative *relative* frequencies merely by dividing cumulative frequencies by the size of the sample. This converts the vertical

Figure 3.5
Weights of adult males

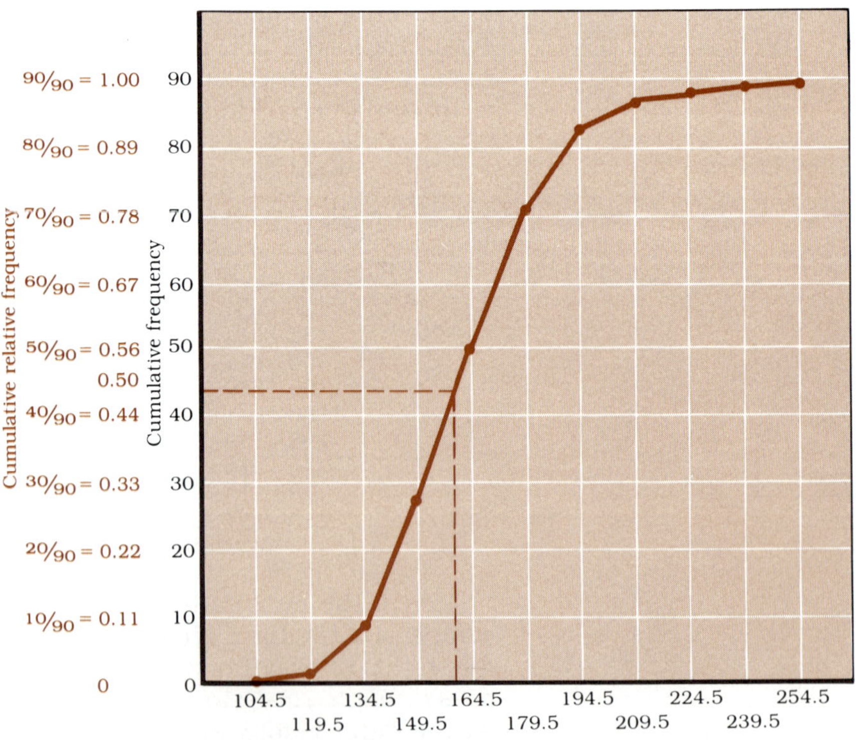

scale to decimals and, when this is done, the graph stretches from 0 to 90/90, or 1 (or 100%). Cumulative relative frequencies are shown in color on the vertical scale in Fig. 3.5. (Again, this is for comparative purposes; normally, they would be shown as the single vertical scale on the graph.)

A graph showing cumulative relative frequencies can be quite handy. For example, if we want to find the weight that falls exactly in the middle of the data—that is, half of the sample had lower weights and half had higher weights—we can use this graph. We find 0.50 on the vertical scale. Since the graph goes from 0 to 1 (100%) as it accounts for the classes, 0.50 is the halfway point (50%) through the data. Now we draw a horizontal line from the vertical scale to the diagonal line, drop straight down to the horizontal scale, and read off 160 pounds, approximately. Within the accuracy of the graph, we have found that half of the sample weighs less than 160 lbs and half weighs more than 160 lbs.

We will come back to the ogive again in Chapter 4 to expand upon this idea. It is a quick, useful tool for finding the position of individual values in a sample and determining the general nature of the sample.

Computer Aids

We do not examine long, involved data lists in this book because they require number crunching and time-consuming tallying; rather, we concentrate on understanding, analysis, and interpretation. Small amounts of data can be used to determine whether or not you understand rules, concepts, and procedures. Such understanding is essential before you attempt to use a computer as a technological tool to assist you.

Statistical computer software packages are available that can take data, construct frequency distributions, graph these distributions, and do many of the calculations called for in this book. It is important for you to understand how these packages can be used, so that you can take advantage of computer speed and capabilities for processing large amounts of data. If you know what the computer can do and how to use it, you will not have to do a lot of tedious work when confronted with large data lists.

EXERCISES/Section 3.3

1. Plot a histogram from the frequency distribution you prepared in Exercises/Section 3.1, no. 5.
2. Plot a histogram from the frequency distribution you made in Exercises/Section 3.1, no. 6.
3. Construct a relative frequency histogram from your work in Exercises/Section 3.1, no. 7.
4. Construct a relative frequency histogram from your work in Exercises/Section 3.1, no. 8.

5. Construct an ogive from your frequency distribution for Exercises/Section 3.1, no. 9.

6. Construct an ogive from your frequency distribution for Exercises/Section 3.1, no. 10.

7. Construct a histogram from your stem-and-leaf diagram for Exercises/Section 3.2, no. 1.

8. Construct a relative frequency histogram from your stem-and-leaf diagram for Exercises/Section 3.2, no. 2.

9. Construct an ogive from your stem-and-leaf diagram for Exercises/Section 3.2, no. 3.

10. Construct an ogive from your stem-and-leaf diagram for Exercises/Section 3.2, no. 4.

Study Notes

KEY TERMS

class ▪ A category or group of data values.

class boundaries ▪ Numbers halfway between consecutive upper and lower class limits. They involve rounding in the measurement process and serve to complete or fill the gaps in a set of classes.

class mark ▪ The midpoint of a class, found halfway between class limits or class boundaries.

$$\frac{\text{Lower limit [boundary]} + \text{Upper limit [boundary]}}{2}$$

class width ▪ The distance between consecutive lower class limits or the distance between class boundaries.

cumulative frequency ▪ The sum of the frequencies in any given class and all preceding classes.

cumulative relative frequency ▪ Cumulative frequency divided by the size of the sample.

frequency ▪ The number of times a particular value occurs in the data in a class under consideration.

frequency distribution ▪ A list of possible values of the variable under consideration and the frequency with which these values occur in the data. A frequency distribution may be grouped or ungrouped.

lower class limit ▪ The smallest possible data value in a class.

ogive ▪ A line graph where cumulative frequency or cumulative relative frequency is plotted on the vertical axis and class boundaries are plotted on the horizontal axis.

relative frequency ▪ Frequency divided by the size of the sample.

upper class limit ▪ The largest possible data value in a class.

OUTLINE

I. Frequency distributions
   A. Ungrouped
      1. To be used when there are only a few possible values.
      2. List values individually in order of size.
      3. Count the number of times that each occurs.
   B. Grouped
      1. There should be 5–12 classes.
      2. The classes should be of equal width.
      3. Every piece of data should fit into one and only one class.

II. Stem-and-leaf diagrams
   A. Classes or stems are the first digit(s) of data values.
   B. Leaves are the last digit of data values.
   C. The number of leaves is the frequency for a stem.

III. Graphs
   A. Should be clear and easy to read.
   B. A graph should be given a title and all parts should be labeled.

C. All intervals along the axes should be equally spaced.

D. Axes displaying quantities or frequencies should start with 0.

E. Histograms

1. Classes go on the horizontal axis.
2. Frequency or relative frequency goes on the vertical axis.

F. Ogives

1. Class boundaries go on the horizontal axis.
2. Cumulative frequency or cumulative relative frequency goes on the vertical axis.
3. Dots go at the upper class boundary of a class, opposite the cumulative frequency or cumulative relative frequency for the class.

REVIEW PROBLEMS

The following numbers, which apply only to Problems 1–4, are the quiz-score averages for 20 students in a math class.

89	67	78	71	46
81	39	58	88	83
87	93	93	99	34
32	89	84	65	63
68	67	88	91	75

1. Using classes 30–39, 40–49, etc., organize these numbers into a frequency distribution and include a cumulative frequency column.

2. For the first class, identify its
a) limits. b) mark. c) boundaries.

3. What is the class width?

4. Draw a histogram and an ogive for this frequency distribution.

5. Use the following table of IQ scores to produce a histogram and an ogive.

Class	f
70–79	2
80–89	10
90–99	15
100–109	25
110–119	17
120–129	9
130–139	2

6. Construct a stem-and-leaf diagram for runs batted in by 40 baseball players.

30	19	32	19	27
25	18	16	37	32
42	27	5	42	51
19	31	27	13	46
18	43	14	12	30
3	9	24	8	25
53	11	13	22	15
21	21	41	31	10

7. Sales records show the following number of cans of a certain brand of cat food sold each day for a 2-month period.

48	36	26	25	22	34
47	42	37	26	42	41
31	34	24	35	40	32
22	28	27	37	33	24
29	46	32	34	28	43
34	40	29	30	32	41
24	37	30	37	51	23
22	41	42	41	19	32
31	41	34	24	33	28
43	32	28	34	27	34

a) Set up a stem-and-leaf diagram.
b) Draw a histogram.
c) Draw an ogive.

8. Use the data from Problem 6 to construct a frequency distribution using a class width of 5.
a) Draw a histogram.
b) Draw an ogive.

9. The average discharge volumes, in thousands of cubic feet per second, at the mouths of major rivers are as follows:

Columbia	262	Ohio	258
St. Lawrence	243	Yukon	240
Atchafalaya	183	Tennessee	64
Red	62.3	Kuskokwim	62
Mobile	61.4	Snake	50
Arkansas	45.1	Copper	43
Tanana	41	Ouachita	40
Susquehanna	37.2	Willamette	35.7
Alabama	32.4	White	32.1
Wabash	30.4	Tombigbee	27.3
Pend Oreille	29.9	Cumberland	26.9
Apalachicola	24.7	Stikine	26
Illinois	22.8	Koyukuk	22
Porcupine	20	Hudson	19.5
Allegheny	19.3	Delaware	17.2

a) Construct a frequency table using a class width of 30.

b) Draw a histogram.

c) Draw an ogive.

10. The drainage areas, in thousands of square miles, of the 30 rivers in Problem 9 are:

258	203.9
302	327.6
95.1	40.9
93.2	49
43.8	109
160.6	24
44	20
27.6	11.2
22.6	28
33.2	20.1
25.8	18.1
19.6	20
27.9	32.4
45	13.4
11.7	11.4

a) Construct a frequency distribution using a class width of 50.

b) Draw a histogram.

c) Draw an ogive.

Descriptive measures—summarizing data

4

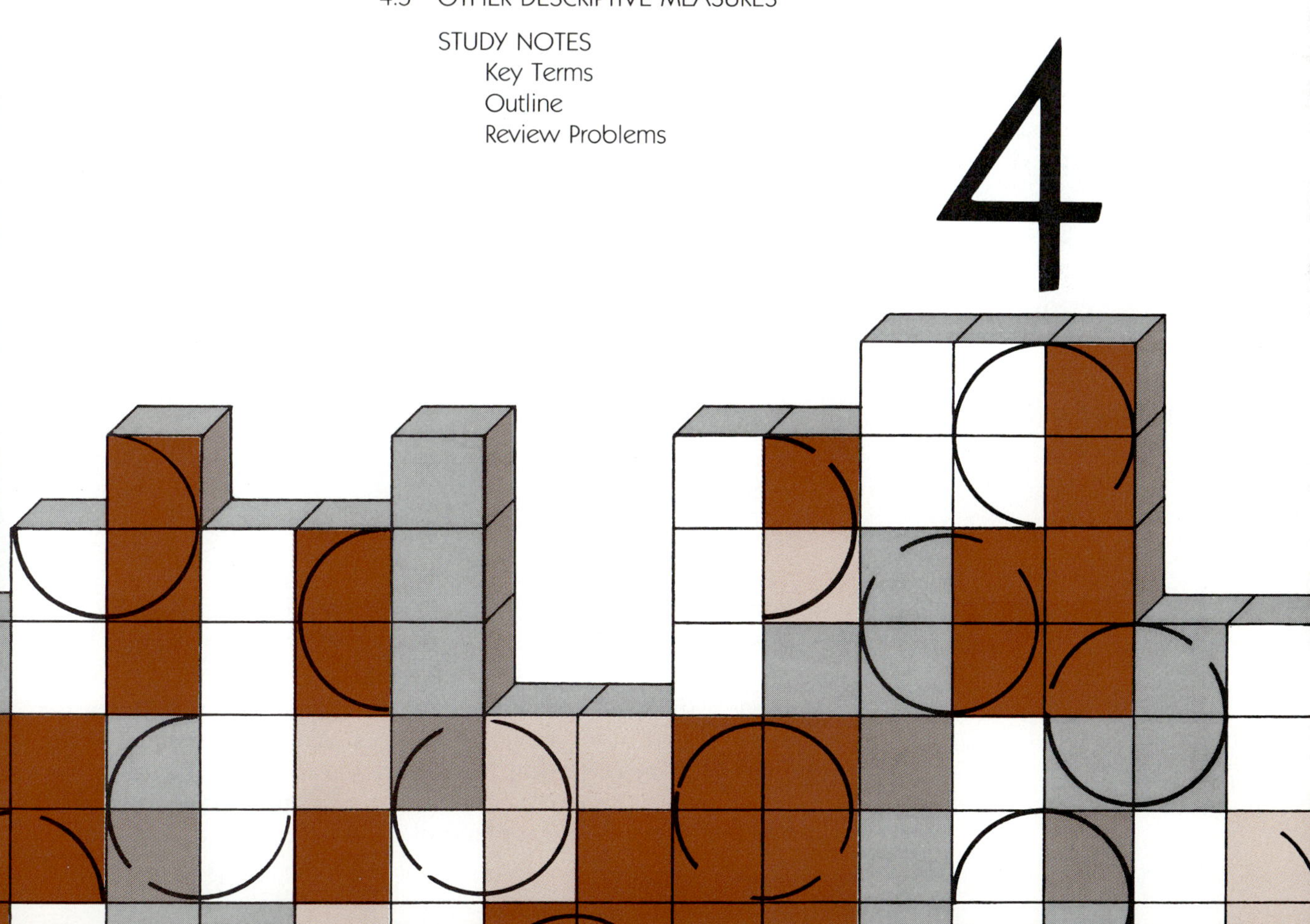

4.1 Types of Data

This chapter deals with arithmetic procedures that can be used to summarize and describe data. First, however, we need to discuss the various types of data because some do not lend themselves to the use of arithmetic. It is important to understand what can and cannot be done to information that you collect.

What is the difference between the responses we would get if we asked someone: What is your height? and What is your eye color? The first response will be a number, but the second response will not be numerical. Those variables or characteristics that do not have numerical values are called **qualitative.**

> ***qualitative data*** ▪ Information that is nonnumerical in nature.

Attributes, such as brown in the case of eye color, are qualitative rather than numerical. Hair color, sex, religious affiliation, political party, off or on, open or closed, success or failure, opinion—these all are examples of qualitative variables. None will involve a numerical response when asked about. Consequently, we cannot add, subtract, multiply, and divide this type of data. We *can* make graphs to display the information because we can count the number of responses that fit into each category.

Sometimes numbers are attached to such variables:

0 Male
1 Female

or

5 Strongly agree
4 Somewhat agree
3 Neutral, or no opinion
2 Somewhat disagree
1 Strongly disagree

But these numbers are artificial, and are usually put there for purposes of computer tally. To understand how artificial these numbers are, try reversing them and see whether it makes any difference. Does the truth of someone's being female change by calling this response a 0 instead of a 1? Determining whether or not we are dealing with qualitative data can best be summarized by the following question: If I asked someone a question about a particular characteristic, would I get a number as a response? If we would not get a number, we are dealing with qualitative data.

Data that is numerical in nature, that is, variables for which numbers are values, are called **quantitative.** Since they are numbers, we can use arithmetic with them.

> ***quantitative data*** ■ Information that is given as numbers. These numbers can be either measurements or counts.

A further breakdown of quantitative data is required. What is the difference between the responses we would get if we asked someone: What is your height? or How many children are there in your family? The first response would give us a measurement that could be any one of an infinitely wide range of possibilities. In theory we could get almost any height as an answer, say 5′8.33214″, although a number like this is usually rounded to 5′8″. Data that are measurements are called *continuous* data. The response to the second inquiry, about family size, would be a count, a whole number, that is one of only a relatively few possibilities. Data that involve counts are called *discrete* data. We have to be careful when performing arithmetic operations—particularly division—with discrete data. Division can result in a decimal answer, and sometimes this is awkward to interpret when only whole numbers have any meaning. For example, the proverbial average family has 2.3 children. More on this later.

Let us check our understanding of the different types of data. What type of data is involved in the statement, There are six people in the room with blue eyes? Remember the generalization: Would we get a numerical response if we asked an individual the color of his or her eyes? No, we would not get a number in response, so we have qualitative data. The *variable* is eye color. The number 6 tells us that there are six people in the room but nothing about the variable and, since it is not a response, it does not determine the type of data.

Before we proceed to analyze quantitative data, we need two more definitions. A **parameter** is a measurable, and thus a numerical, characteristic of the whole population. It is the true value of the characteristic under study. Thus the number of people who actually voted for the Democratic candidate in the last election is a parameter. The estimates projected by polling organizations before the election are not. They are guesses of the true value, the parameter, based on sample results.

parameter ■ A numerical characteristic of the population. The true, fixed value of some characteristic of interest, *as opposed to a*

statistic ■ A calculated sample result; a measurable characteristic of a sample. Its value may change from sample to sample.

We will be working with sample results in the rest of Chapter 4; then, starting in Chapter 8, we will learn how to use these results to make estimates or judgments about parameters. In order to know that we are talking about a calculated sample result, a statistic, we will denote sample results by italic letters, just as variables in algebra are designated by italic letters. To distinguish collected data values, say $X = 5'8''$ for height, from calculated summarizing sample statistics, we will attach an extra symbol such as ˆ or ¯ to the letter. Thus $\bar{X} = 5'8''$ means that this number was obtained by calculation, not by asking an individual. Parameters will be labeled with Greek letters to distinguish them from sample results. The standard Greek letters used will be introduced as we discuss various parameters.

Analyzing qualitative data requires the use of methods different from those used for analyzing quantitative data. These methods are usually nonparametric in nature; that is, they do not deal with single numerical descriptors of the population, with the exception of those variables that have only two values, such as success or failure and on or off. We will look at nonparametric methods for dealing with qualitative data in Chapters 13 and 14 and handle the case of a variable with only two nonnumerical values in Chapters 6 and 10.

EXERCISES/Section 4.1

Are the following types of data qualitative or quantitative and, if quantitative, are they continuous or discrete?

1. Grades on English compositions.
2. Grades on chemistry tests.
3. The number of speeding tickets written by Officer Jones in May.
4. Religion, coded

1	Catholic	4	Moslem
2	Protestant	5	Other
3	Jewish	6	None

5. Temperature.
6. Job title.
7. IQ.
8. Social Security number.
9. Blood pressure.
10. Cost of hamburger.
11. Unemployment rate.
12. Number of women bank executives.
13. Brand of cigarettes.
14. Number of years of schooling.
15. Percent voting in the last election by precinct.

16. Reaction time to alcohol.

17. Salary.

18. Daily water usage.

19. Number of cases of child abuse.

20. Opinion on national-brand versus house-brand grocery item.

4.2 Measures of Central Tendency

When confronted with large amounts of data, most people find it difficult to absorb much of the information quickly. We need some descriptive measures that will summarize information and tell us something about it. We begin with measures of central tendency, that is, numbers that describe the center or centering tendencies of data, which typically are called *averages.* The word average is used in several ways and thus is imprecise, so we will introduce and carefully define more technical alternatives to it.

Mean

For discussion purposes, let us consider a variable X whose values are 2, 3, 5, 3, and 7. We write this as follows:

$$X\text{: } 2, 3, 5, 3, 7.$$

The first measure of central tendency is the **mean.** It is what is usually meant by "average." The mean is found by adding all of the data values and dividing by the number of values. As such, it is the arithmetic average.

$$\frac{2 + 3 + 5 + 3 + 7}{5} = \frac{20}{5} = 4.$$

We just calculated a sample statistic, a number characterizing the sample, so we label it with an italic letter. The letter chosen is that for the variable, X; to indicate that we have calculated the mean, we put a bar over it.

$$\bar{X} = 4.$$

> ***mean*** ▪ The arithmetic average found by adding the data values and dividing by the number of data values.

There are many situations in which the addition of numbers is required. Thus it will be helpful to use a shorthand method called *summation notation* to indicate the operation of addition. This notation is represented by the capital Greek letter sigma, Σ.

$$\bar{X} = \frac{\Sigma X}{n},$$

where n is the sample size.

For those who are not familiar with the summation notation, Appendix B presents procedures using Σ.

If we could add all the numbers in a population, we could calculate the population mean, a parameter. It is designated by the Greek letter μ, read "mu."

$$\mu = \frac{\Sigma X}{N},$$

where N is the size of the population.

At this point let us make a comment about rounding. When division is performed, answers are often rounded. However, the results of calculations will frequently be needed for further work in drawing conclusions about the data, so it is advisable to keep two decimal places more in an answer than existed in the original data.

What happens to the mean if one of the numbers is extreme? Let us revise our data list to form a new one:

$$Y\text{: } 2, 3, 5, 3, 7, 1000.$$

Now the mean is

$$\bar{Y} = \frac{2 + 3 + 5 + 3 + 7 + 1000}{6} = \frac{1020}{6} = 170.00.$$

Figure 4.1

The mean has been drastically affected by this extreme; it has been pulled toward it. This illustrates one of the properties of the mean: It is the balance point of the numbers. If the X data were arranged as in Fig. 4.1, the mean is where the line would balance ($20/5 = 4$). Thus if an extreme shows up in the list, the balance point is pulled in that direction.

Median

When we have a skewed distribution such as in the Y list, we should consider a second measure of central tendency: the **median.** The median is the number that divides a list of data into two equal parts; half the data will have values less than the median and half will have values greater.

> ***median*** ■ The middle number; the number that divides a list or distribution of data into two equal parts.

Thus before we can find a median, we have to *put the numbers in order:*

$$X\text{: } 2, 3, 3, 5, 7.$$

By examining this list, we see that 3 is the middle number and divides the list into two equal parts.

$$2, 3, \underset{\text{Median}}{\overset{\downarrow}{3}}, 5, 7$$

What happens if there is not just one middle number but two, as in the case of a list with an even number of entries?

$$Y\text{: } 2, 3, \overbrace{3, 5}, 7, 1000$$

The median is the mean of the two middle numbers, or halfway between them:

$$\frac{3 + 5}{2} = 4.00.$$

This also illustrates that extremes do not radically affect the median. When we added 1000 to the list, the median only went from 3 to 4. Finding the median merely requires counting numbers on each side of the middle number. It does not take into account how far those numbers are from the middle in value.

News & Views 4.1 provides an example of the use of the median. The article is correctly explicit by not referring just to average income or average house price.

Stem-and-leaf diagrams are quite handy for use in calculating the median. They put a large list of data roughly in order; when we determine which stem contains the median, it is simply a matter of ordering the leaves of the stem to pick out the median. In Example 3.4, we obtained the following stem-and-leaf diagram for 20 pieces of data.

		f
12	9 9	2
13	1 8 4 9 9 8	6
14	4 6 4 1 8 5	6
15	4 0 1 0	4
16	1	1
17	0	1
		20

NEWS & VIEWS 4.1

Median house price gains

House prices soared to a median $72,000 last year and "cast a pall over the American dream of home ownership," savings and loan officials said yesterday. The $72,000 price tag meant a typical down payment of $16,100 and monthly housing payments of $816, including taxes and insurance, the U.S. League of Savings Associations said in a new study. First-time buyers typically paid $58,900 for a house and agreed to monthly housing expenses of $721 after making an $8,600 down payment. . . .

The median should occur halfway between the tenth and eleventh numbers. If we find cumulative frequencies, we can determine which stem contains the median.

Stem	Leaves	f	Σf
12	9 9	2	2
13	1 8 4 9 9 8	6	8
14	4 6 4 1 8 5	6	14
15	4 0 1 0	4	18
16	1	1	19
17	0	1	20

The first stem gives us two pieces of data. The second stem gives us six more for a total of eight. Adding the third stem gets us to 14, so it must contain the tenth and eleventh pieces. Ordering the leaves of the 14 stem,

1	4	4	5	6	8
ninth	tenth	eleventh			

we find that the tenth and eleventh data values are each 4, so the median is 144. If the two leaves had been different in magnitude, we would have calculated their mean to give us the answer.

Mode

The third measure of central tendency is the **mode.** It is the number that occurs most frequently. The mode for the X and the Y values we have been using is 3.

> ***mode*** ■ The most frequently occurring value.

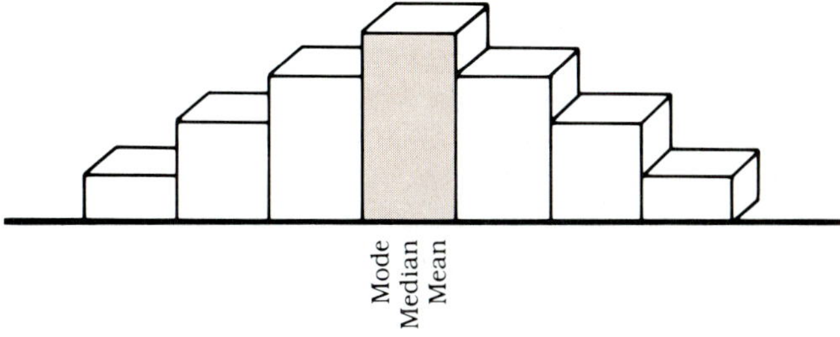

Figure 4.2

If all numbers occur with equal frequency, there is no mode. If two numbers predominate, the list is *bimodal.* (We have already seen a bimodal histogram in Chapter 3.) Of the measures of central tendency, only the mode will always take the same form as the data, because it is one of the actual data numbers; the mean will and the median may require division, and extra decimal places may result from these calculations. If we want a number that is easily explainable in terms of the characteristics of the data, the mode should be used. The 2.3 children in the average family is a mean, and the 0.3 resulted from division. Perhaps a better, more easily interpreted figure might be one that says the average family has 2 or 3 children, using modal numbers.

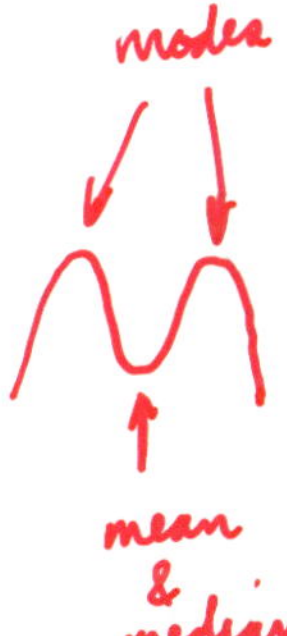

When the distribution is symmetric, as in Fig. 4.2, the mean, median, and mode all occur at the same place. The highest frequency (mode) is also the balance point (mean) and divides the diagram in half (median). In a skewed distribution (Fig. 4.3) the mean is pulled in the direction of the tail. The median also shifts in that direction but not so far. The mode remains at the high point.

Let us consider the following situation in a labor negotiation. The local municipal orchestra is out on strike, and labor and management are both trying to get the public on their side. Man-

Figure 4.3

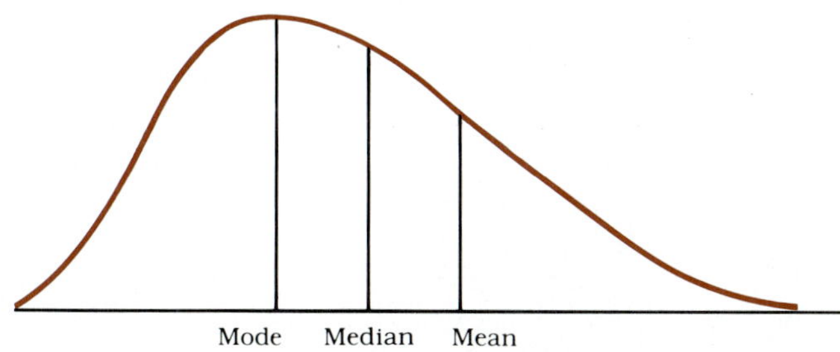

agement claims that the average salary is $27,300; labor claims that it is $17,000. Is one side deliberately lying? No. Each is using the meaning for average that suits its own purposes. Management has chosen the mean, which includes the large salaries earned by the principal players. Labor has chosen the mode, which is the beginning salary of all the young players. Be careful when you hear the word *average* thrown about.

The article in News & Views 4.2 provides an example of the use of large and impressive numbers that are applicable to an entire population in order to make a point about increasing costs of health care. The final paragraph includes the ambiguous usage of "averages". Are these numbers means? Were they calculated by taking into account hospital size based on the number of beds, by merely adding across hospitals, or what?

NEWS & VIEWS 4.2

Health costs mounting

In 1980, we Americans spent the staggering total of $217.9 billion—almost 10% of the gross national product—on personal health care.

We paid physicians $46.6 billion, a rise of 14.5% from 1979. Hospital costs soared to $99.6 billion, up 16.2%. And the cost of drugs rose to $19.2 billion.

The 1981 statistics are not yet available, but we probably spent even more on health care. The best medicine in the world undoubtedly is practiced in the U.S.—but we surely pay for it.

Fortunately, an estimated 88.3% of U.S. employees work for companies that offer private health insurance. There are 26.6 million other Americans, however, who have no such coverage. For them, a prolonged hospital stay or serious illness can spell endless debt and inevitable bankruptcy.

Lucky for White House press secretary James S. Brady, 41, that he's covered by federal employees' workmen's compensation. According to a spokeswoman at George Washington University Medical Center, "The hospital charges for Brady's stay from March 30 to Nov. 23, 1981, range from $150,000 to $200,000—and that's without the doctor's charges for surgery."

Brady was shot in the brain during the assassination attempt on President Reagan last March. He underwent four serious operations, "the cost of which," the spokeswoman says, "I do not yet have." She adds, "Nor do I have any estimate on what his outpatient physical therapy—four hours a day, five days a week—will cost." One physician says Brady's total will be at least $250,000.

Doctors predict that in time Brady will be able to walk with a cane. The fact that he is alive is regarded as a near miracle and ample evidence of the excellence of the medicine, care and surgery available in this country. Their cost, however, is a separate but most relevant matter.

Incidentally, the Equitable Life Assurance Society recently reported that the most expensive city in the nation in which to be hospitalized is Washington, D.C. A semi-private room averages $219 per day, a private room $246, and an intensive care room $617.

Source: Lloyd Shearer in *Parade*, January 24, 1982.

Using a Grouped Frequency Distribution

We complete the discussion of measures of central tendency by examining the use of a frequency table to derive the mean, median, and mode. Let us use the frequency table developed in Section 3.1, in which X is the class mark.

Class	f	X
105–119	2	112
120–134	7	127
135–149	19	142
150–164	23	157
165–179	19	172
180–194	13	187
195–209	4	202
210–224	1	217
225–239	1	232
240–254	1	247
	90	

The mode is the easiest to obtain, so we look for it first. We start by finding the modal class.

modal class ■ The class with the largest frequency.

In this case it is the 150–164 class. We have to pick one value to represent that class. Remember that the class mark is the midpoint of the class and as such is a rough average of the numbers in the class. The most representative number then is the class mark or, in this case, 157.

mode (from a grouped frequency distribution) ■ The class mark of the modal class.

The mean is also found by using class marks. Normally, we would add 90 data values and then divide by 90. But, in this case, we do not have 90 individual numbers with which to work; therefore we want each class to contribute to the sum in proportion to the number of data values contained in the class, that is, its frequency. The first class contains two numbers, but we do not know what they are; we represent them by the class mark—making the assumption that the data values are evenly distributed throughout the class. Then we multiply the frequency by the class mark, using two 112's, or 2 · 112. The next class contributes seven 127's to the sum, or 7 · 127, and so on.

Class	f	X	$f \cdot X$
105–119	2	112	2 · 112 = 224
120–134	7	127	7 · 127 = 889
135–149	19	142	19 · 142 = 2698
150–164	23	157	23 · 157 = 3611
165–179	19	172	19 · 172 = 3268
180–194	13	187	13 · 187 = 2431
195–209	4	202	4 · 202 = 808
210–224	1	217	1 · 217 = 217
225–239	1	232	1 · 232 = 232
240–254	1	247	1 · 247 = 247

We add the results in the last column and divide by n, or 90.

$$\bar{X} = \frac{\Sigma fX}{n} = \frac{14{,}625}{90} = 162.50.$$

$$\boxed{\bar{X} = \frac{\Sigma fX}{n}.}$$

The median is found by a process called interpolation. As with the stem-and-leaf diagram, we begin by finding the class that contains the median. We need a cumulative frequency column, just as we did to plot an ogive.

Class	f	Σf
105–119	2	2
120–134	7	9
135–149	19	28
150–164	23	51
165–179	19	70
180–194	13	83
195–209	4	87
210–224	1	88
225–239	1	89
240–254	1	90

Since there are 90 pieces of data, the median is the forty-fifth. Through the third class, we have 28 pieces of data. Going through the fourth class gets us past the forty-fifth to 51. Thus the median is in the 150–164 class. At the lower class boundary, 149.5, we have 28 data values. How many more do we need?

$$45 - 28 = 17$$

We need 17 of the 23 in the next class (the 150–164 class). That class has a width of 15. If we need 17 of the 23 data values, we also need 17/23 of the width or

$$\frac{17}{23} \cdot 15 = 11.$$

Thus we go 11 units into this class to find the median: 149.5 + 11 = 160.5

In general, we find the position of the cumulative frequency median (45 in this case) and subtract the cumulative frequency of the class preceding the median class (28 in this case). Let us call this result F. It represents the number of data values in the median class that we need in order to get to the median. If f is the frequency in the median class and w is the class width, then

$$\text{Median} = \text{Lower class boundary} + \frac{F}{f} \cdot w.$$

EXAMPLE 4.1 Find the mode, mean, and median using the following frequency distribution; $n = 70$.

Class	f	Σf	X
0– 9	6	6	4.5
10–19	8	14	14.5
20–29	24	38	24.5
30–39	15	53	34.5
40–49	17	70	44.5

Solution The mode is the class mark for the 20–29 class: 24.5.

To find the mean requires that we add an fX column.

fX
$6 \cdot 4.5 = 27$
$8 \cdot 14.5 = 116$
$24 \cdot 24.5 = 588$
$15 \cdot 34.5 = 517.5$
$17 \cdot 44.5 = 756.5$
2005.0

$$\bar{X} = \frac{2005}{70} = 28.64.$$

TABLE 4.1

Year	Median Age	Year	Median Age
1820	16.7	1900	22.9
1830	17.2	1910	24.1
1840	17.8	1920	25.3
1850	18.9	1930	26.5
1860	19.4	1940	29.0
1870	20.2	1950	30.2
1880	20.9	1960	29.5
1890	22.0	1970	28.1

The median class contains the thirty-fifth number (70/2). The cumulative frequency column tells us that it is in the 20–29 class. We have 14 data values at the class boundary of 19.5; we need $35 - 14 = 21$ more from the 24 in the median class.

$$\text{Median} = 19.5 + \frac{21}{24}(10) = 28.75. \quad \square$$

Table 4.1 shows the changes in median age for the population of the United States from 1820 to 1970. The changes have occurred in general because of increased longevity. The aberrations in 1950 may be greater than the median for later years because of the deaths of young men in World War II.

Conclusion

We have already mentioned some of the characteristics and proper use of each of the measures of central tendency. We conclude this section with some further comments about them.

The mode is useful when qualitative data are being considered. The most popular color for automobiles has meaning, but a mean or median color does not. The mode, however, may not give us any information about trends in the data. It is not necessarily unique or even in the middle of a set of data; in fact it may exhibit no central tendency at all. It does not take into account data-value size, reflecting merely how many data values there are.

The median does consider all the data values but only their ranking—that is, where they fall—not how big or small each is. The median is unique and is in the center of the data.

The mode and the median are subject to greater fluctuation from one sample to the next from the same population. The mean uses all the data values, is the most stable (it varies less from one sample to another), and thus is the measure used most often. It most consistently best approximates the true population mean for normal populations. Again, we see in News & Views 4.3 results expressed in means because of the decimal values.

NEWS & VIEWS 4.3

Tooth decay found to drop in children

The prevalence of tooth decay in American schoolchildren has declined 32 percent in the last 10 years, a major health advance brought about by better dental care, wider use of fluoride and perhaps changing dietary habits.

Government researchers from the National Institute of Dental Research yesterday said that results of a national survey of 38,000 schoolchildren show a steady drop in the number of decayed, missing and filled teeth detected in examinations.

Dr. Janet A. Brunelle told a session at the annual meeting of the American Association for the Advancement of Science that the prevalence of tooth decay in children aged 5 to 17 dropped from 7.1 to 4.8 problems per child in the decade ending in 1980.

Tooth decay costs the nation billions of dollars a year. "This reduction represents an enormous savings in costs of needed dental treatment," said Dr. Brunelle.

The study found that not only were children getting fewer cavities and other forms of decay, but increasing numbers were totally free of tooth decay.

Source: *Baltimore Sun,* January 7, 1982.

EXERCISES/Section 4.2

1. Find the mean, median, and mode of the following data.

X: 1, 7, 9, 4, 3, 7, 6, 2, 8, 1, 9, 7

2. Find the mean, median, and mode of the following data.

Y: 51, 63, 21, 51, 72, 81, 66, 43, 31

3. Wages at the ABC Corporation are shown in Table 4.2. Find the **(a)** mean, **(b)** median, and **(c)** modal wages.

4. The price of lettuce on 5 consecutive Mondays was $0.99, $0.99, $1.19, $1.09, and $0.79.

a) Find the mean price charged for lettuce during this period.

b) What is the median?

c) What is the mode?

5. The ages at death of a sample of 10 people were as follows.

63, 72, 81, 82, 55, 49, 63, 71, 77, 79

Find the **(a)** mean, **(b)** median, and **(c)** modal ages.

TABLE 4.2

President	$250,000	Production-line worker	20,000
Vice-president	100,000	Production-line worker	20,000
Vice-president	100,000	Production-line worker	20,000
Comptroller	50,000	Production-line worker	20,000
Accountant	50,000	Production-line worker	20,000
Production-line worker	20,000	Production-line worker	20,000
Production-line worker	20,000	Secretary	19,000
Production-line worker	20,000	Secretary	19,000
Production-line worker	20,000	Janitor	10,000

6. The rainfall per month, in inches, for one year is as follows.

4.21	8.62	1.42	3.82
5.03	6.14	1.78	4.79
8.91	5.21	3.71	5.11

Find the (a) mean, (b) median, and (c) modal inches of rainfall.

7. Ten babies in a sample first walked at the following ages (in months).

9, 11, 12, 12, 14, 16, 18, 10, 9, 9

Find the (a) mean, (b) median, and (c) modal ages.

8. Using the following frequency table

Class	*f*
80– 99	6
100–119	12
120–139	19
140–159	24
160–179	15
180–199	4

find the (a) mean, (b) median, and (c) mode.

9. Using this frequency table

Class	*f*
850– 899	36
900– 949	42
950– 999	17
1000–1049	20
1050–1099	5

find the (a) mean, (b) median, and (c) mode.

10. The following frequency table shows information about the cost of homes, in thousands of dollars, and the number sold in the various classes.

Class	*f*	Class	*f*
30– 44	3	105–119	10
45– 59	8	120–134	9
60– 74	10	135–149	6
75– 89	18	150–164	4
90–104	11	165–179	1

Find the (a) mean, (b) median, and (c) modal costs of homes sold.

11. The number of gallons of gasoline used per month is given in the following frequency table; *f* represents the number of drivers in each class.

Class	*f*
20– 39	2
40– 59	6
60– 79	15
80– 99	21
100–119	14
120–139	2

Find the (a) mean, (b) median, and (c) modal number of gallons used.

12. The following frequency table is based on batting averages and the number of baseball players in each class.

Class	*f*
0–0.059	1
0.060–0.119	4
0.120–0.179	10
0.180–0.239	15
0.240–0.299	12
0.300–0.359	8

Find the (a) mean, (b) median, and (c) modal batting averages.

13. The number of hours spent doing homework per week by the 30 students in Professor Smith's class were:

0, 7, 8, 19, 23, 6, 41, 32, 17, 18, 24,
27, 32, 40, 0, 17, 21, 24, 35, 6, 9,
11, 41, 7, 16, 17, 14, 14, 22, 28

Make a stem-and-leaf diagram and find the median number of hours from it.

14. Determine the median from your stem-and-leaf diagram in Exercises/Section 3.2, no. 2.

15. Determine the median from your stem-and-leaf diagram in Exercises/Section 3.2, no. 3.

4.3 Measures of Variability or Dispersion

Knowing the mean helps us to summarize data, but this is often insufficient for a satisfactory characterization of a data set. If you could not swim but someone told you that, on the average, the river you want to cross is only 2-feet deep, would you calmly wade across? We hope not. A mean of 2 feet tells you nothing about the 10-foot hole in front of you. Means have a leveling effect. They tell us nothing about variations or extremes that may exist in the data. Thus we need to discuss *measures of variability or dispersion*, which describe how numbers vary or are spread out.

Range

The simplest measure of variability is the **range.**

> ***range*** ■ The difference between the largest data value and the smallest data value (largest − smallest).

Let us consider again the X set of data from Section 4.2:

$$X\text{: } 2, 3, 5, 3, 7.$$

The range is $7 - 2 = 5$.

For the Y set of data

$$Y\text{: } 2, 3, 5, 3, 7, 1000,$$

the range is $1000 - 2 = 998$.

In general, *the larger the calculated value is for the measure of variability, the more spread out the data is.* This should be obvious from the two data sets used but later, when we make more complicated calculations, it would be a good idea for us to remember what these measures indicate. *The larger the measure of variability, the more spread out the data is.*

The range has some of the same weaknesses as the mode and the median. It varies greatly from one sample to another of the same population. It uses only two of the data values and does not tell us anything about clustering. The range of the Y data is very large compared to that of the X data because of one extreme value, but both lists contain clusters of small values.

Variance

We need another measure of variability that depends on all the data values, uses the mean as a reference point for those values, and measures the distance of the values from the mean. To determine how far apart numbers are, we subtract; if we subtract the mean from each number in the X list, we get

$$\begin{array}{ccccc} 2-4 & 3-4 & 5-4 & 3-4 & 7-4 \\ -2 & -1 & 1 & -1 & 3 \end{array}$$

These results are called *deviations* from the mean. If we now want to find the average or mean deviation, we add the deviations and divide by the number of deviations:

$$\frac{-2+(-1)+1+(-1)+3}{5} = \frac{0}{5} = 0.$$

It seems that we have gotten nowhere. A calculation that measures the spread of a set of data values and comes out 0 is not too helpful. In general, if we were to calculate the deviations from the mean and add them, we would always get 0. Why? The negative numbers and the positive numbers sum to zero because the mean is the balance point of the data. The sum of the distances to numbers larger than the mean must exactly balance the sum of the distances to numbers smaller than the mean to maintain the balance.

We do not really care about which side of the mean the numbers fall on, but how far they are from the mean. Thus the minus signs are getting in the way. We can get rid of them by using absolute values, which discard the minus signs and make all numbers positive or 0. However, a more frequently used method is to square (multiply by itself) each of the deviations before adding them:

$$\begin{array}{ccccccccc} (-2)^2 & + & (-1)^2 & + & 1^2 & + & (-1)^2 & + & 3^2 \\ \downarrow & & \downarrow & & \downarrow & & \downarrow & & \downarrow \\ 4 & + & 1 & + & 1 & + & 1 & + & 9 \end{array}$$

So far we have subtracted $\bar{X}$ from each X, squared the results, and added the squares.

$$X - \bar{X}, \quad (X - \bar{X})^2, \quad \text{and} \quad \Sigma(X - \bar{X})^2.$$

Next, we divide this result by $n - 1$:

$$\frac{\Sigma(X - \bar{X})^2}{n - 1}.$$

This expression is called the **variance** and is labeled s^2. Thus

$$\text{Variance} = s^2 = \frac{\Sigma(X - \bar{X})^2}{n - 1}$$

The s is squared to remind us that we have squared the deviations. Using $n - 1$ (rather than n) yields an s^2 from a sample that better approximates the population variance.

variance ■ A measure of variability, using the mean as a reference point and found by the equation

$$s^2 = \frac{\Sigma(X - \bar{X})^2}{n - 1}. \qquad \text{(1a)}$$

Still using the numbers from our X list, we can now calculate the variance.

$$s^2 = \frac{\Sigma(X - \bar{X})^2}{n - 1}$$
$$= \frac{(2-4)^2 + (3-4)^2 + (5-4)^2 + (3-4)^2 + (7-4)^2}{5-4}$$
$$= \frac{(-2)^2 + (-1)^2 + 1^2 + (-1)^2 + 3^2}{4}$$
$$= \frac{4+1+1+1+9}{4} = \frac{16}{4} = 4.$$

If we could collect data from an entire population, we could calculate the population variance, which is a parameter. It is labeled with the Greek letter σ^2, read "sigma squared."

$$\sigma^2 = \frac{\Sigma(X - \mu)^2}{N}$$

Note that we use the population mean, μ, to calculate the deviations and then divide by the population size, N. For an entire population, we do not subtract 1 from N in the denominator.

Standard Deviation

A related measure of variability is needed because the variance presents some problems in interpretation. By squaring the deviations, we changed the units of the data. If we have 5 pounds and subtract 4 pounds, we get 1 pound; when we square this we get 1 pound squared, which has no meaning. Similarly, \$5 − \$4 = \$1, but what is 1 dollar squared? In order to eliminate squaring of the units and to aid in interpreting the results, we take the square root, $\sqrt{\ }$, of the variance. The result is called the **standard deviation.**

$$s = \sqrt{\frac{\Sigma(X - \bar{X})^2}{n - 1}} \qquad \text{(1b)}$$

Thus the standard deviation from our X list is $s = \sqrt{4} = 2$. If we are dealing with the population variance σ^2, the square root produces the population standard deviation.

$$\sigma = \sqrt{\frac{\Sigma(X - \mu)^2}{N}}$$

standard deviation ■ The square root of the variance.

There is an alternative method, which is algebraically equivalent, for determining the variance and the standard deviation.

$$s^2 = \frac{n\Sigma X^2 - (\Sigma X)^2}{n(n - 1)} \quad \textbf{(2a)}$$

and thus

$$s = \sqrt{\frac{n\Sigma X^2 - (\Sigma X)^2}{n(n - 1)}}. \quad \textbf{(2b)}$$

Equations (2a) and (2b) are usually more accurate and easier to use. They are especially convenient for use with calculators. Equations (1a) and (1b) require that you calculate the mean and subtract it from every data value, square each result, and then add. This can be very time-consuming for a large amount of data. Perhaps a more important problem arises if $\bar{X}$ is not exact but needs to be rounded before further calculations are made. This rounding introduces error that is compounded as calculations proceed. Equations (2a) and (2b) require no rounding until the end and each requires only one subtraction in order to complete the numerator.

To review, let us recalculate the variance and standard deviation for the X data.

X	X^2
2	4
3	9
5	25
3	9
7	49
20	96

$$s^2 = \frac{n\Sigma X^2 - (\Sigma X)^2}{n(n - 1)} = \frac{5 \cdot 96 - (20)^2}{5 \cdot 4}$$

$$= \frac{480 - 400}{20} = \frac{80}{20} = 4,$$

and

$$s = \sqrt{4} = 2.$$

Using a Grouped Frequency Distribution

Before leaving standard deviation and variance, we need to discuss how to calculate them using a frequency table. The equation for variance using a frequency table is:

$$s^2 = \frac{n\Sigma X^2 f - (\Sigma X f)^2}{n(n-1)} \quad \text{(3a)}$$

We need an Xf column just as we did to find the mean. We also need a column that comes from multiplying $X \cdot X \cdot f$; we merely multiply each number of the Xf column by X and get X^2f.

The standard deviation is the square root of the variance:

$$s = \sqrt{\frac{n\Sigma X^2 f - (\Sigma X f)^2}{n(n-1)}}. \quad \text{(3b)}$$

EXAMPLE 4.2 Find the variance and standard deviation using the frequency table from Section 3.1.

Solution We begin by adding Xf and X^2f columns to the table:

Class	f	X	Xf	X^2f	
105–119	2	112	224	112 · 224 =	25,088
120–134	7	127	889	127 · 889 =	112,903
135–149	19	142	2,698	142 · 2698 =	383,116
150–164	23	157	3,611	157 · 3611 =	566,927
165–179	19	172	3,268	172 · 3268 =	562,096
180–194	13	187	2,431	187 · 2431 =	454,597
195–209	4	202	808	202 · 808 =	163,216
210–224	1	217	217	217 · 217 =	47,089
225–239	1	232	232	232 · 232 =	53,824
240–254	1	247	247	247 · 247 =	61,009
	90		14,625		2,429,865

$$s^2 = \frac{90 \cdot 2{,}429{,}865 - (14{,}625)^2}{90 \cdot 89}$$

$$= \frac{4{,}797{,}225}{8010} = 598.90.$$

$$s = \sqrt{598.90} = 24.47. \quad \square$$

EXERCISES/Section 4.3

For Exercises 1 through 12, refer to Exercises/Section 4.2, nos. 1–12, respectively.

1. a) Find the range.
   b) Find the variance and standard deviation using Eqs. (1a) and (1b).
2. a) Find the range.
   b) Find the variance and the standard deviation using Eqs. (1a) and (1b).
3. a) Find the range.
   b) Find the variance and the standard deviation using Eqs. (2a) and (2b).
4. a) Find the range.
   b) Find the variance and the standard deviation using Eqs. (2a) and (2b).
5. Find the standard deviation using **(a)** Eq. (1b) and **(b)** Eq. (2b).
6. Find the standard deviation using **(a)** Eq. (1b) and **(b)** Eq. (2b).
7. Find the standard deviation using **(a)** Eq. (1b) and **(b)** Eq. (2b).
8. Find the variance and the standard deviation using Eqs. (3a) and (3b).
9. Find the variance and the standard deviation using Eqs. (3a) and (3b).
10. Find the variance and the standard deviation using Eqs. (3a) and (3b).
11. Find the variance and the standard deviation using Eqs. (3a) and (3b).
12. Find the variance and the standard deviation using Eqs. (3a) and (3b).
13. Compare the standard deviations for the X values and Y values in the following data sets.
    a) X: 1, 2, 3, 4, 5
       Y: 101, 102, 103, 104, 105
    b) X: 1, 3, 5, 7, 9
       Y: 999, 1002, 1005, 1008, 1011
    c) X: 4, 5, 6, 7, 8
       Y: 0.01, 0.10, 0.50, 0.90, 1.00

4.4 Interpreting Descriptive Measures and Distributions; Chebyshev's Theorem

So far, you may not be able to see the connections between the things we have been learning; much of the initial work in elementary statistics is that way. We still need to learn more definitions and skills for use in drawing conclusions and making estimates about a population. However, we can begin to put together some of what we learned about distributions and graphing with the descriptive measures we have just studied.

Histograms of sample results by their very nature are jagged, but if we made smaller and smaller classes and took larger and larger samples we would see a smoothing effect that, ultimately, can be represented by a curve, as in Figs. 4.4 and 4.5.

We have already said that the mean, the median, and the mode occur at the same place in a symmetric distribution; that in a skewed distribution, the mode is still at the highest point, but

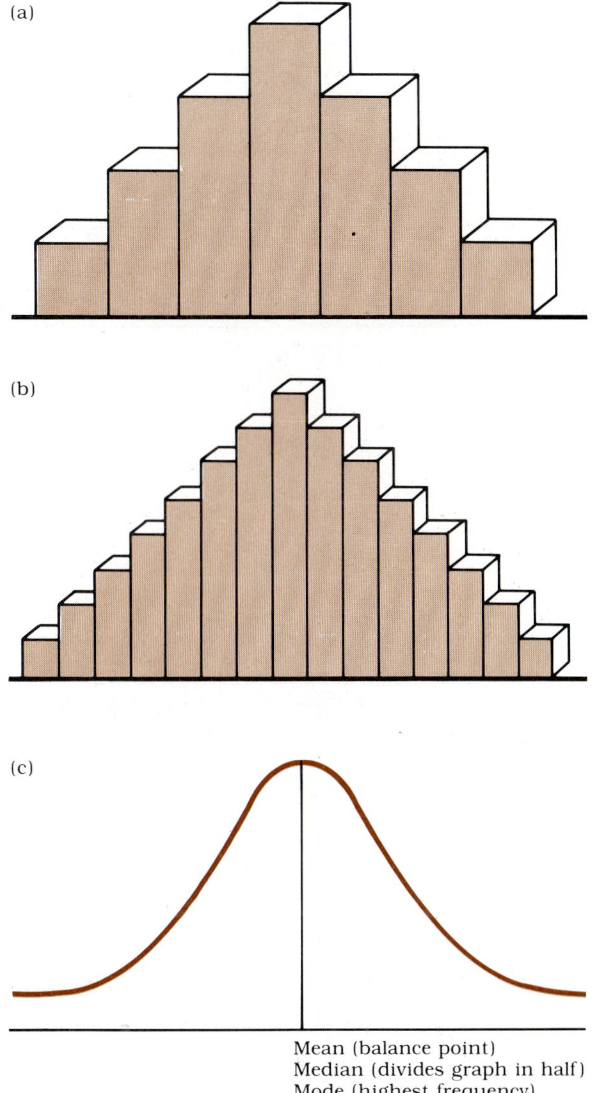

Figure 4.4

the mean is pulled toward the tail to maintain balance. The median is pulled as well but not so much as the mean. These relationships are shown in Fig. 4.4(c) and Fig. 4.5(c).

The standard deviation can be related to these diagrams in several ways. A special distribution that we will use a lot is called

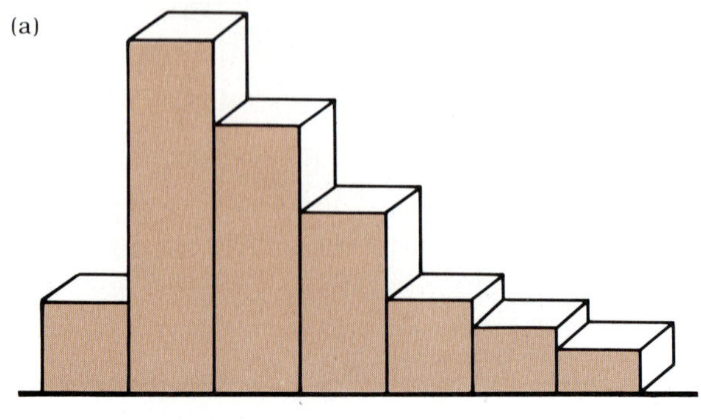

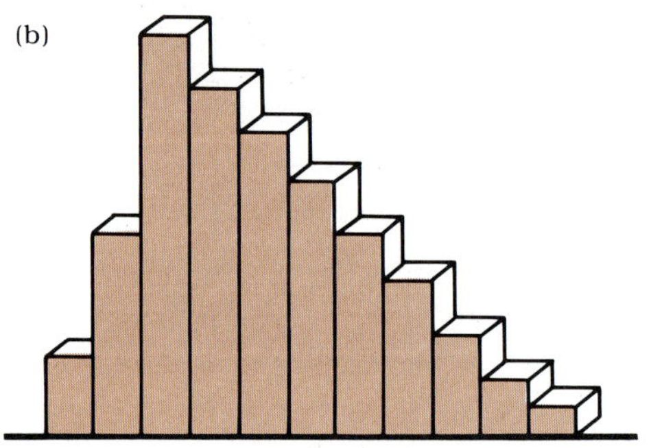

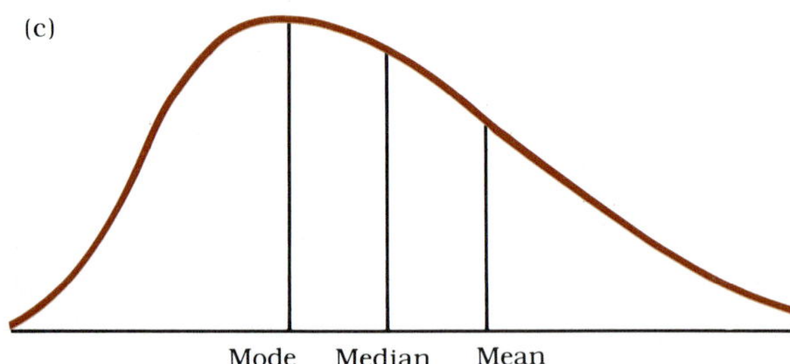

Figure 4.5

the **normal distribution.** It is symmetric and can be almost entirely described by counting three standard deviations on both sides of the mean, as shown in Fig. 4.6.

If we know the mean and the standard deviation, we can determine the limits between which almost all values lie in a normal distribution. For example, if the mean height of adult males is 5′9″ and the standard deviation is 3″, we can determine that

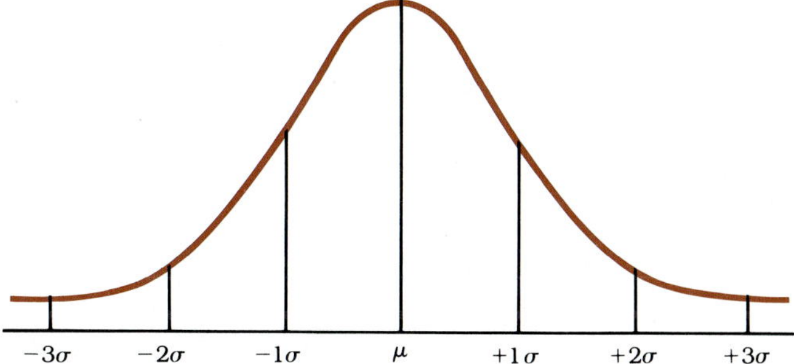

Figure 4.6

nearly all adult males are between 5′ and 6′6″ in height (Fig. 4.7). This type of distribution receives further, detailed attention, starting in Chapter 6.

If we do not know whether we have a normal distribution, we can still make some judgments about a population, using the mean and the standard deviation. To do this we use Chebyshev's theorem.

> ***Chebyshev's theorem*** ■ For any specific numerical value k that exceeds 1, the proportion of a population that lies within k standard deviations of the mean is at least
>
> $$1 - \frac{1}{k^2}.$$

Again, we start at the mean and count the number of standard deviations. If $k = 2$, we go out two standard deviations from the mean and find that at least

$$1 - \frac{1}{2^2} = 1 - \frac{1}{4} = \frac{3}{4},$$

Figure 4.7

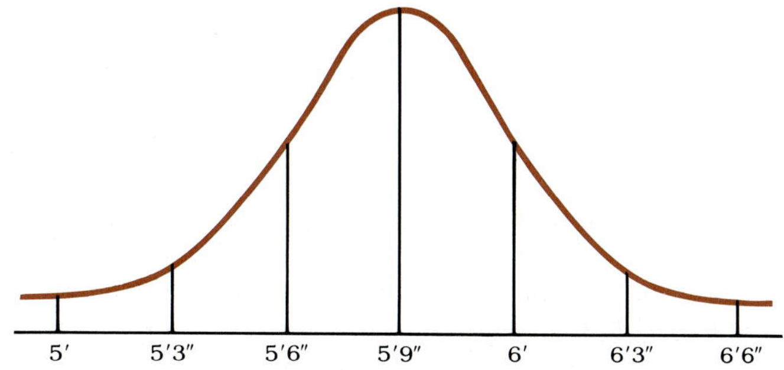

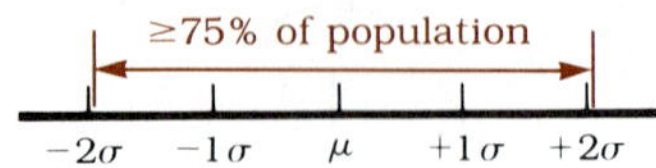

Figure 4.8

or 75%, of the population is contained in this interval, as shown in Fig. 4.8.

If we go out 3 standard deviations, we will account for at least

$$1 - \frac{1}{3^2} = 1 - \frac{1}{9} = \frac{8}{9},$$

or about 89%, of *any* population. The shape of the distribution makes no difference. Chebyshev's theorem is valid for any real number greater than one, such as $k = 2.25$, 3.7, and so on.

We found that our data on weights in Table 3.1, from which we constructed the frequency distribution and graph, provided $\bar{X} = 162.5$ lbs and $s = 24.5$ lbs. Let us apply Chebyshev's theorem for $k = 2$, as shown in Fig. 4.9. Counting the data values in the interval 113.5–211.5, we find that there are 85. Chebyshev's theorem tells us that this should be at least 3/4 or 75% of our data; we actually have 85/90 = 94%.

Figure 4.9

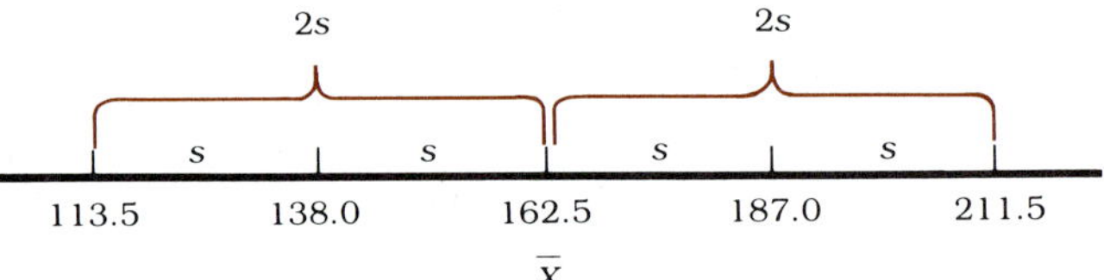

EXERCISES/Section 4.4

1. What percent of the data, at least, should lie between the data values 7.5–22.5 for a distribution having a mean of 15 and a standard deviation of 3?

2. Use Exercises/Section 3.1, nos. 4–10, to
   a) calculate $\bar{X}$ and s from your frequency tables.
   b) construct a diagram similar to Fig. 4.8, using your calculated $\bar{X}$ and s.
   c) count the number of data values between $\bar{X} - 2s$ and $\bar{X} + 2s$ and find the percentage that this number is of your whole sample. How does this compare with the *at least* 3/4 predicted by Chebyshev's theorem?
   d) count the number of data values between $\bar{X} - 3s$ and $\bar{X} + 3s$, and find the percentage that this number is of your whole sample. How does this compare with the *at least* 8/9 predicted by Chebyshev's theorem?

4.5 Other Descriptive Measures

Fractiles

If you received a graded test paper back from your professor with only the class mean $\bar{X}$ and the class standard deviation s on the top of the paper, would you be satisfied? Of course not. You would want to know how you performed in absolute terms and relative to the rest of the class.

When determining the place of an individual data value in a list, we are applying measures of *position.* We have already become familiar with one of these: the median. The median is the position in an ordered list that divides the list in half. We can easily generalize beyond the median by dividing a data list (or a population) into any fractional parts we wish, not just in half. The numbers that do this are called **fractiles.**

> ***fractile*** ▪ A number that cuts off the bottom portion of an ordered data list (the portion with the smallest values); this portion is expressed as a fraction, such as the bottom 1/10.

Fractiles expressed as percentages are called **percentiles.** The tenth percentile, written P_{10}, is the number that divides a list into two parts: 10% of the list is smaller than or equal to P_{10}, and, at most, 90% of the list is greater than P_{10}.

> ***percentile (P_m)*** ▪ A number that divides an ordered data list into two parts: m% of the list is smaller than or equal to P_m and, at most, $(100 - m)$% of the list is greater than P_m.

EXAMPLE 4.3 We return to the list of 90 weights in Table 3.1. Find: **(a)** P_{10}, **(b)** P_{25}, **(c)** P_{70}, and **(d)** P_{90}.

Solution

a) We are working with 90 numbers and want to divide the list into two parts: one with 10% of the data, one with 90%. Ten percent of the data is

$$0.10 \cdot 90 = 9 \text{ numbers.}$$

We always start at the smallest value and count up 9 values. The ninth value is 134, so $P_{10} = 134$.

b) $0.25 \cdot 90 = 22.5$ numbers. Since we are counting numbers in the data list, we must go to the twenty-second or the twenty-third number. In order to guarantee that we will have 25% in the lower portion on the list, we will round up to the twenty-third. It is 146, so $P_{25} = 146$.

c) $0.70 \cdot 90 = 63$ numbers. The sixty-third number is 174, so $P_{70} = 174$.

d) $0.90 \cdot 90 = 81$. The eighty-first number is 190, so $P_{90} = 190$. □

NOTE

1. Be sure to put the numbers in order (from lowest to highest) before you determine percentiles.
2. Percentiles never go above P_{99}. There is no P_{100}. By definition P_m divides the data list into two groups; P_{100} would not do this.

One other set of fractiles occurs in statistics, especially in nonparametric discussions. These are the **quartiles.**

quartiles ■ Values in a data list which divide the list into four equal parts.

There are three quartiles: $Q_1 = P_{25}$; $Q_2 = P_{50}$ = median; and $Q_3 = P_{75}$.

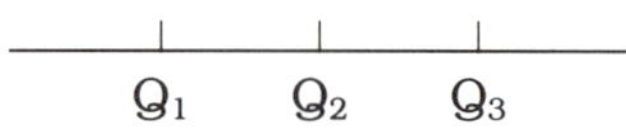

Quartiles can be used to provide a measure of central tendency: the **midquartile.**

midquartile ■ A measure of central tendency, found by

$$\frac{Q_1 + Q_3}{2}.$$

They also provide a measure of dispersion: the **interquartile range.**

interquartile range ■ A measure of dispersion, found by

$$Q_3 - Q_1.$$

EXAMPLE 4.4 Use the following data to find

a) Q_1, Q_2, Q_3; **b)** the midquartile; **c)** the interquartile range.

4	19	35	48
7	20	36	50
9	22	39	50
10	23	39	56
11	24	40	58
16	27	40	58
17	27	40	62
18	31	40	69
18	31	41	70
18	32	47	80

Solution There are 40 numbers, so the quartiles should be every $40/4 = 10$th number.

a) $Q_1 = 18$, $Q_2 = 32$, $Q_3 = 47$.

b) $\dfrac{Q_1 + Q_3}{2} = \dfrac{18 + 47}{2} = \dfrac{65}{2} = 32.5$.

c) $Q_3 - Q_1 = 47 - 18 = 29$. □

To determine fractiles (as either percentiles or quartiles) from grouped data, we employ the ogive shown in Fig. 3.5. Recall that the vertical scale is a cumulative relative frequency scale. To determine a percentile P_m, go up the vertical scale to the proper height as determined by m. Then draw a horizontal line at this height until it intersects the diagonal line. Drop down vertically at this point and P_m will be the value on the horizontal axis.

EXAMPLE 4.5 Use Fig. 4.10 to find **(a)** P_{70} and **(b)** Q_1.

Figure 4.10
Weights of adult males

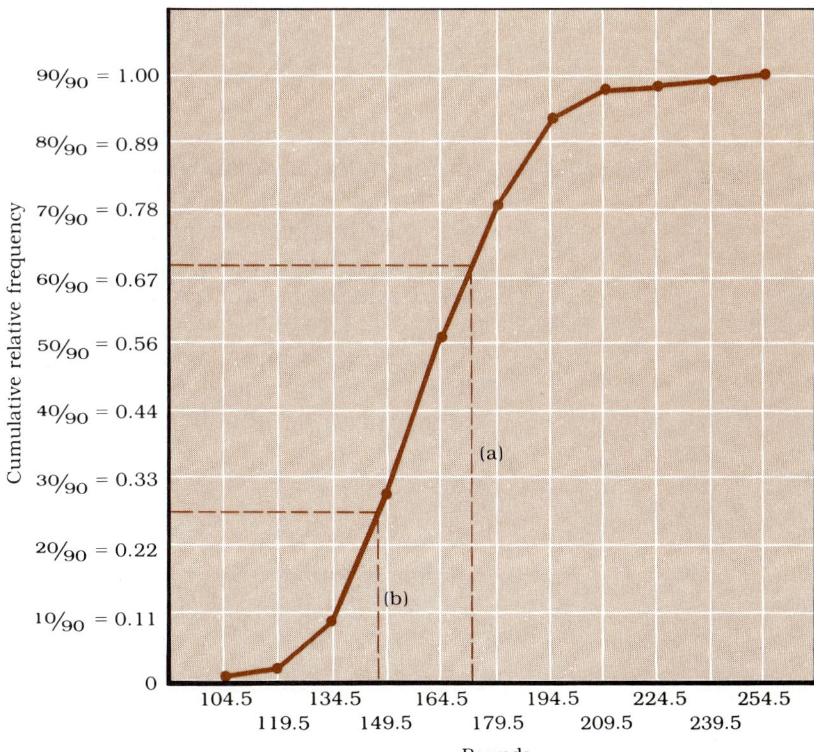

Solution

a) P_{70} means that we go up the vertical axis to 0.70. We draw a dashed line across to the diagonal line and drop down to the horizontal axis. This gives us a reading of about 173, so P_{70} = 173.
b) $Q_1 = P_{25}$, so we go up the vertical axis to 0.25. We draw a line across and drop down to the horizontal axis. This yields a reading of about 147, so Q_1 = 147. □

This method of finding percentiles and quartiles depends on the accuracy of the graph and how careful we are in using it. We should always use graph paper when employing this method.

Scores

A frequently used measure of position is the **z score.** It gives the position of an individual data value in relation to the mean in standard deviations. When discussing Chebyshev's theorem, we marked off intervals in standard deviations. We will do the same for z scores.

> ***z score*** ■ The position of an individual data value in relation to the mean, expressed in standard deviations.

If the mean height for adult males is 5′9″ with a standard deviation of 3″, we say that someone with a height of 6′ is one standard deviation above the mean, or has a z score of +1. Someone with a height of 5′3″ is two standard deviations below the mean, or has a z score of −2. The z score of the mean is 0. In general, we are transforming the scale from one involving units of measurement to one having no units of measurement, as shown in Figure 4.11.

Using z scores allows us to make comparisons at this stage. Later they will enable us to analyze a normal distribution. Consider a mean weight of 160 lbs with a standard deviation of 20 lbs

Figure 4.11

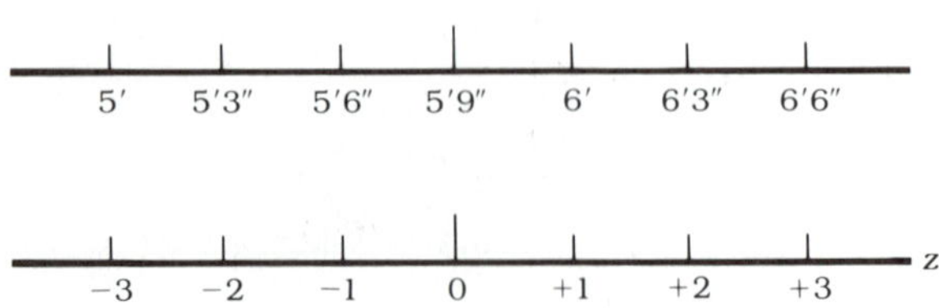

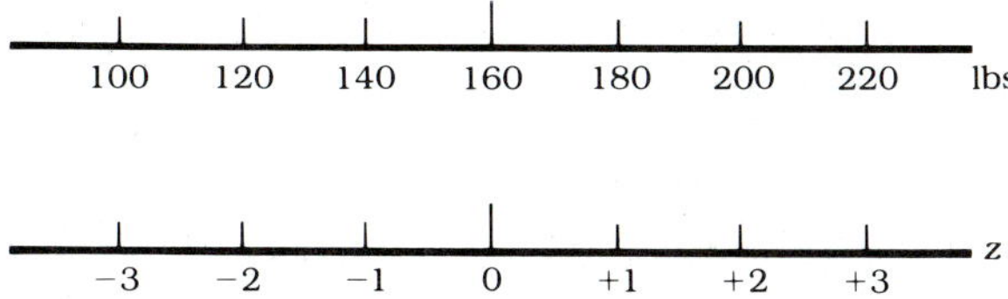

Figure 4.12

(Fig. 4.12). The person who weighs 180 lbs and is 6′ tall has a z score of 1 for both characteristics. A person who weighs 200 lbs ($z = 2$) and is 5′9″ tall ($z = 0$) is well above the mean in weight but not in height.

EXAMPLE 4.6 We can even compare the proverbial apples and oranges. Which is larger *relatively:* an apple that is 2.8″ in diameter, with a mean diameter of 2.6″ and a standard deviation of 0.1″ for this variety of apples, or an orange that is 4″ in diameter, with a mean diameter of 3.5″ and a standard deviation of 0.2″ for this variety of oranges?

Solution From Figure 4.13 we see that the apple has a z score of 2, and the orange has a z score between 2 and 3 (actually, 2.5). Thus the orange is larger, relative to the group from which it comes. □

Many times, the z scores for specific data values are not so obvious, especially when they are not whole numbers. In general, we want to find the distance from the mean; we do this using subtraction:

$$X - \bar{X} \text{ for a sample} \qquad \text{or} \qquad X - \mu \quad \text{for a population.}$$

Figure 4.13

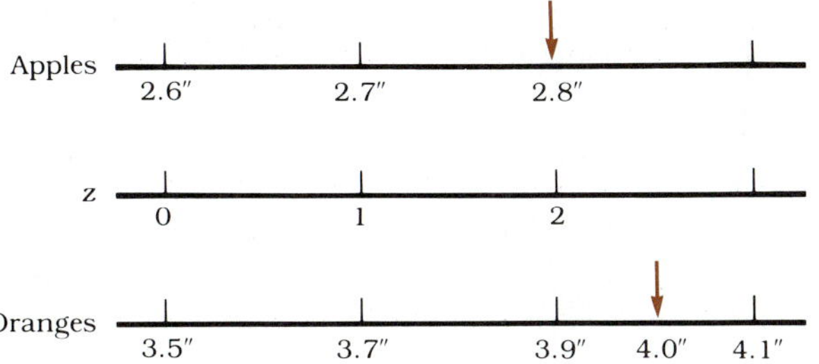

To convert this difference into standard deviations, we divide by the standard deviation.

$$z = \frac{X - \bar{X}}{s} \quad \text{for a sample}$$

$$z = \frac{X - \mu}{\sigma} \quad \text{for a population}$$

Perhaps the easiest way for you to remember the z-score equations and to start to notice a pattern that will be maintained in more complex situations is to state the basic equation in words:

$$z = \frac{\text{Individual data value} - \text{Mean}}{\text{Standard deviation}}.$$

EXAMPLE 4.7 What is the z score of an individual who received an 86 on a test that had a mean of 73 and a standard deviation of 11?

Solution

$$z = \frac{X - \bar{X}}{s} = \frac{86 - 73}{11} = \frac{13}{11} = 1.18.$$

It is a good idea to carry z scores out to two decimal places. This degree of accuracy will be necessary later. □

EXAMPLE 4.8 **(a)** What test score goes with a z score of 1.40 in Example 4.7? **(b)** With a z score of −2.10?

Solution The z-score equation requires some rearrangement before we can find the individual data value. We want to find X in the equation

$$z = \frac{X - \bar{X}}{s}.$$

To get X by itself on one side of the equation we do the algebra to solve for X. First, we multiply both sides by s:

$$z \cdot s = \frac{X - \bar{X}}{s} \cdot s.$$

Then we add $\bar{X}$ to both sides:

$$\bar{X} + z \cdot s = X - \bar{X} + \bar{X},$$

or

$$X = \bar{X} + zs$$

To solve the problem, we substitute numbers for $\bar{X}$, z, and s.

(a) $X = 73 + (1.40)(11) = 88.40$.

(b) $X = 73 + (-2.10)(11) = 49.90$. □

EXERCISES/Section 4.5

1. The fuel economy, in miles per gallon, was measured for 36 vehicles.

13.2	17.8	20.1	22.5	23.8	27.4
13.8	18.0	21.0	22.9	24.3	27.9
13.9	18.1	21.0	23.1	24.6	28.0
14.5	18.6	21.1	23.1	25.0	28.7
15.2	19.1	21.7	23.4	25.3	29.1
16.2	19.9	22.5	23.7	27.0	29.8

Find: a) P_{20} b) P_{60} c) the quartiles

What is the d) midquartile? e) interquartile range?

2. The following data show annual income, in thousands of dollars, for 35 families.

7	11	12	17	21
9	11	14	18	22
9	11	14	18	22
9	12	14	19	25
10	12	15	20	26
10	12	16	20	28
11	12	17	21	30

Find: a) P_{40} b) P_{80} c) Q_1.

3. The following are the number of runs batted in by 60 baseball players.

3	14	19	27	37
7	14	21	29	41
8	15	21	30	42
9	15	21	30	42
10	16	22	31	46
11	17	24	31	46
12	18	25	32	51
13	18	25	32	53
13	19	27	34	54
13	19	27	35	54
13	19	27	35	54
14	19	27	36	55

Find: a) P_{56} b) P_{30} c) the quartiles

What is the d) midquartile? e) interquartile range?

4. The following are weekly starting salaries, in dollars, for 30 jobs in data processing.

115	130	150	175	200	225
120	135	150	180	200	250
125	140	160	190	200	275
127	150	175	192	225	275
130	150	175	200	225	300

Find: a) P_{70} b) P_{90} c) Q_3

5. Forty cities reported the following average temperatures, in °F, in July.

60.1	68.1	74.6	81.7
62.3	68.9	75.1	81.9
62.4	69.0	76.5	82.3
62.4	70.1	76.8	82.4
64.6	70.6	77.2	83.1
56.4	71.2	77.7	83.6
65.5	72.3	79.4	86.4
67.2	73.4	79.4	86.4
67.5	73.6	80.6	87.1
67.8	73.9	81.2	88.2

Find: a) P_{84} b) P_{10} c) the quartiles

What is the d) midquartile? e) interquartile range?

6. Using the ogive from Exercises/Section 3.3, no. 6, find
a) P_{40} b) Q_1

7. Using the ogive from Exercises/Section 3.3, no. 5, find
a) P_{60} b) Q_2 c) Q_3 d) P_{70}

8. Using the ogive from Exercises/Section 3.3, no. 10, find
a) P_{10} b) P_{20} c) Q_1 d) Q_3

9. Using the ogive from Exercises/Section 3.3, no. 9, find

a) P_{30} b) P_{80} c) Q_1 d) Q_3

10. Standard 100-watt light bulbs have a mean life of 850 hrs with a standard deviation of 50 hours.

a) Find the z scores of bulbs that last 810 hrs, 930 hrs, 835 hrs, and 990 hrs.
b) What is the life of a light bulb with a z score of 1.2? −2.1? 1.75? 1.96?

11. According to the label, a particular brand of ketchup contains 28 oz with a standard deviation of 0.3 oz.

a) What is the z score of a bottle that contains 28.7 oz? 26.9 oz? 27.6 oz? 29.1 oz?
b) What is the weight of the contents of a bottle that has a z score of 1.65? −2.05? 2.33? −1.4?

12. The mean number of keypunch errors made by operators when transcribing data is 15 for every 75 cards punched with a standard deviation of 4.

a) What is the z score of an operator who made 12 errors? 17? 11? 27?
b) What is the number of errors made (to one decimal place) by an operator with a z score of 1.55? −2.14? 1.28? −0.65?

13. Random errors are always introduced when people make measurements; the values obtained will vary from one measurement to the next. On the average these random errors should cancel each other. Assume that the mean random error is 0 and the standard deviation is 0.03 foot in measurement of a line.

a) Find the z score of a measurement that is 0.048 foot above the true value of the measurement; 0.045 foot below; 0.012 above; 0.006 below.
b) What is the random error of a measurement having a z score of 1.45? 2.50? −1.41? −2.03?

14. Mean weekly sales by new cashiers at the U-Save Discount Food Store is $8200 with a standard deviation of $2250.

a) Find the z scores of new employees who checked out $7300, $6200, $8750, and $9300 worth of groceries.
b) What are the weekly sales figures for cashiers with z scores of 1.58, 2.20, −1.48, and −1.65?

15. A company that hired high school dropouts with learning disabilities found that their average weekly production rate was 23 items assembled with a standard deviation of 2.9 items.

a) What are the z scores of employees with weekly production rates of 27, 32, 15, and 20 items assembled?
b) What are the weekly production rates of employees who have z scores of 2.1, 3.1, −1.5, and −0.5.

Study Notes

KEY TERMS

fractile ■ A number that cuts off the bottom portion of an ordered data list (the portion with the smallest values); this portion is expressed as a fraction, such as the bottom 1/10.

interquartile range ■ A measure of dispersion, found by $Q_3 - Q_1$.

mean ■ The arithmetic average found by adding the data values and dividing by the number of data values.

median ■ The middle number; the number that divides a list or distribution of data into two equal parts.

midquartile ■ A measure of central tendency, found by

$$\frac{Q_1 + Q_3}{2}.$$

modal class ■ The class with the largest frequency.

mode ■ The most frequently occurring value. From a grouped frequency distribution, the class mark of the modal class.

parameter ■ A numerical characteristic of the population. The true, fixed value of some characteristic of interest. Parameters are labeled with Greek letters.

percentile (P_m) ■ A number that divides an ordered data list into two parts: m% of the list is smaller than or equal to P_m and, at most, $(100 - m)$% of the list is greater than P_m.

qualitative data ■ Information that is nonnumerical in nature.

quantitative data ■ Information that is given as numbers. These numbers can be either measurements or counts.

quartiles ■ Values in a data list which divide the list into four equal parts.

range ■ The difference between the largest data value and the smallest data value (largest − smallest).

standard deviation ■ The square root of the variance.

statistic ■ A calculated sample result; a measurable characteristic of a sample. Its value may change from sample to sample. Statistics are labeled with italic letters.

variance ■ A measure of variability, using the mean as a reference point and found by the equations

$$s^2 = \frac{\Sigma(X - \bar{X})^2}{n - 1} \quad \text{or} \quad s^2 = \frac{n\Sigma X^2 - (\Sigma X)^2}{n(n - 1)}.$$

z score ■ The position of an individual data value in relation to the mean, expressed in standard deviations.

$$z = \frac{\text{Individual data value} - \text{Mean}}{\text{Standard deviation}}.$$

OUTLINE

I. Types of data
   A. Qualitative—attributes, nonnumerical responses.
   B. Quantitative
      1. Measurements.
      2. Counts.

II. Measures of central tendency
   A. Mean
      1. From a list of data values,

$$\bar{X} = \frac{\Sigma X}{n}.$$

      2. From a frequency distribution:

$$\bar{X} = \frac{\Sigma Xf}{n}.$$

   B. Median
      1. Put the numbers in order.
      2. The middle number or the mean of the two middle numbers.
      3. Can be determined from a stem-and-leaf diagram.
      4. From a frequency distribution:

$$\text{Lower class boundary} + \frac{F}{f}(w)$$

   C. Mode
      1. From a list: the most frequently occurring number.
      2. From a frequency distribution: the class mark of the modal class.

III. Measure of variability or dispersion
   A. Range.
   B. Variance.
      1. From a list:

$$s^2 = \frac{\Sigma(X - \bar{X})^2}{n - 1}$$

or

$$s^2 = \frac{n\Sigma X^2 - (\Sigma X)^2}{n(n - 1)}.$$

      2. From a frequency distribution:

$$s^2 = \frac{n\Sigma X^2 f - (\Sigma Xf)^2}{n(n - 1)}.$$

   C. Standard deviation: the square root of the variance.

IV. *Chebyshev's theorem* For any specific numerical value k that exceeds 1, the proportion of a population that lies within k standard deviations of the mean is at least

$$1 - \frac{1}{k^2}.$$

V. Fractiles
   A. Percentiles
      1. Divide the data into predetermined percentages.

2. Found by counting.
3. Determined from an ogive.

B. Quartiles—divide the data into quarters.

VI. z scores

A. The position of an individual data value relative to the mean, expressed in standard deviations.

B. $z = \dfrac{\text{Individual data value} - \text{Mean}}{\text{Standard deviation}}$.

C. To find the individual data value when the z score is known:

$$X = \bar{X} + zs$$

REVIEW PROBLEMS

For each of the following data sets, determine the mean, median, mode, range, variance, and standard deviation.

1. Time, in minutes, required to jog 10 laps by members of a fitness class:

   15 12 9 11 12 10 13 15 14 15 10 12 11

2. Amount of money, in dollars, spent on books by 8 randomly selected students:

   45 58 95 72 68 62 53 110

3. Amount of time, in minutes, spent by 9 students on a statistics homework assignment:

   55 42 68 40 52 48 45 42 40

4. Mileage ratings for 8 new cars:

   32 28 35 30 38 30 31 34

5. Price per gallon, in cents, for unleaded gasoline at 7 local gas stations:

   110.9 122.9 115.9 120.9
   130.9 132.9 130.9

6. Assessed value, in dollars, of 5 randomly selected houses:

   45,000 41,500 53,000 42,000 50,000

7. Number of children in 11 randomly chosen families:

   2 1 2 2 3 2 4 1 5 2 3

8. Amount of time, in minutes, spent on a two-way commute by 8 workers in a certain office:

   30 42 65 20 40 35 30 38

For each of the frequency distributions in Problems 9–12, determine the mean, median, mode, variance, and standard deviation.

9. A statistics teacher's grade record disclosed the following grade distributions:

Grade Intervals	f
30–39	1
40–49	0
50–59	3
60–69	3
70–79	5
80–89	12
90–99	9

10. The department of motor vehicles and police records for a certain local jurisdiction indicate that the number of accidents for a year was distributed according to the drivers' ages as follows:

Age	f
16–20	25
21–25	16
26–30	10
31–35	5
36–40	3
41–45	2
46–50	0
51–55	1
56–60	0
61–65	2

11. A certain automobile dealer kept records of the price paid by each customer and organized the results for a recent month into the

following distribution table:

Price	f
\$5,000– 5,999	12
6,000– 6,999	18
7,000– 7,999	10
8,000– 8,999	8
9,000– 9,999	3
10,000–10,999	0
11,000–11,999	5

12. The manager of the meat department of a local supermarket recorded the weights of frozen turkeys received on a recent day. The results are tabulated as follows:

Weight	f	Weight	f
11.00–11.99	10	16.00–16.99	20
12.00–12.99	16	17.00–17.99	18
13.00–13.99	19	18.00–18.99	30
14.00–14.99	25	19.00–19.99	28
15.00–15.99	8	20.00–20.99	5

13. The admissions office of a local college reports that the average score on a standardized entrance exam for incoming freshmen is 550 with a standard deviation of 50. What is the z score corresponding to each of the following:
a) $X = 620$ **c)** $X = 550$ **e)** $X = 690$
b) $X = 500$ **d)** $X = 460$

14. The quality control officer at an auto assembly plant reports that the average weight of front bumpers being installed on a certain model car is 12 kg with a standard deviation of 0.015 kg. Determine the z score for:
a) $X = 11.9$ **d)** $X = 11.75$
b) $X = 12.01$ **e)** $X = 12.015$
c) $X = 12.025$

15. An English teacher reports that the mean grade on a remedial vocabulary test worth 140 points was 110 with a variance of 144. What is the grade corresponding to the following z scores:
a) $z = 1.5$ **c)** $z = -2.5$ **e)** $z = 0$
b) $z = 0.75$ **d)** $z = -0.25$

16. A seedsman notes that the germination time for a variety of annuals is 120 hrs (5 days) with a standard deviation of 8 hrs. What is the germination time for each of the following:
a) $z = -2$ **c)** $z = 0$ **e)** $z = 2.5$
b) $z = -3.5$ **d)** $z = 1.25$

17. As a result of a strike, the Cost of Living Council allowed steel workers a contract with a mean salary of \$17,000 and a standard deviation of \$800. This next year each worker will receive an increase of \$500. What will be the new mean and the new standard deviation?

18. A tourists' guide to a major eastern city lists the following minimum prices, in dollars, for one night's lodging in 32 hotels and motels.

42	48	60	54
65	100	56	50
99	49	52	80
80	30	84	56
36	88	58	42
75	96	92	95
79	52	52	55
96	56	60	56

Determine:
a) Q_1, Q_2, Q_3.
b) P_{40}, P_{65}, P_{20}, P_{90}.
c) midquartile.

19. The following lists the grades earned by 18 students on a math quiz:

81	81	70	62	74	34
45	72	87	87	28	100
51	96	81	23	91	98

Determine:
a) Q_1, Q_2, Q_3.
b) P_{10}, P_{60}, P_{84}.
c) midquartile, interquartile range.

20. The following birth weights, in kilograms, were recorded for 12 babies in the pediatric section of a local hospital:

3.4	3.0	2.8	3.9	3.2	3.6
4.1	3.5	4.4	3.5	3.1	4.0

Determine:
a) Q_1, Q_2, Q_3.
b) P_{30}, P_{70}, P_{85}.
c) midquartile, interquartile range.

Probability

5

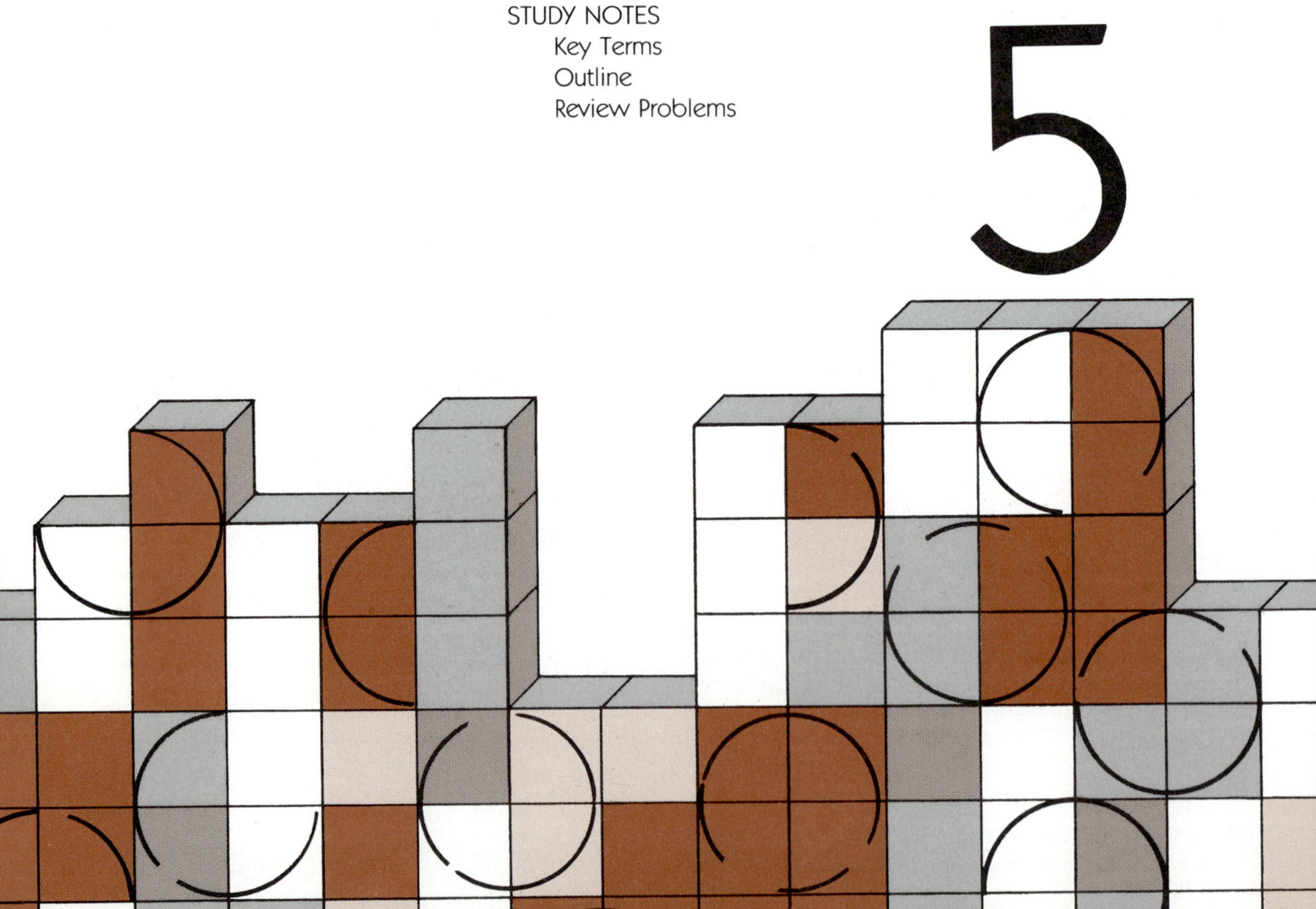

5.1 Introduction

At this point in our discussion we need to digress. So far, we have learned how to describe a sample with graphs and numerical measures, but this is not an end in itself. We will want to try to draw conclusions about this information and apply them to a population as a whole. The leap from concrete data to an unknown population requires that we take some risks. To help minimize these risks and to help us estimate their size, we need to study probability. We have already studied samples from which we may wish to generalize to populations. The aim of this chapter and the next is to study some small, ideal populations and generate possible samples that we can study and generalize to real-life situations. We will also study some simple, known populations and try to draw some generalizations and find laws that will help us with statistical experimental situations. Chapter 5 provides the ideas necessary for development of the essential probability distributions encountered in Chapter 6. In News & Views 5.1, we see how probabilities can impact on various facets of our lives.

5.2 Simple Probability

To begin the discussion of probability, we need to define some terms. The word **experiment** will be used in its broadest sense to refer to any activity that generates data. The result of an individual trial of an experiment is called an **outcome;** the single outcome or a set of outcomes of an experiment constitute an **event.**

> ***experiment*** ▪ Any activity that generates data.
>
> ***outcome*** ▪ The result of an individual trial of an experiment.
>
> ***event*** ▪ An outcome or a set of outcomes of an experiment.

Work with probability also requires that we be able to count outcomes. To do this accurately, we use the concept of **sample space.**

> ***sample space*** ▪ The complete, nonoverlapping list of all possible outcomes of an experiment.

NEWS & VIEWS 5.1

Mr. Wizard comes to court

New scientific evidence is helping to show jurors whodunit

Did Wayne Williams murder Nathaniel Cater and Jimmy Ray Payne? No one saw either crime, and there were no fingerprints. But there is plenty of circumstantial evidence in the extraordinary Atlanta case, including carpet fibers found on the victims and bloodstains in Williams' station wagon. So prosecutors are placing their faith in test tubes, microscopes and forensic specialists; in hour upon hour of testimony, experts have said that all the scientific evidence points to Williams. Last week the defense fought back. Kansas State University Professor Randall Bresee claimed that the prosecution's fiber analysis was too imprecise. In fact, said Bresee, he had examined fibers from a carpet in Defense Attorney Mary Welcome's office and found them "microscopically similar" to those from Williams' home.

No matter who prevails, the trial is highlighting a major development in the criminal courtroom. With the help of a variety of technical advances, more and more silent evidence is being turned into loudly damning testimony. FBI Laboratory Chief Thomas Kelleher (whose technicians handle half a million pieces of evidence a year) reports that forensic science is growing so fast that even the most sophisticated researchers cannot keep up. The granddaddy of scientific evidence is the fingerprint, introduced in 1901. Because a person's print is unique, there is still no better physical evidence. But now there are a number of new ways of linking a criminal to a crime that are nearly as clear-cut. Suspects are being asked not only for fingerprints but for footprints, blood samples and pieces of hair.

Over the past ten years, no area has developed faster than the examination of bloodstains. "Before, we used to be satisfied with identifying a blood sample as type A, B, AB or O. Now we have 13 or more different antigen and enzyme systems we can pick out," says Gary Howell, 34, director of the Kansas City regional crime lab. The probability that any two people will share the same assortment of these blood variables is .1% or less. Because of that, Howell was recently able to use two tiny bloodstains to help convict a double murderer. . . .

Source: *Time,* March 1, 1982, p. 90.

Let us look at some examples in order to explore ways of generating sample spaces. The easiest way to illustrate the sample-space concept is with coin and dice tosses.

EXAMPLE 5.1 Toss one coin one time. What is the sample space of this experiment?

Solution We toss the coin and get either heads (H) or tails (T). There are two possible outcomes, so the sample space is H,T. □

Consider the sobering views expressed in News & Views 5.2. We like to think that the results of psychiatric, medical, scientific, and social theory and practice are based on more than pure chance—but sometimes they aren't.

EXAMPLE 5.2 Toss 2 coins one time each. What is the sample space?

Solution To be sure that we include all possible outcomes, the technique employed in a trial involving more than one item is to pretend that the items are distinguishable from each other. Assume that we have a penny and a dime. The sample space consists of four possible outcomes, or two outcomes for one item for each of the two outcomes of the other item.

Penny	Dime
H	H
H	T
T	H
T	T

□

NEWS & VIEWS 5.2

Insanity studies called no better than coin flip

Psychiatric studies are no more reliable than the flip of a coin in determining whether a person found legally insane by a jury may safely be returned to society, medical experts told a Senate panel yesterday.

Two officials who head state mental-illness programs said there is no available data to support the idea that clinicians can predict future violent behavior in patients who are examined for as long as a month.

"We could stand here and flip coins, or a judge could do the same thing, and be as accurate as clinicians who give elaborate presentations of patients based on 30-day studies," said Dr. Henry J. Steadman, head of special projects research at the New York State Department of Mental Hygiene.

"There's never been a study that shows a better than 50-50 chance for anyone to predict behavior of someone in a mental facility," Dr. Steadman told the Senate judiciary subcommittee on criminal law.

Dr. Stuart B. Silver, Maryland's assistant secretary for forensic services, said statistics on mental patients he compiled for his state show that predictions of recovery are unreliable.

"We cannot predict future acts of violence. We can only make unreliable judgments based on past experience," Dr. Silver testified.

The subcommittee, headed by Senator Arlen Specter (R, Pa.), is examining ways to make it harder for a defendant to use the insanity defense in criminal cases.

Mr. Specter opened hearings on the issue last month after John W. Hinckley, Jr., was found not guilty by reason of insanity on charges of trying to assassinate President Reagan on March 30, 1981.

Source: *Baltimore Sun,* July 15, 1982.

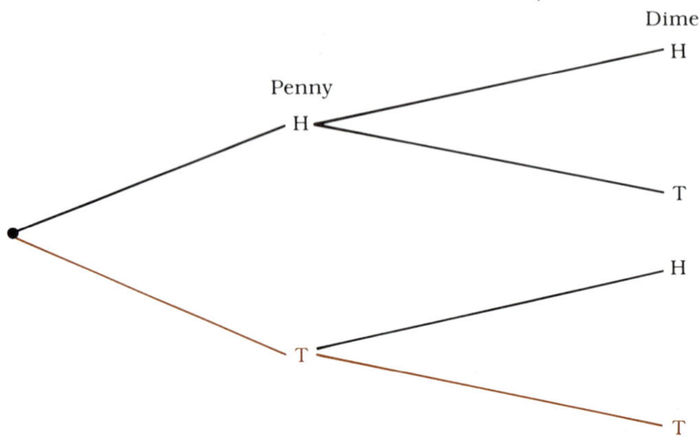

Figure 5.1

A more systematic and easily understood method of listing outcomes is known as a *tree diagram.* Tree diagrams begin with a dot, which is connected by a line (branch) to each possible outcome for an individual item, as shown for the penny (Example 5.2) in Fig. 5.1. We then add a set of branches off these outcomes to connect with each possible outcome for the next item, as for the dime as in Fig. 5.1. An outcome for this two-item trial, therefore, consists of the entire limb from the beginning dot on the left to one of the final endpoints on the right, as shown in color in Fig. 5.1. Thus there are as many outcomes as there are endpoints. If these outcomes are all equally likely to occur, probabilities are determined by counting.

EXAMPLE 5.3 Toss three coins one time each. What is the sample space?

Solution We have a penny, a nickel, and a dime. Construct a tree diagram, to find the sample space, as in Fig. 5.2. The eight outcomes are listed on the right of the tree diagram. □

Now consider whether tossing three coins once has the same sample space as tossing one coin three times. Complete the partial tree diagram in Fig. 5.3 to convince yourself.

EXAMPLE 5.4 Toss a die one time. What is the sample space?

Solution 1, 2, 3, 4, 5, and 6 are all possible outcomes. For an unweighted, fair die all six faces have an equal chance of occurring and are said to be *equally likely* to occur. □

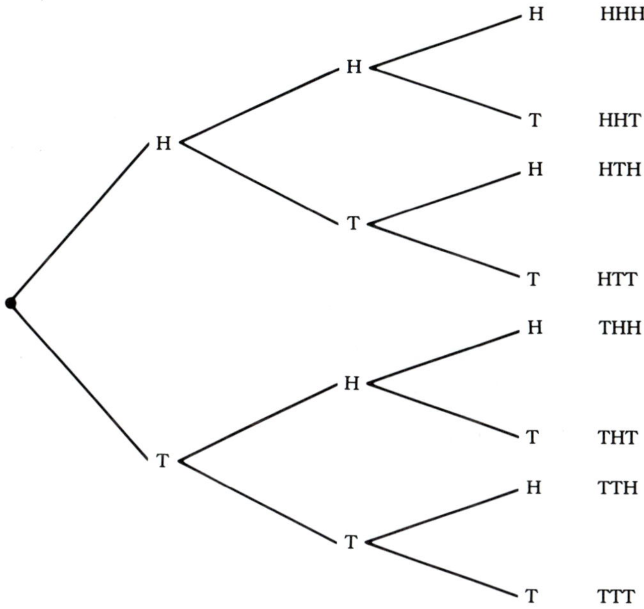

Figure 5.2

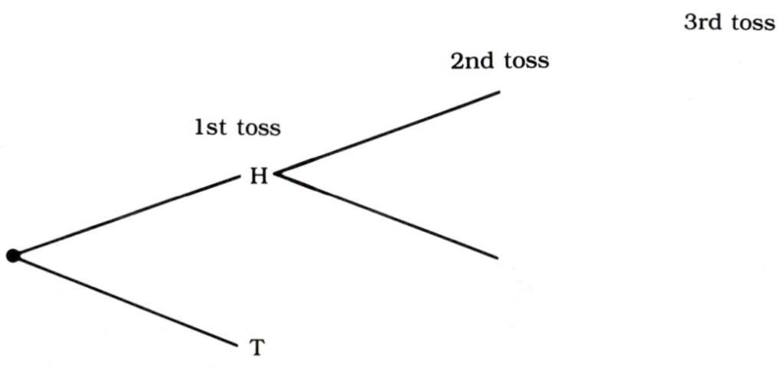

Figure 5.3

EXAMPLE 5.5 Toss two dice one time. What is the resulting sample space?

Solution We construct another tree diagram (Fig. 5.4). The 36 outcomes are listed to the right of the tree diagram. □

We are accumulating sample spaces with which to work in order to discuss probabilities.

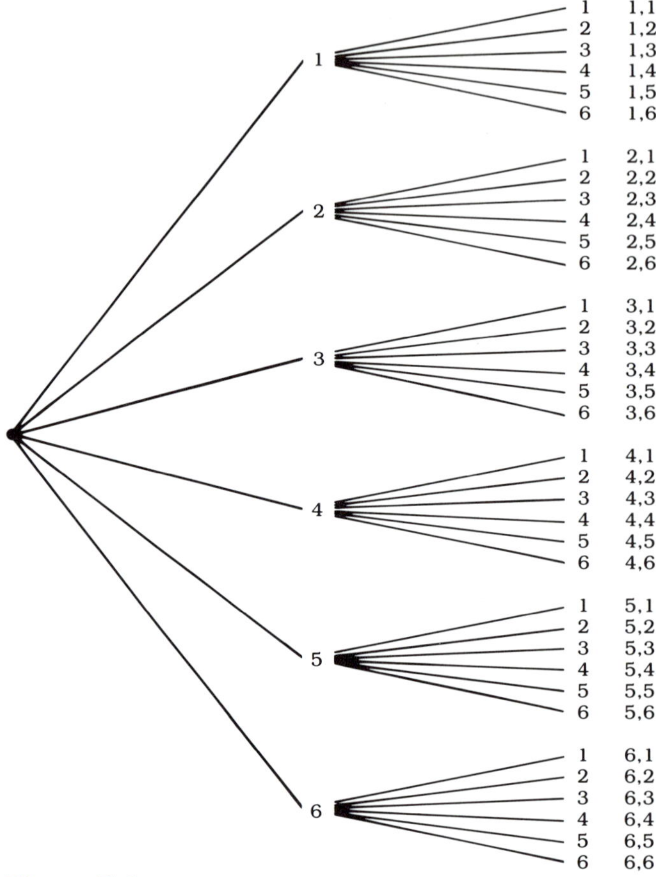

Figure 5.4

If you have ever pitched pennies with a friend, you will have no trouble understanding the statement: The probability of getting heads is ½. Since there are two faces on a penny, heads and tails, you have one chance out of two of getting a head. Now, what does it mean to say: The probability of rain is 20%? Obviously, this is not so simple. The penny example requires that you merely think of all the possibilities (heads, tails) and then count, an illustration of classical probability. The rain example also involves counting but, in order for us to have anything to count, records of weather conditions must have been kept. Based on such records, we could say that in the past—on 100 days with weather conditions exactly like those predicted for tomorrow—rain occurred on 20 such days. Stated in probability terms, this means that there is a 20% chance of rain under those particular conditions. This is an **experimental** or **observational probability.**

> ***probability*** ■ The relative frequency with which an event A
>
> 1. actually occurs (experimental)
>
> *or*
>
> 2. is expected to occur (classical).
>
> $$P(A) = \frac{\text{The number of occurrences of } A}{n}$$

At first glance, it appears that these are two very different situations, and that we have two different definitions of probability. This cannot be the case, if we are to develop laws and rules governing the behavior of probabilities. To show that these definitions are compatible, we can take a simple example of classical probability—tossing coins—and make an experiment out of it. Table 5.1 gives the results obtained when 30 people each tossed four coins at the same time five different times and observed the number of heads for each toss. Thus the column labeled 1 represents five tosses of four coins by one person, with the result that one head was observed one time and two heads were observed four times.

By totaling the results across for all 30 people, we find four occurrences of 0 heads, thirty-six occurrences of 1 head, fifty-eight occurrences of 2 heads, forty-two occurrences of 3 heads, and ten occurrences of 4 heads. The definition of probability tells

TABLE 5.1

	Person															
Result	**1**	**2**	**3**	**4**	**5**	**6**	**7**	**8**	**9**	**10**	**11**	**12**	**13**	**14**	**15**	**16**
0H													1			
1H	1	2	3		1		1	2	2		1	2			1	2
2H	4		1	2	2	4	3	2		2	2		2	3	2	1
3H		2		2	2		1	1	3	3	2	3	2	2	1	2
4H		1	1	1		1									1	

	Person													
Result	**17**	**18**	**19**	**20**	**21**	**22**	**23**	**24**	**25**	**26**	**27**	**28**	**29**	**30**
0H		1	1					1						
1H	2	1	1	2	2		1	2	1	2			1	3
2H	2	2	1	2	3	1	4	2	2	1	4	2	2	
3H	1			1		3			2	2	1	2	2	2
4H		1	2			1						1		

us to use relative frequencies (fractions), so

$$P(0H) = 4/150 = 0.027;$$

$$P(1H) = 36/150 = 0.24;$$

$$P(2H) = 58/150 = 0.387;$$

$$P(3H) = 42/150 = 0.28;$$

$$P(4H) = 10/150 = 0.067.$$

Note that when we have a sample space, the probabilities add to $150/150 = 1$. Since this is an experiment, we could have chosen to terminate it at any point short of 150 tosses. Similarly, it could be extended beyond the 150 tosses.

We need to make one other generalization from this experiment. Let us follow one row, the 2H row, across and treat it as a cumulative record of information for the coin tosses, as shown in Table 5.2. Person 1 had five tosses. At that point

$$P(2H) = 4/5 = 0.80.$$

Person 2 had ten tosses without a heads, so we now have

$$P(2H) = 4/10 = 0.40.$$

Adding in the results from Person 3, we get

$$P(2H) = 5/15 = 0.33.$$

TABLE 5.2

	Person									
	1	**2**	**3**	**4**	**5**	**6**	**7**	**8**	**9**	**10**
***P*(2H)**	$\frac{4}{5}$	$\frac{4}{10}$	$\frac{5}{15}$	$\frac{7}{20}$	$\frac{9}{25}$	$\frac{13}{30}$	$\frac{16}{35}$	$\frac{18}{40}$	$\frac{18}{45}$	$\frac{20}{50}$
	.80	.40	.33	.35	.36	.43	.46	.45	.40	.40
	11	**12**	**13**	**14**	**15**	**16**	**17**	**18**	**19**	**20**
***P*(2H)**	$\frac{22}{55}$	$\frac{22}{60}$	$\frac{24}{65}$	$\frac{27}{70}$	$\frac{29}{75}$	$\frac{30}{80}$	$\frac{32}{85}$	$\frac{34}{90}$	$\frac{35}{95}$	$\frac{37}{100}$
	.40	.37	.37	.39	.39	.38	.38	.38	.37	.37
	21	**22**	**23**	**24**	**25**	**26**	**27**	**28**	**29**	**30**
***P*(2H)**	$\frac{40}{105}$	$\frac{41}{110}$	$\frac{45}{115}$	$\frac{47}{120}$	$\frac{49}{125}$	$\frac{50}{130}$	$\frac{54}{135}$	$\frac{56}{140}$	$\frac{58}{145}$	$\frac{58}{150}$
	.38	.37	.39	.39	.39	.38	.40	.40	.40	.39

NEWS & VIEWS 5.3

Stereotyping

We all "play the odds" when we indulge in stereotyping. Asking for *Mr.* Bank Executive may indicate prejudice on the part of the speaker, but it may also mean that there are so many more male than female bank executives that the speaker is relying on the greater probability of being correct. Ask your friends who are women in management positions whether callers assume that they are the secretary when they answer their own telephones. What stereotyped generalizations do *you* make? Are they based on data—and hence probabilities—or on prejudice that is blind to individual capabilities?

The probability of getting 2H fluctuates drastically at first, but continuing on in Table 5.2 we see that $P(2H)$ begins to fluctuate less and from persons 9–30 varies by no more than 0.03. As we generated more data, the probability became more stable. If we continued the experiment for larger and larger numbers of tosses, the fluctuations at two decimal places would disappear; the experimental probability would closely approach the classical probability of obtaining two heads when tossing four coins. We should be able to determine the classical probability using a tree diagram. Try it.

We have just demonstrated the **law of large numbers.**

> ***law of large numbers*** ■ As the number of trials increases, the experimental probability of a particular event stabilizes and approaches the classical or theoretical probability of the event occurring on one trial.

We can represent this schematically as follows:

$$\text{Experimental probability} \xrightarrow[\text{As } n \text{ gets large}]{} \text{Classical probability.}$$

By studying some simple examples, we can now develop some general rules and principles governing the behavior of probability. Then we can transfer these rules to more complex experimental cases.

EXAMPLE 5.6 If we toss a penny, a nickel, and a dime, one outcome could be H on the penny, T on the nickel, and T on the dime; this is usually written as HTT. This outcome is part of the event 1H. Also included in event 1H are the outcomes THT and TTH. How many outcomes are in the event 0H? Only TTT. Which event has a higher probability: 1H or 0H? The answer is 1H, since the event 1H contains more outcomes. What is $P(1\text{H})$?

Solution Referring to Fig. 5.2, we can see that

$$P(1\text{H}) = 3/8$$

and can determine the rest of the probabilities:

$$P(0\text{H}) = 1/8;$$

$$P(1\text{H}) = 3/8;$$

$$P(3\text{H}) = 1/8. \quad \square$$

EXAMPLE 5.7 Using Fig. 5.4, find P(doubles), P(sum of 3), P(sum of 7), and P(sum of 10).

Solution Note that there are 36 outcomes. We count the number of occurrences for each combination required in the list of outcomes.

$$P(\text{doubles}) = 6/36.$$

$$P(\text{sum of } 3) = 2/36.$$

$$P(\text{sum of } 7) = 6/36.$$

$$P(\text{sum of } 10) = 3/36.$$

Be sure to count the combinations for yourself to help you become familiar with lists of outcomes. □

EXAMPLE 5.8 Sales records for a certain week indicate that the Ace Record Shop had the following sales: 150 rock, 120 country-western, 80 light classics, and 50 classical records. If a record were chosen at random from this list, what is the probability that it would be (a) rock? (b) country-western? (c) light classical? (d) classical?

Solution Since the total number of records sold was 400,

a) P(rock) = 150/400.
b) P(country-western) = 120/400.
c) P(light classical) = 80/400.
d) P(classical) = 50/400. □

EXAMPLE 5.9 For an ordinary deck of cards, find the following probabilities:

a) P(ace)
b) P(club)
c) P(queen of hearts)
d) P(red card)
e) P(red jack)
f) P(face card)

Solution

a) There are four aces, one in each suit:

$$P(\text{ace}) = 4/52.$$

b) There are 13 cards that are clubs:

$$P(\text{club}) = 13/52.$$

c) There is only one queen of hearts:

$$P(\text{queen of hearts}) = 1/52.$$

d) There are two red suits, each of which has 13 cards:

$$P(\text{red}) = 26/52.$$

e) There are two red jacks:

$$P(\text{red jack}) = 2/52.$$

f) Each suit has three face cards:

$$P(\text{face card}) = 12/52. \quad \square$$

EXAMPLE 5.10 A hat contains one white tennis ball, one orange tennis ball, and one yellow tennis ball. Draw two of the tennis balls from the hat. What is the sample space of possible color combinations?

Solutions Although at first glance this problem may seem clear, it is somewhat ambiguous. Do we draw one ball, record its color, put it back, and draw again? If so, we find the sample space using the tree diagram in Fig. 5.5.

Figure 5.5
Outcomes with replacement

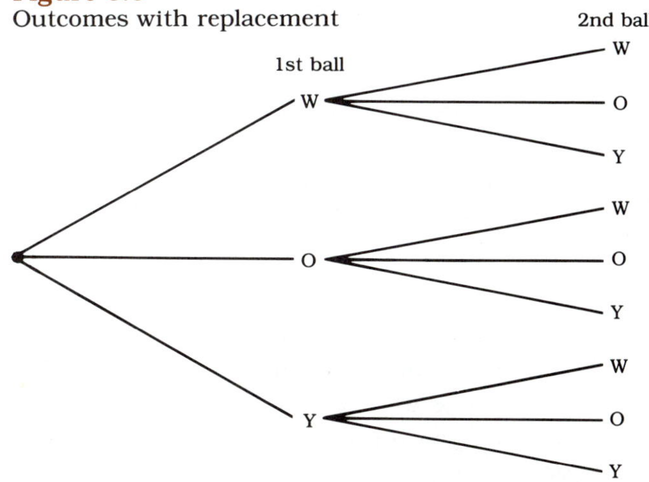

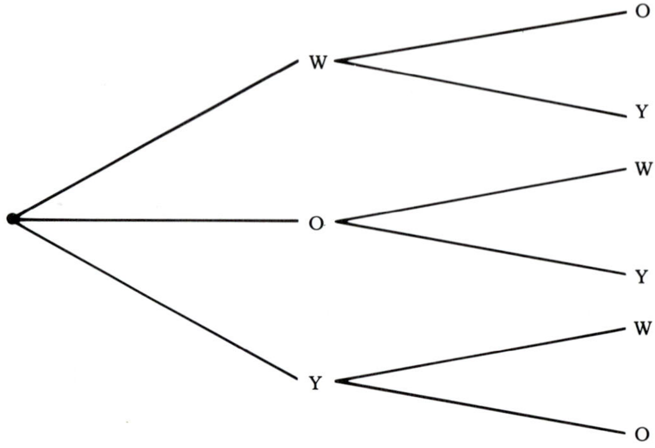

Figure 5.6
Outcomes without replacement

If we draw one ball and then a second one without replacing the first, the tree diagram of outcomes (Fig. 5.6) is different because all cases of *doubles* must be eliminated. Probabilities are likewise affected.

	With Replacement	Without Replacement
P(doubles)	3/9	0/6
P(1W)	4/9	4/6

Be sure that you can find all four of the outcomes in the last row. List outcomes as WW, WO, WY, etc., if necessary. Which tree diagram is used to display outcomes for the situation when two balls are drawn simultaneously? □

Let us now state some generalizations about the properties of probabilities. By definition, probabilities are always fractions. Is there any smallest value? Example 5.10 gave us a probability of 0. Since we are counting, negatives are not possible. Is there any largest value? If we included all on the list in our count, we would have $n/n = 1$. Thus we can generalize by saying that

$$0 \leq P(X) \leq 1.$$

At either extreme we have interesting cases. If $P(A) = 0$, then A is *impossible;* it cannot happen. If $P(B) = 1$, then B is *certain;* it must occur in every trial. Consider a five-card poker hand. What is the probability that the hand contains at least two cards of the same suit? Before you start worrying about counting all such hands, think for a moment. If the first four cards in the hand are each of a different suit, the fifth *must* repeat one of those suits. Therefore P(at least 2 cards of the same suit) = 1.

Examine Examples 5.2 and 5.3, as well as the coin-tossing experiment. What did we get when we added the probabilities of the complete, nonoverlapping list of events in the experiment?

$$P(\text{OH}) \quad P(1\text{H}) + P(2\text{H}) + P(3\text{H}) + P(4\text{H}) = 150/150 = 1.$$

Our second generalization, then, is

$$\Sigma P(X) = 1,$$

where the values of X represent a complete, nonoverlapping list of events, or the sample space.

PROPERTIES

1. $0 \leq P(X) \leq 1$.
2. $\Sigma P(X) = 1$, where the values of X represent a sample space.

These two properties are important in the solution of problems where counting is impossible. Probabilities can never be negative, regardless of how they are obtained. Any set of numbers that lies between 0 and 1 and adds to 1 can be treated as a set of probabilities. It will be important to remember this later when we deal with percentages of a population, another example of numbers between 0 and 1 which add to 100% = 1.

Probabilities can be used to support the findings of research. The *Federalist Papers* written after the American Revolution were published anonymously. However, the authors of most were known to be either Madison or Hamilton. By analyzing the writing styles in those articles for which the authors are known, scholars have determined with a high degree of probability the authorship of the others. Another version of such research that was reported recently appears in News & Views 5.4.

By one hand?

Computers reread Genesis

Jewish tradition holds that Moses wrote the first five books of the Bible. Conservative Christians agree. But 19th century Protestant critics emphasized the Pentateuch's diversity rather than its unity. They deemed it a scissors-and-paste job on materials from different centuries by four anonymous authors: the Jahwist ("J"), Elohist ("E"), Priestly writer ("P") and Deuteronomist ("D"). Though traditionalists rejected it, this J.E.P.D. theory hardened into liberal orthodoxy.

Now the four-author thesis has come under a powerful new attack. The ancient view, it seems, is supported by that most modern deity of omniscience, the computer. Bible Scholar Yehuda Radday of Haifa's Israel Institute of Technology reports that a five-year computer study of *Genesis* shows that it is the work of a single writer and that the J.E.P.D. theory must be "rejected or at least thoroughly revised."

Radday and three aides studied "the only unquestionable data," the words of the Hebrew text, and concentrated on 56 criteria of "language behavior" (such as use of conjunctions and word length) that are outside the conscious control of an author. The key finding: a remarkably high 82% probability that the same person wrote the supposed J and E passages. The P passages were as distinct as the critics have long maintained, but Radday contends that the difference can be explained totally by the formalistic content. Says he: "My love letters to my wife are completely different from my scholarly articles."

None of this proves that the single writer was Moses, but Radday is already hard at work on *Exodus* and will then search *Leviticus, Numbers* and *Deuteronomy*. Radday's challenge will be fiercely resisted by scholars who favor the entrenched J.E.P.D. theory, partly because he deals exclusively with linguistic criteria and ignores stylistic variations. Radday is no right-wing ideologue, however. Earlier he earned wide acclaim when his whirring computers supported the conventional theory that multiple authors produced the books of *Judges, Zechariah* and *Isaiah*. Says Radday: "These scholars can't have it both ways, approving a method when it suits them, and repudiating it when it does not."

EXERCISES/Section 5.2

1. For a throw of two dice, what are the following probabilities?

a) P(sum of 4)
b) P(sum of 9)
c) P(sum larger than 9)
d) P(sum smaller than 4)
e) P(even sum)
f) P(odd sum)
g) P(sum of 10 or more)
h) P(one die is larger than the other)
i) P(at least one die 5 or larger)
j) P(both dice even)
k) P(sum of 2, 3, 11, or 12)
l) P(a 5 shows)
m) P(a 2 shows)

2. A card is drawn from a 52-card deck. Find the following probabilities. (*Hint:* Count jack and king as odd.)

a) P(club)
b) P(10)
c) P(club and 10)
d) P(club or 10)
e) P(odd number)
f) P(below 8)
g) P(space and below 8)
h) P(red face card)
i) P(ace of spades)
j) P(9 or 10)

3. In preparing for the annual family picnic, Mr. Olechewski packed the following in the family cooler: 12 cans of light beer, 24 cans of regular beer, 18 cans of cola, and 12 cans of sugar-free diet soda. At the picnic, one of the family reached into the cooler and pulled out a can at random. What is the probability that the can was

a) light beer? c) cola? e) beer?
b) regular beer? d) diet soda?

4. Two cards are chosen from a well-shuffled deck, with the first card not replaced. Find the following probabilities:
   a) the first card drawn is not a jack; P(jack on the second card)
   b) the first card is a 4; P(6 or lower on the second card)
   c) the first card is a 7; P(even number on the second card). *Hint:* Count queen as even.
   d) the first card is a 10; P(face card on the second card)
   e) the first card is a diamond; P(diamond on the second card)

5. A designer of children's games included the following decision mechanism in one of his recent games. A player's next move is determined by the outcome of the spin of a pointer (see Fig. 5.7) and the subsequent toss of a die. Draw a tree diagram to illustrate the possible outcomes of this decision sequence. Use the tree to determine the probability of:
   a) W1 c) R4 e) W odd
   b) Y6 d) G even

6. The numbers 1–365 are each written on a piece of paper and put in a bowl. One is drawn. Find the following probabilities.
   a) *P*(128 is drawn)
   b) *P*(even number is drawn)
   c) *P*(number below 200 is drawn)
   d) *P*(number between 100 and 199 inclusive is drawn)
   e) *P*(your age is drawn)

7. A roulette wheel has segments numbered 0–36 on it; the numbers are alternately red and black in color, except for 0. A ball is dropped on the spinning wheel and comes to rest on one of the numbers. Find the following probabilities.
   a) *P*(even) (*Hint:* Do not count 0.)
   b) *P*(red)
   c) *P*(number between 1 and 18 inclusive)
   d) *P*(number between 19 and 36 inclusive)
   e) *P*(21) f) *P*(0)

8. Toss a die 60 times and record your results. From your record, find the following probabilities.
   a) *P*(1) c) *P*(3) e) *P*(5)
   b) *P*(2) d) *P*(4) f) *P*(6)
   Are these the results you expected?

9. A box contains three white balls, four red balls, and two blue balls. You draw two balls. What is the probability that both are the same color
   a) if you put the first one back before drawing the second?
   b) if you do not put the first one back?

10. A first grade class contains 11 girls and 9 boys. The honor of leading the class in the daily Pledge of Allegiance is determined by drawing the name of a student from a hat. What is the probability that the first name drawn will be a girl's? Assuming that, indeed, the first name drawn is a girl's and that her name is not returned to the hat, what is the probability that the name drawn on the following day will be that of another girl? A boy?

11. On the first day of class, a math instructor discovered that, of the 32 students enrolled in her class, 4 are repeating the course, 20 already have a calculator, 22 are female, 9 are new to the college, and 10 are planning to graduate within the next six months. If a student is chosen at random from this class, what is the probability that the student is
   a) repeating the course.
   b) a calculator owner.
   c) female.
   d) new to the college.
   e) not repeating the course.
   f) not a calculator owner.
   g) male.
   h) not new to the college.
   i) planning to graduate in 6 months.
   j) not planning to graduate in 6 months.

Figure 5.7

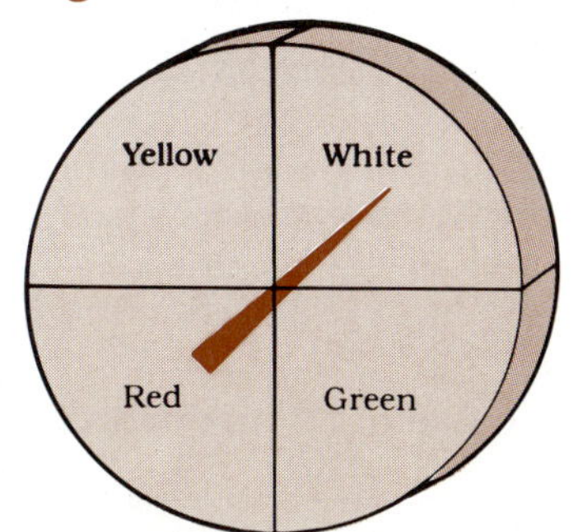

5.3 Rules for Combining Events

There are many instances when determining probabilities by counting is difficult, if not impossible, to do. Thus we need rules to help us calculate probabilities. We develop these rules using examples in which we can count to verify the results and then generalizing them to any situation.

Complementary Events

The notion of complement requires that we divide a sample space into two nonoverlapping groups; the groups do not need to be the same size. This idea should be a familiar one if you think of even and odd, success and failure, or face cards and nonface cards in a deck. The word *not* can frequently trigger the use of complements.

> ***complement of event*** **A** ▪ The set of all outcomes in the sample space *not* in event A. The complement of A is written as $\bar{A}$.

If we have A and $\bar{A}$, what is $P(A) + P(\bar{A})$? Since every outcome is in either A or $\bar{A}$, we have included all the possible outcomes. Thus

$$P(A) + P(\bar{A}) = 1.$$

Subtracting $P(\bar{A})$ from both sides of the equation, we get

RULE

$$P(\bar{A}) = 1 - P(A).$$

EXAMPLE 5.11 When you roll two dice, what is the probability of *not* getting doubles?

Solution Let us choose to count the smaller group of outcomes, the event *doubles*, and calculate to find the probability of its complement.

$$P(\text{not doubles}) = 1 - P(\text{doubles}).$$

$$P(\text{doubles}) = 6/36;$$

$$P(\text{not doubles}) = 1 - 6/36 = 30/36. \quad \square$$

The phrase *at least* can also trigger the use of complements.

EXAMPLE 5.12 When you toss two dice, what is the probability of getting a sum of at least 3?

Solution What is the complement of *a sum of at least 3?* It is a sum less than 3. This can only occur as a sum of 2, or (1, 1) from the tree diagram in Fig. 5.4. Thus

$$P(\text{sum of at least } 3) = 1 - P(\text{sum of } 2)$$
$$= 1 - 1/36 = 35/36.$$

We again used complements to count few outcomes and calculate the larger probability. □

We will utilize this subtraction property with probability distributions in Chapter 6. It will help us use information we have in order to find probabilities we need.

Events Formed Using the Word OR

New events can be formed by combining outcomes or other events using the word *or*. The probabilities for these new events can be found by using a simple rule involving arithmetic. We can formulate this rule by looking at some examples. For the examples in the rest of this section, consider the outcome of drawing one card from a deck; let

A = drawing a queen;

B = drawing a 6;

C = drawing a spade;

D = drawing a red card.

EXAMPLE 5.13 Find $P(A \text{ or } B)$.

Solution There are four queens and four 6's, so

$$P(A \text{ or } B) = P(\text{queen or } 6) = 8/52.$$

Now, $P(A) = 4/52$ and $P(B) = 4/52$. If we add $4/52 + 4/52$, we get 8/52; thus

$$P(A \text{ or } B) = P(A) + P(B) \quad □$$

EXAMPLE 5.14 Find $P(B \text{ or } C)$.

Solution We are asked to find $P(6 \text{ or spade})$. If you get out a deck of cards and actually count, you will find that

$$P(B \text{ or } C) = 16/52.$$

Now, $P(B) = 4/52$ and $P(C) = 13/52$, but

$$P(B \text{ or } C) \neq P(B) + P(C).$$

Why not? Because you counted the 6 of spades twice, both as a 6 and as a spade. Therefore in order to make the arithmetic work, you must subtract out one of the times you counted it.

$$\begin{aligned} P(6 \text{ or spade}) &= P(6) + P(\text{spade}) - P(\text{both 6 and spade}) \\ &= 4/52 + 13/52 - 1/52 \\ &= 16/52. \end{aligned}$$ □

We must be careful with events that overlap, that is, which have outcomes in common. Events such as A and B that do not overlap are called **mutually exclusive;** they mutually exclude each other.

> ***mutually exclusive events*** ■ Two or more events that have no outcomes in common.

EXAMPLE 5.15 Are A and C mutually exclusive? Find $P(A \text{ or } C)$.

Solution $P(A) = 4/52$, $P(C) = 13/52$, but the queen of spades occurs in both A and C. Thus A and C are not mutually exclusive and

$$P(A \text{ or } C) = 4/52 + 13/52 - \underset{\substack{\uparrow \\ \text{For the queen of spades}}}{1/52} = 16/52.$$ □

EXAMPLE 5.16 Are A and D mutually exclusive? Find $P(A \text{ or } D)$.

Solution $P(A) = 4/52$ and $P(D) = 26/52$. There are two red queens, the queen of hearts and the queen of diamonds, so A and D are not mutually exclusive.

$$P(A \text{ or } D) = 4/52 + 26/52 - \underset{\substack{\uparrow \\ \text{For the 2 red queens}}}{2/52} = 28/52$$ □

EXAMPLE 5.17 Are C and D mutually exclusive? Find $P(C \text{ or } D)$.

Solution $P(C) = 13/52$ and $P(D) = 26/52$. There are no red spades; C and D are mutually exclusive.

$$P(C \text{ or } D) = 13/52 + 26/52 = 39/52.$$ □

We can state, in general, that

RULE

$$P(A \text{ or } B) = P(A) + P(B) - P(A \text{ and } B)$$
$$= P(A) + P(B) \quad \text{only if } A \text{ and } B \text{ are mutually exclusive.}$$

If A and B are mutually exclusive, they have no outcomes in common. Therefore $P(A \text{ and } B) = 0$.

In general, are A and $\bar{A}$ mutually exclusive? By definition $\bar{A}$ contains all outcomes not in A, so A and $\bar{A}$ are mutually exclusive.

In Chapter 6, we will need to be careful about subtracting out probabilities when events overlap. We will be deriving probabilities from tables, which means that we *must* be aware of possible overlapping events and deal with them accordingly.

EXERCISES/Section 5.3

1. What is the probability of not throwing two 6's when rolling two dice?
2. What is the probability of not getting a face card when drawing a card from a deck?
3. Are the following events mutually exclusive? Find $P(A \text{ or } B)$.
   a) When throwing two dice,

      A = getting a sum of 6;

      B = rolling a 4 on one die.
   b) When drawing a card,

      A = drawing a jack;

      B = drawing a spade.
   c) When throwing two dice,

      A = rolling a sum of 10;

      B = rolling a sum of 6.
   d) When tossing two coins,

      A = getting at least one H;

      B = getting at least one T.
   e) When throwing two dice,

      A = getting at least one 6;

      B = getting a sum of 10.
4. When tossing two dice, what is the probability of getting
   a) a sum of 7 or 11?
   b) doubles, or a sum of 7?
   c) doubles, or a sum of 8?
   d) a sum of 9 or an even sum?
   e) a 1 on the first die or a sum of 7?
5. Draw a card from a 52-card deck. What is the probability of getting
   a) a jack or a spade?
   b) a face card or a club?
   c) a jack or an even number? (*Hint:* Count the queen as even.)
   d) a number below 10 or a queen? (*Hint:* Count the ace as 1.)
   e) a number below 10 or a club?

5.4

Independence and Conditional Probabilities

Events Formed Using the Word AND

Now we need to consider events that are formed by combining outcomes or other events, using the word *and*. The event A and B consists of all those outcomes contained in both A and B at the same time, that is, all outcomes in the overlap of A with B. In Section 5.3, we worked with examples that had overlapping events but were able to determine their joint probability by counting. We will continue to use examples that have countable events and develop principles and rules to be applied for $P(A \text{ and } B)$ in general.

EXAMPLE 5.18 If one card is drawn from a deck of 52, what is the probability that that card will be both red and a jack, $P(A \text{ and } B)$? Let

A = the event *a red card is drawn.*

B = the event *a jack is drawn.*

Solution Since there are two red jacks, the jack of hearts and the jack of diamonds, $P(A \text{ and } B) = 2/52 = 1/26$. As before, let us analyze this in parts:

$$P(A) = 26/52 = 1/2$$

$$P(B) = 4/52 = 1/13$$

How can we combine the last two fractions to produce the answer $1/26$? What arithmetic process should we use? If we multiply ($1/2 \cdot 1/13 = 1/26$), we find that

$$P(A \text{ and } B) = P(A) \cdot P(B). \quad \square$$

However, lest we be too hasty in using this generalization, let us consider another example.

EXAMPLE 5.19 If one card is drawn from a deck of 52, what is the probability that it will be both red and a spade? Let C = the event *a spade is drawn*. What is $P(A \text{ and } C)$?

Solution If we try to apply the generalization from Example 5.18, we find that

$$P(A) = 26/52 = 1/2.$$

$$P(C) = 13/52 = 1/4.$$

But is $P(A \text{ and } C)$ = P(drawing a red and a spade at the same time) = $\frac{1}{2} \cdot \frac{1}{4} = \frac{1}{8}$? No, it is impossible to draw a red spade, so

$$P(A \text{ and } C) = 0. \quad \square$$

How is Example 5.18 different from Example 5.19? When we wanted to add probabilities, we had to consider the property of mutual exclusivity; in order to multiply probabilities, we must consider the property of independence.

> ***independence*** ■ Two events, A and B, are said to be independent if the occurrence of one does not affect the *probability* of the occurrence of the other. If A occurs, $P(B)$ does not change because of that fact. Similarly, if B occurs, $P(A)$ remains constant.

Determining whether or not A and B are independent requires that we ask ourselves some questions about what happens to a sample space when we are told that some event has occurred in the sample space. Let us reexamine Examples 5.18 and 5.19. We know that

$$P(A) = 1/2;$$

$$P(B) = 1/13;$$

$$P(C) = 1/4.$$

We want to find $P(A \text{ and } B)$ and $P(A \text{ and } C)$. Are A and B independent? What is the probability of B if we pretend that A has occurred? If we know that a red card has been drawn, we know that there are 26 possibilities to consider. We can *limit our sample space* to these 26 red cards. Now, what is the probability that the drawn card is a jack? There are 2 jacks in our limited sample space, so

$$P(B \text{ knowing } A \text{ has occurred}) = 2/26 = 1/13.$$

Our knowing that A has occurred has not affected $P(B)$ because it is still $\frac{1}{13}$. Thus A and B are independent. If we reserve the roles of A and B, will we come to the same conclusion? If we know that a jack has been drawn, we know that we have one of four cards: the jack of hearts, the jack of diamonds, the jack of spades, or the jack of clubs. What is the probability that we have a red card? There are 2 possibilities out of 4:

$$P(A \text{ knowing } B \text{ has occurred}) = 2/4 = 1/2,$$

and again $P(A)$ is unaffected by B. Thus A and B are independent.

In general, when we know (or pretend) that one event has occurred and wish to find the probability of a second event, we write

$$P(A/B),$$

which is read "the probability of A given B has occurred." We are looking for the probability of A using B as additional information to limit the sample space and the amount of counting required. This type of probability is called **conditional probability.**

> ***conditional probability*** ■ A probability determined by using one event as a condition for limiting the sample space of the desired event.

Using this approach for Example 5.19, we find that $P(A) = 1/2$ and $P(C) = 1/4$. If we assume that A has occurred (we have drawn a red card), what is the probability of drawing a spade, $P(C/A)$. There are no red spades, so $P(C/A) = 0/26 = 0$. Thus A and C are not independent. Using the knowledge of A's occurrence requires us to change C's probability to 0.

Let us reverse the order in which we consider A and C to find out whether this makes a difference in our conclusion. If we assume that C has occurred (we have drawn a spade), what is the probability of drawing a red card, $P(A/C)$?

$$P(A/C) = 0/13 = 0$$

Again, the conditional probability for A is different from the unconditional $P(A)$, so A and C are not independent.

Let us go back to our generalization that calls for multiplying probabilities when events are joined by the word *and.* We saw in Example 5.18 that, when A and B were independent, $P(A \text{ and } B) = P(A) \cdot P(B)$. When events are not independent, as in Example 5.19, multiplication will also work, provided we use conditional probabilities.

$$P(A \text{ and } C) = P(A) \cdot P(C/A) = 1/2 \cdot 0 = 0.$$

or

$$P(A \text{ and } C) = P(A/C) \cdot P(C) = 0 \cdot 1/4 = 0.$$

We can now generalize as follows:

RULE

$$\begin{aligned} P(A \text{ and } B) &= P(A) \cdot P(B/A) \\ &= P(A/B) \cdot P(B) \\ &= P(A) \cdot P(B) \quad \text{if } A \text{ and } B \text{ are independent.} \end{aligned}$$

Before we leave this rule, let us recall the definition of independence: It tells us that knowing the occurrence of one event, and using it as a condition, makes no difference in the probability of the other event. Stated in symbols, this gives us

$$\left.\begin{array}{c} P(A/B) = P(A) \\ \text{or} \\ P(B/A) = P(B) \end{array}\right\} \text{if } A \text{ and } B \text{ are independent.}$$

Conditional probability is a difficult concept, so let us consider several examples. These examples indicate that independence is a useful property and that, if it is not guaranteed, probabilities of joint events will be affected. In general, *and* should trigger *multiply* just as *or* meant *add*, but care is essential whenever independence is in doubt.

EXAMPLE 5.20 Consider the tossing of two dice; let

A = the event *obtaining an even sum;*
B = the event *obtaining doubles;*
C = the event *obtaining a sum of 6.*

Find: (a) $P(A/B)$, (b) $P(B/A)$, (c) $P(B/C)$, and (d) $P(C/A)$.

Solution

a) $P(A/B)$ tells us that B has occurred: we have rolled doubles. We need to consider only the six cases of doubles. How many of these are even sums? All of them.

$$P(A/B) = 6/6 = 1.$$

b) $P(B/A)$ tells us to assume that A has occurred: we have an even sum. There are 18 even sums in the 36 outcomes when two dice are tossed. Now how many of those 18 are doubles? All doubles are even sums, so there are 6.

$$P(B/A) = 6/18 = 1/3.$$

Note that $P(A/B) \neq P(B/A)$.

c) $P(B/C)$ tells us that we have a sum of 6. There are five such cases, one of which is a double, (3, 3).

$$P(B/C) = 1/5.$$

d) $P(C/A) = 5/18$, as there are 18 even sums of which five are six. □

EXAMPLE 5.21 In a random sample of 1000 registered voters, people were asked to state their family income and opinion regarding the president's foreign policy. The following table shows the results.

		Income Level ($)				
		0–10,000	10,001–25,000	25,001–100,000	Over 100,001	Total
Opinion	Favor	75	100	150	100	425
	Oppose	175	175	75	40	465
	No opinion	50	25	25	10	110
	Total	300	300	250	150	1000

Let

A = choosing someone with an income of $0–10,000;
B = choosing someone who opposes the president's foreign policy.

Find: **(a)** $P(A)$, **(b)** $P(B)$, **(c)** $P(A/B)$, and **(d)** $P(B/A)$.

Solution

a) $P(A) = 300/1000$.

b) $P(B) = 465/1000$.

c) We are limited to those who oppose the president's foreign policy:

$$P(A/B) = 175/465.$$

d) We are limited to those with incomes of $0–10,000:

$$P(B/A) = 175/300. \quad \square$$

EXAMPLE 5.22 A box contains two green and five yellow balls. Consider the drawing of two balls.

a) If the first ball is replaced before the second is drawn, find
   1) $P(2G)$.
   2) P(one yellow and one green).
   3) P(two the same color).

b) If the first ball is not replaced before the second is drawn, find
   1) $P(2G)$.
   2) P(one yellow and one green).
   3) P(two the same color).

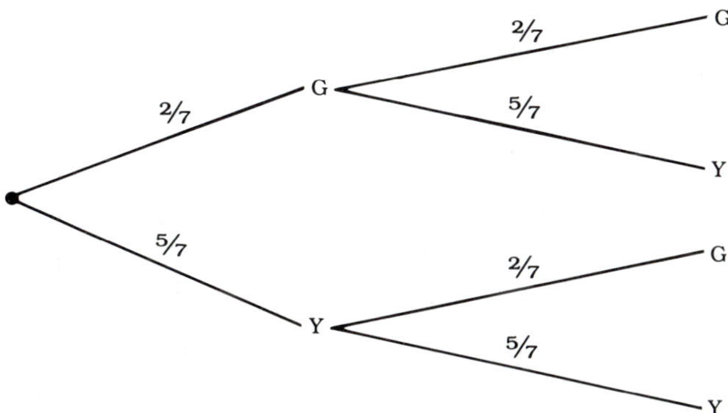

Figure 5.8

Solution

a) We begin by drawing the tree diagram of the sample space (Fig. 5.8). A helpful device is to label branches of the diagram with appropriate probabilities. Note that the probabilities associated with the second set of branches are identical to those of the first set. Why? The first limb represents a green ball followed by a second green ball or "G and G" in order, and so on.

1) $P(2G) = 2/7 \cdot 2/7 = 4/49$.

2) P(one yellow and one green) $= P$(Y and G or G and Y) $= P$(Y and G) $+ P$(G and Y) $= 5/7 \cdot 2/7 + 2/7 \cdot 5/7 = 20/49$.

3) P(two the same color) $= P(2Y \text{ or } 2G) = P(2Y) + P(2G) = 5/7 \cdot 5/7 + 2/7 \cdot 2/7 = 29/49$.

We can generalize to say that: *we multiply along limbs to obtain probabilities for the entire limb; and, if an event utilizes more than one limb, we add probabilities of those limbs required for the event.*

b) We now have different probabilities, as shown on tree diagram in Fig. 5.9. Note that the probabilities at each branching point add to 1, and that the probabilities associated with the second set of branches are different from those for the first set.

1) $P(2G) = 2/7 \cdot 1/6 = 2/42$.

2) P(one yellow and one green) $= 2/7 \cdot 5/6 + 5/7 \cdot 2/6 = 20/42$.

3) P(two the same color) $= 2/7 \cdot 1/6 + 5/7 \cdot 4/6 = 22/42$. □

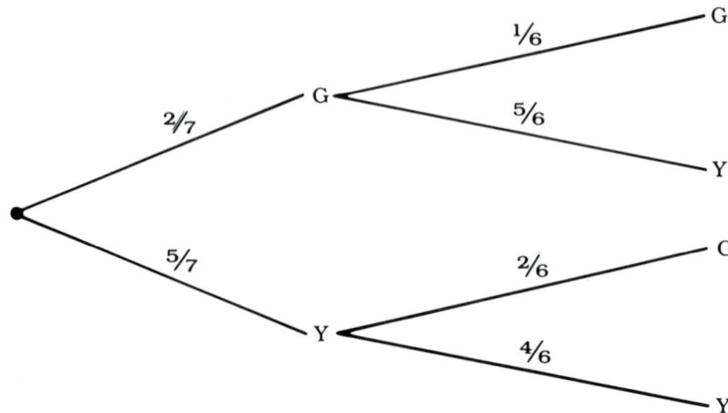

Figure 5.9

EXAMPLE 5.23 An auto mechanic kept records of characteristics of car repair jobs he did during a recent two-week period. Of the 64 cars brought in for repair, 24 were imports, 35 had \$50 or less worth of repair work done, 12 had to be kept overnight, and 17 required work on the electrical system; 52 of the jobs were paid for by credit cards.

a) If one of these jobs is chosen at random, what is the probability that
   1) it involved a foreign car.
   2) the car had to be kept overnight.
b) If two job invoices are chosen at random without replacement, what is the probability that
   1) they both are for domestic cars.
   2) neither car had more than \$50 work done on it.

Solution

a) Since there are 64 cars,
   1) $P(\text{foreign car}) = 24/64$.
   2) $P(\text{overnight}) = 12/64$.
b) 1) $P(\text{both domestic cars}) = P(\text{1st is domestic}) \cdot P(\text{2nd is domestic/1st is domestic}) = 40/64 \cdot 39/64 = 1560/4096$.
   2) $P(\text{neither had more than \$50 work done}) = P(\text{1st did not}) \times P(\text{2nd did not/1st did not}) = 35/64 \cdot 34/64 = 1190/4096$. □

We will utilize the concept of independence in Chapter 6, as we broaden our notion of probability in order to allow us to relate it to distributions. The analysis of data such as that in Example

5.21 will also require an understanding of independent events and the resulting multiplication of probabilities.

Sometimes independence of events is obvious. Coins and dice do not have memories; therefore consecutive tosses provide independent events. Repetitive draws from a population with replacement returns the population to its original state before each new draw, so events defined for that population are independent. In polling situations, people are not returned to a population after they have been questioned and thus given an opportunity to be repolled. However, when populations are large, the resulting change in the denominator of any calculated probabilities (one less to choose from for the second draw, etc.) is so small that changes in the resulting probabilities are minimal.

NEWS & VIEWS 5.5

The use of probabilities in the courtroom

Trials in which the prosecution builds a case based on circumstantial evidence are based on the hope that the jury will intuitively use joint probabilities. The Federal Rules of Evidence define relevance in terms of probability: Evidence is relevant if it has "any tendency to make the existence of any fact that is of consequence to the determination of the action more probable or less probable than it would be without the evidence." What is the likelihood that the crime was committed by the accused when evidence A *and* evidence B *and* evidence C, etc., all point to him? The prosecutor hopes that the jury will think that this joint probability is very large, and thus the probability that the accused did commit the crime is large. Probability theory tells us that we should multiply the probabilities of events joined by the word *and*. However, research seems to indicate that most people intuitively add component probabilities when combining probabilities and thus lean toward the prosecutor's view.

The most famous case of this type involved the mugging of an elderly person in Los Angeles. A couple described as a man with beard and mustache and a blond girl with ponytail were seen leaving the scene in a yellow car. Later a couple resembling this description was arrested. The prosecutor brought in a mathematics professor to testify; he suggested that the following conservative probabilities could be assigned to the characteristics noted by the witnesses:

Yellow auto	1/10
Man with mustache	1/4
Girl with ponytail	1/10
Blond girl	1/3
Bearded man	1/10

The prosecutor multiplied these probabilities, although the events are far from independent. The jury convicted the couple on the basis of the resulting small probability that there could be another such couple. Fortunately, this misuse of mathematics was spotted on appeal and the decision overturned.

The great mathematician Leibniz remarked that trial lawyers practice the art of logic based on contingencies, and Justice Oliver Wendell Holmes said in 1897: "For the rational study of the law, . . . the man of the future is the man of statistics."

EXERCISES/Section 5.4

1. Are the following events independent?
   a) Drawing numbers in the draft lottery.
   b) The sex of your first child and the sex of your second child.
   c) Getting an even number on the roll of one die and getting a 5 on the same roll.
   d) Drawing two cards at the same time from a deck and getting a club on the first and a club on the second.
   e) Drawing two cards, replacing the first before drawing the second, and getting a club followed by another club.
2. Two dice are used. What is the probability of getting the following?
   a) A sum of 7 three times in a row.
   b) At least one 6 on one toss.
   c) Doubles, followed by a sum of 7, and then doubles again.
   d) Doubles twice in three tosses.
   e) A sum of 7 only once in three tosses.
3. Draw two cards from a deck, putting the first card back and reshuffling before drawing another.
   a) What is the probability of getting two aces?
   b) What is the probability that both are hearts?
   c) What is the probability of getting a jack and a spade as your two cards?
   d) What is the probability of getting a 6 and a spade as your two cards?
4. When you draw two cards, what is the probability of getting a spade and a heart (a) with replacement of the first before drawing the second and (b) without replacement?
5. When you draw two cards, what is the probability of getting a number below 6 and a face card (a) with replacement and (b) without replacement?
6. When you draw two cards, what is the probability of getting at least one diamond (a) with replacement and (b) without replacement?
7. What is the probability of getting the following on ten tosses of a coin?
   a) HHHHHTTTTT
   b) HTHTHTHTHT
   c) THTHTHTHTH
   d) TTTTTHHHHH
   e) HHHTTTHHTT
8. Are mutually exclusive events ever independent?
9. Consider the drawing of one card from a deck. Find
   a) P(red/diamond).
   b) P(diamond/red).
   c) P(6/numbered card).
   d) P(6/even numbered card).
   e) P(even numbered card/6).
10. Consider the tossing of two dice, one green and one red. Find
    a) P(sum of 10/doubles).
    b) P(doubles/sum of 10).
    c) P(red die is even/green die is even).
    d) P(green/6).
11. The following breakdown shows political affiliation by sex for a sample of 100 men and 100 women.

	Men	Women
Democrat	55	45
Republican	40	50
Independent or other	5	5

    Find:
    a) P(Woman/Republican).
    b) P(Democrat/Man).
    c) P(Other/Man).
    d) P(Other).
    e) Are the events *choosing a man* and *choosing an independent or other voter* independent?
12. A manufacturing concern gathered the following information about production of defective items.

Day of Week	Number of Defective Items	Number of Nondefective Items
M	62	1230
TWTh	84	4561
F	45	1018

Find:

a) P(defective/Monday).
b) P(nondefective/TWTh).
c) P(Friday/defective).
d) P(Friday).
e) Are the events *choosing Friday* and *choosing a defective item* independent?

13. A random sample of 25 cars chosen from the parking lot of a nearby high school disclosed that 5 were a shade of red, 7 were a shade of blue, and 3 were a shade of yellow. Let R, B, and Y denote the choice of a car of each color shade respectively. Determine:

a) $P(\bar{R})$ d) P(R or B) g) $P(\overline{\text{R or B}})$
b) $P(\bar{B})$ e) P(R or Y) h) $P(\overline{\text{R or Y}})$
c) $P(\bar{Y})$ f) P(B or Y) i) $P(\overline{\text{B or Y}})$

14. A political action committee surveyed the electorate in a certain state on the question of a tax deduction proposal to be decided in an upcoming general election. The summary of its results is as follows:

	Support	Oppose	Undecided
Male	27	20	3
Female	23	20	7

Designate O = oppose, S = support, U = undecided, M = male, and F = female. Determine:

a) P(O/M) c) P(U/F) e) P(U)
b) P(S/F) d) P(O) f) P(S)
g) Does P(S/M) = P(S)? What conclusion do you draw from your answer?

15. In a box of ten light bulbs, two were damaged in shipping. You need to change bulbs in three fixtures. What is the probability that you will choose three undamaged bulbs on your first attempt?

Study Notes

KEY TERMS

complement of event A ■ The set of all outcomes in the sample space *not* in event *A*.

conditional probability ■ A probability determined by using one event as a condition for limiting the sample space of the desired event.

experiment ■ Any activity that generates data.

event ■ An outcome or a set of outcomes of an experiment.

independence ■ Two events, *A* and *B*, are said to be independent if the occurrence of one does not affect the *probability* of the occurrence of the other.

mutually exclusive events ■ Two or more events that have no outcomes in common.

outcome ■ The result of an individual trial of an experiment.

probability ■ The relative frequency with which an event *A* occurs.

$$P(A) = \frac{\text{The number of occurrences of } A}{n}$$

sample space ■ The complete, nonoverlapping list of all possible outcomes of an experiment.

OUTLINE

I. Simple probability
  A. Experiments generate sample spaces.
  B. Probability—a concept originating as a relative frequency.
  C. Law of large numbers—allows us to study simple classical problems and generalize to experimental ones.
  D. $0 \le P(X) \le 1$.
  E. $\Sigma P(X) = 1$, where values of X represent a sample space.

II. Rules
  A. Complements: $P(\bar{A}) = 1 - P(A)$.
  B. $P(A \text{ or } B) = P(A) + P(B) - P(A \text{ and } B)$

C. $P(A \text{ and } B) = P(A) \cdot P(B/A)$
$= P(B) \cdot P(A/B)$
$= P(A) \cdot P(B)$, if A and B are independent.

D. Conditional probabilities—found by using the given information to limit the sample space.

E. When using tree diagrams, multiply along limbs and add when more than one limb is required.

REVIEW PROBLEMS

1. The records office at a local community college disclosed that students planning to graduate declared majors in the following numbers: 300 in general studies, 75 in nursing, 125 in engineering, 210 in business, and 250 in arts and sciences. If a student is chosen at random from this list (assuming that no student has more than one declared major), what is the probability that the student's major is
   a) general studies?
   b) nursing?
   c) engineering?
   d) business?
   e) arts and sciences?
2. Preparing for school the other day, Sue Meyers packed two English notebooks, one math notebook, three science notebooks, and one foreign language notebook. When she got to school, she reached into her book bag and picked out a notebook at random. What is the probability that it was for
   a) English?
   b) math?
   c) science?
   d) foreign language?
3. Two cards are chosen from a well-shuffled deck; find the following probabilities if the first card is replaced before the second is drawn.
   a) P(jack on the second card).
   b) P(6 or lower on the second card).
   c) P(even number on the second card). (*Hint:* Count queen as even.)
   d) P(face card on the second card).
   e) P(diamond on the second card).
4. While gazing out of her office window the other day, an office manager observed that, of all the cars in the first two rows of the parking lot nearest her office, 10 were subcompacts, 8 were compacts, 12 were midsized, and 15 were full-sized. She also observed a fellow employee approach that part of the parking lot. Assuming that the employee's car is in the first two rows, what is the probability that the employee's car is
   a) subcompact?
   b) compact?
   c) midsized?
   d) full-sized?
5. For the toss-four-coins experiment done in Section 5.2, list all of the outcomes in the sample space using a tree diagram. Find the classical probabilities.
   a) P(0H)
   b) P(1H)
   c) P(2H)
   d) P(3H)
   e) P(4H)
6. In a game, a player's next move is determined by the outcome of a coin toss, followed by a pointer spin (Fig. 5.10). Draw a tree diagram to illustrate all possible outcomes of this decision sequence; use the diagram to determine the probability of:
   a) TW b) HG c) TY d) HR
7. A box contains four balls: one yellow, one green, one red, and one blue. You are to draw three balls. Construct a tree diagram that lists outcomes when
   a) each ball is replaced before another is drawn.
      Find:
      1) P(1Y)
      2) P(2G)
      3) P(3R)
      4) P(2Y)
      5) P(3G)
      6) P(3B)
      7) P(3Y)
      8) P(YRG)

Figure 5.10

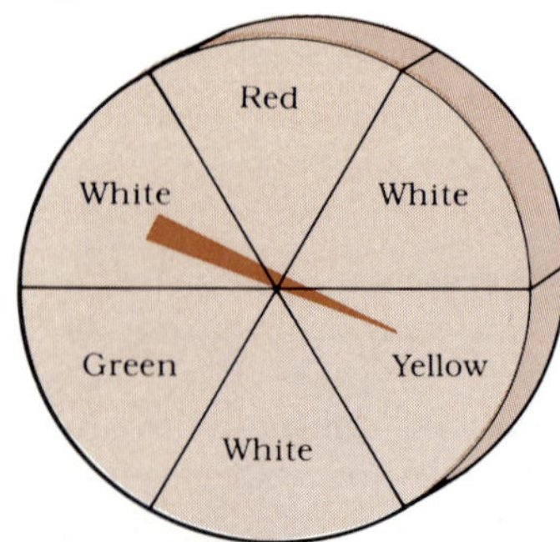

b) a ball is not replaced before drawing another.

Find:

1) $P(1Y)$ 2) $P(2G)$ 3) $P(3R)$ 4) $P(2Y)$ 5) $P(3G)$ 6) $P(3B)$ 7) $P(3Y)$ 8) $P(YRG)$

8. An auto mechanic kept records of characteristics of car repair jobs over a two-week period. Of the 64 cars brought in for repair, 24 were imports, 35 had $50 or less worth of repair work done, 12 had to be kept overnight, and 17 required work on the electrical system; 52 of the jobs were paid for by credit cards.

a) If one of these jobs is chosen at random, what is the probability that the job
   1) involved $50 or less in repair work.
   2) required electrical work.
   3) was paid for by credit card.
   4) was not on a foreign car.
   5) required more than $50 worth of repair work.
   6) did not involve keeping the car overnight.
   7) did not require electrical work.
   8) was not paid for by credit card.

b) If two job invoices are chosen at random without replacement, what is the probability that
   1) both were paid for by credit card.
   2) neither was for a car kept overnight.
   3) both were for cars kept overnight.
   4) neither job required electrical work.

9. Consider the drawing of two cards from a deck without replacement; that is, one card and then the other is drawn without returning the first to the deck. Let

A = the event *a ten is drawn on the first draw;*

B = the event *a ten is drawn on the second draw;*

C = the event *a jack is drawn on the second draw.*

Find: (a) $P(A \text{ and } B)$ (b) $P(A \text{ and } C)$

10. A box contains two red, three blue, and five yellow marbles. One is chosen at random. Determine:

a) $P(\bar{R})$ b) $P(\bar{B})$ c) $P(\bar{Y})$ d) $P(R \text{ or } B)$ e) $P(R \text{ or } Y)$ f) $P(B \text{ or } Y)$ g) $P(\overline{R \text{ or } B})$ h) $P(\overline{R \text{ or } Y})$ i) $P(\overline{B \text{ or } Y})$

11. A box contains ten black counters, numbered 0 through 9, inclusive, and ten white counters, also numbered 0 through 9. A counter is withdrawn at random. Let E represent even number; D, odd number; B, black counter; and W, white counter. Determine the probability that the counter drawn is

a) even or black.
b) even and black.
c) 3 or 5.
d) 2, 8, or 9.
e) neither white nor even.
f) not divisible by 3.

12. A *face deck* consists only of the jacks, queens, kings, and aces of the regular bridge or poker deck of playing cards. A card is drawn at random from such a deck. What is the probability that the card drawn is

a) not a king?
b) jack or ace?
c) red jack?
d) not a heart?
e) red or ace?
f) neither black nor king?

13. A *number deck* consists of the cards numbered 2–10 from a regular bridge or poker deck of playing cards. A card is drawn at random from such a deck. What is the probability that the card drawn is

a) not an even red card?
b) black or even?
c) red and divisible by 3?
d) larger than 5 and black?
e) less than 2 and red?
f) red or black?
g) red or 5 or 6?
h) black or 2 or 3?

14. While driving through Canada on vacation recently, Sara Montague got into the habit of accumulating loose change in her change purse. By the end of her trip, she had accumulated 25 pennies, 10 nickels, 15 dimes, and 10 quarters in Canadian currency; 20 pennies, 10 nickels, 5 dimes, and 5 quarters in U.S. currency. She reached into her change purse and drew out a coin at random. What is the probability that the coin she chose is

a) a dime?
b) not a quarter?
c) not a penny?
d) a U.S. coin or penny?
e) a nickel or dime?
f) a U.S. coin worth more than a penny?
g) a Canadian coin worth less than a dime?

15. A political pollster surveyed a sample of voters on their opinions regarding a nuclear freeze. The results, broken down by age groups, are summarized as follows:

	29 and Under (Y)	30–55 (M)	56+ (E)
Support (S)	65	140	95
Oppose (O)	35	110	55

Determine:

a) P(Y) d) P(S) g) P(S/Y) j) P(E/S)
b) P(M) e) P(O) h) P(M/S) k) P(S/E)
c) P(E) f) P(Y/S) i) P(S/M)

16. A random sample of 100 cars in the student parking lot was characterized according to age and type of tire. The results of this breakdown are:

	A	B	C
Radial	9	25	21
Nonradial	6	20	19

Where:

A = current model year;
B = 2–5 yrs old;
C = 6 or more years old.

Determine:

a) P(A) d) P(R) g) P(R/A) j) P(R/C)
b) P(B) e) $P(\bar{R})$ h) P(B/R) k) P(C/R)
c) P(C) f) P(A/R) i) P(R/B)

17. A box contains six red counters numbered 0–5 and four blue counters numbered 0–3. A counter is withdrawn; its number and color are recorded. Determine:

a) P(R) e) P(0) i) P(4) m) P(5/B)
b) P(B) f) P(1) j) P(5) n) P(B/5)
c) P(4/R) g) P(2) k) P(2/B)
d) P(R/4) h) P(3) l) P(B/2)

18. The experiment described in Problem 17 is extended. The first counter is not replaced and a second counter is drawn. Determine the probability that
a) the first is blue and the second red.
b) the first is red and the second red.
c) the first is 5 and the second blue.
d) the first is 0 and the second red.

19. A particular missile has a 98% chance of hitting the target. If four missiles are fired independently, find:
a) P(all four hit their targets).
b) P(none hits its target).
c) P(at least one hits its target).
d) P(exactly 1 hits its target).

20. If there is a 5% chance that a vending machine will not work properly—either in making change or dispensing the proper item—find the probability that you can get the snack you want (drink, crackers, and candy bar) from three different machines.

21. Component *X* on the space shuttle has a 2% chance of failure. Since it is critical to safety, it has three backups *A*, *B*, and *C* that respond in case of failure. *A* responds if *X* fails; *B* responds if *X* and then *A* fail; *C* responds if *X*, then *A*, then *B* fail. If the backups also have a 2% failure rate, what is the probability that
a) all will fail?
b) the component will fail but its backups will not? (*Hint:* Do we need to worry about *B* and *C* if *A* does not fail?)

Probability distributions

6

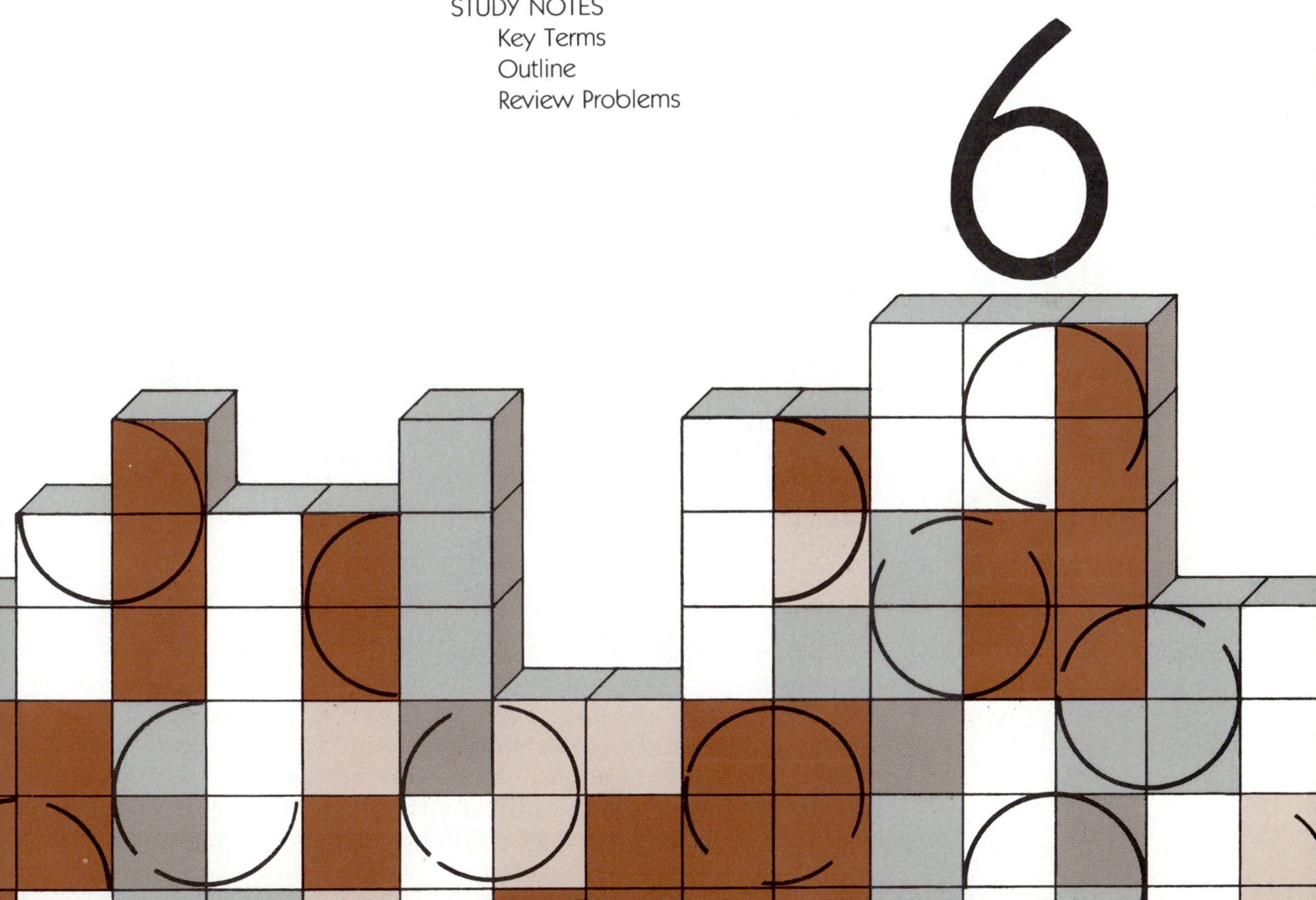

6.1 Probability Distributions: An Introduction

In Chapters 3 and 4, we studied frequency distributions and graphs and then went on to analyze these distributions with various types of descriptive measures. As you will recall, a distribution is a list of all possible values of a variable; we associated frequencies or relative frequencies with these values. Then in Chapter 5, using the law of large numbers, we associated relative frequencies with the concept of probability. Thus we are now ready to put our work with frequency distributions together with our work in probability and develop **probability distributions.**

> ***probability distribution*** ■ A set of all possible values of a variable together with their associated probabilities.

We use probability distributions for two types of variables in this chapter. Initially, we examine probability distributions using discrete variables and counting to take advantage of the simple examples we studied in Chapter 5. This leads us to a frequently used discrete probability distribution: the binomial probability distribution. Then we take up the most frequently used example of a continuous probability distribution: the normal distribution. By studying these two techniques, we will know how to deal with others as we encounter them later in our work. Probability distributions are idealized populations; we need to understand them before we can use them properly to generalize to a population from sample results. We will use such probabilistic models to help us make decisions and draw conclusions when we study statistical inference, beginning in Chapter 8.

The definition of probability distribution requires that we have values of *a* variable. Therefore we cannot necessarily consider an experiment in terms of its outcomes (HH, HT, etc.), but in terms of one variable. It is called a **random variable** when its values are determined by some random procedure, such as coin tossing or those used in selecting a random sample.

> ***random variable*** ■ A variable having values that are determined by a random process.

The random variables used in this section and Section 6.2 are discrete; that is, they always involve counting something. We will now think of events in terms of X = the number of. . . . Because we are able to determine probabilities associated with values of random variables, much of statistical theory is based on using random samples. It is also possible to determine probabilities associated with samples chosen other than randomly, and we will examine some of the ways we can use such information later in the text. Now let us look at some examples.

EXAMPLE 6.1 **Consider the experiment: toss two coins. Determine a probability distribution associated with this experiment.**

Solution Let the random variable be one that counts heads; that is, X = number of heads. Then the values of X are 0, 1, and 2. 0H represents TT, etc. From our work in Chapter 5, we can see that

X	$P(X)$
0	1/4
1	2/4
2	1/4

is the probability distribution we desire. This can also be expressed as a histogram, as frequency distributions were. Since we have discrete data, we will center the bars over the discrete values and plot probabilities on the vertical axis, as in Fig. 6.1. □

This example allows us to examine some properties of probability distributions. First, let us find the areas in the rectangles. Each rectangle has width = 1 and height = $P(X)$. Thus the areas are $1 \cdot 1/4 = 1/4$, $1 \cdot 2/4 = 2/4$, and $1 \cdot 1/4 = 1/4$. These are numbers between 0 and 1 that sum to 1 and thus may be thought of as probabilities.

Figure 6.1

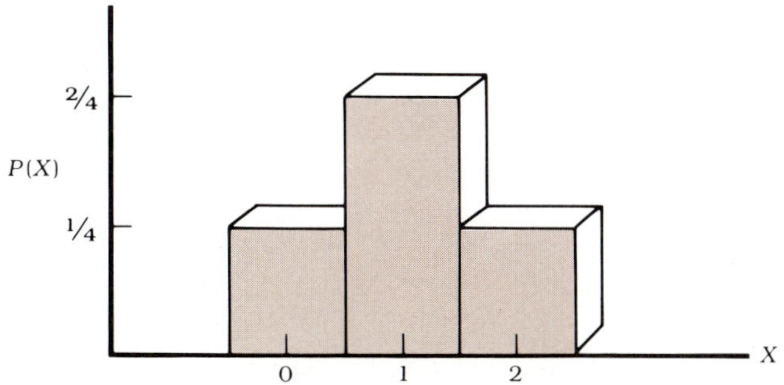

RULE

The areas of the bars are probabilities and

$$\Sigma \text{Areas} = 1.$$

We will generalize this rule to graphs of probability distributions associated with continuous variables (represented by curves) where we do not have rectangles to allow us to calculate areas easily. It is this property—that areas represent probabilities—that allows us to draw conclusions about populations and attach to them probabilities of our being correct.

Since a probability distribution graph includes all possible values of X, and thus the associated $P(X)$'s sum to 1, we are dealing with a population, not a sample. Thus there should be a mean, μ, and a standard deviation, σ, that describe this population. If we recall that μ represents the balance point of a population, it is clear that $\mu = 1$. We can calculate μ if we transform the equation for finding the mean from a frequency distribution,

$$\bar{X} = \frac{\Sigma Xf}{n}.$$

Now

$$\frac{\Sigma Xf}{n} = \sum \frac{(X \cdot f)}{n} = \sum \left(X \cdot \frac{f}{n}\right),$$

where f/n is a relative frequency; in terms of an experiment such as coin tossing, it becomes $P(X)$. Thus

$$\mu = \Sigma[X \cdot P(X)] \quad \text{for a discrete random variable } X.$$

For σ we use

$$\sigma = \sqrt{\Sigma[(X - \mu)^2 \cdot P(X)]}$$

or

$$\sigma = \sqrt{\Sigma[X^2 \cdot P(X)] - \{\Sigma[X \cdot P(X)]\}^2}.$$

Let us try these equations with the values from Example 6.1.

X	$P(X)$	$X \cdot P(X)$	$X^2 \cdot P(X)$
0	1/4	0	0
1	1/2	1/2	1/2
2	1/4	1/2	1
		1	1½ = 3/2

$$\mu = \Sigma X \cdot P(X) = 1.$$

$$\sigma = \sqrt{3/2 - 1^2} = \sqrt{3/2 - 1} = \sqrt{1/2}.$$

EXAMPLE 6.2 Consider the experiment: toss two dice. What is the probability distribution associated with X = number of dots showing on the top two faces? Graph this distribution and find μ and σ.

Solution From Example 5.5, we get:

X	$P(X)$	$X \cdot P(X)$	$X^2 \cdot P(X)$
2	1/36	2/36	4/36
3	2/36	6/36	18/36
4	3/36	12/36	48/36
5	4/36	20/36	100/36
6	5/36	30/36	180/36
7	6/36	42/36	294/36
8	5/36	40/36	320/36
9	4/36	36/36	324/36
10	3/36	30/36	300/36
11	2/36	22/36	242/36
12	1/36	12/36	144/36
		252/36	1974/36

Figure 6.2 shows the probability distribution for these data.

$$\mu = \frac{252}{36} = 7; \qquad \sigma = \sqrt{\frac{1974}{36} - 7^2} = 2.42. \quad \square$$

Figure 6.2

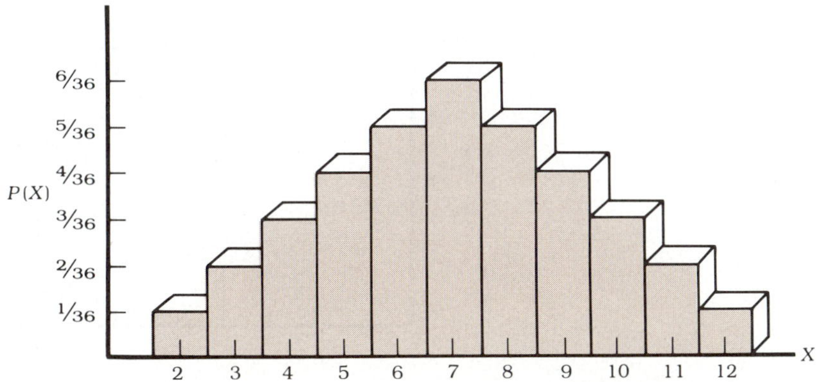

EXAMPLE 6.3 Consider the variable X with values of 3, 4, and 5. The probabilities associated with values of X are determined by an equation called a *function*,

$$P(X) = \frac{(X - 1)}{9}.$$

Find the probability distribution using this information, graph it, and find μ and σ.

Solution

X	$P(X) = (X - 1)/9$	$X \cdot P(X)$	$X^2 \cdot P(X)$
3	$(3 - 1)/9 = 2/9$	6/9	18/9
4	$(4 - 1)/9 = 3/9$	12/9	48/9
5	$(5 - 1)/9 = 4/9$	20/9	100/9
		38/9	166/9

We are sure that this relationship is a probability distribution because the sum of the probabilities is 1. Figure 6.3 shows this distribution.

$$\mu = \frac{38}{9} = 4.22. \qquad \sigma = \sqrt{\frac{166}{9} - \left(\frac{38}{9}\right)^2} = 0.79. \quad \square$$

Figure 6.3

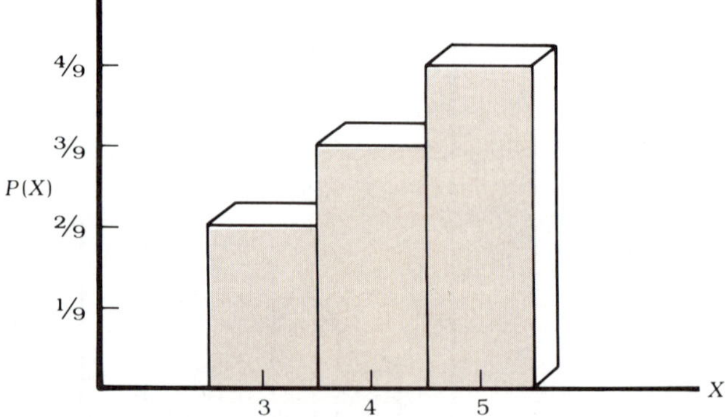

EXERCISES/Section 6.1

1. Consider the tossing of three coins. Let X = the number of tails that land up.
   a) What values can X assume?
   b) Find $P(X)$ for each of these values.
   c) Draw the histogram of this probability distribution.
   d) Calculate μ and σ.
2. Consider a hat containing one white tennis ball, one orange tennis ball, and one yellow tennis ball. Draw two balls from the hat with replacement. Let X = the number of white balls drawn.
   a) What values can X assume?
   b) Find $P(X)$ for each value of X. (A tree diagram may be helpful here.)
   c) Draw a histogram of this probability distribution.
   d) Find μ and σ.
3. A wallet contains five \$1 bills, two \$5 bills, a \$10 bill, and a \$20 bill. One bill is selected. Let X = the value of the bill chosen.
   a) What values can X take on?
   b) Find $P(X)$ for each value of X.
   c) Find μ and σ.
4. Consider the drawing of a card from a well-shuffled deck. Let X = the numerical value of the card, with ace counting as 1, jack as 11, queen as 12, and king as 13.
   a) List the values for X.
   b) Find $P(X)$ for these values.
   c) Draw a histogram of this probability distribution.
   d) Find μ and σ. For this exercise, how does $\mu = \Sigma X \cdot P(X)$ compare with $\mu = \Sigma X/n$?
5. Given

X:	1	2	3	4	5
$P(X)$:	0.1	0.4	0.2	0.2	0.1

   a) Determine μ and σ.
   b) Construct a histogram for this probability distribution.

For each of the following (a) find the probability distribution, (b) construct a histogram, and (c) find μ and σ.

6. $P(X) = X/15$ for $X = 1, 2, 3, 4, 5$.
7. $P(X) = (6 - X)/18$ for $X = 0, 1, 2, 3$.
8. $P(X) = (X^2 + 1)/6$ for $X = 0, 1, 2$.
9. $P(X) = (X^2 - 2)/n$ for $X = 2, 3, 4, 5$. (*Hint:* Find n first.)
10. $P(X) = 2X/n$ for $X = 1, 2, 3, 4$. (*Hint:* Find n first.)

6.2 Binomial Distribution

The binomial probability distribution is frequently used with discrete values of the random variable X. Let us examine its properties using an example.

EXAMPLE 6.4 A recently issued poll reported that 60% of all voters favor the president's foreign policy and 40% do not. Three randomly selected voters are then asked to state their opinions on the same issue. If X = the number favoring the policies, X can have values 0, 1, 2, or 3 in this case. Find the sample space of possible responses from the three voters and determine the associated probability distribution.

Solution The sample space is found using a tree diagram (Fig. 6.4). If we put probabilities on the tree diagram, we can find the probabilities associated with each outcome.

FFF	(3)	$0.6 \cdot 0.6 \cdot 0.6 = (0.6)^3$
$FF\bar{F}$	(2)	$0.6 \cdot 0.6 \cdot 0.4 = (0.6)^2 \cdot 0.4$
$F\bar{F}F$	(2)	$0.6 \cdot 0.4 \cdot 0.6 = (0.6)^2 \cdot 0.4$
$F\bar{F}\bar{F}$	(1)	$0.6 \cdot 0.4 \cdot 0.4 = (0.6) \cdot (0.4)^2$
$\bar{F}FF$	(2)	$0.4 \cdot 0.6 \cdot 0.6 = (0.6)^2 \cdot 0.4$
$\bar{F}F\bar{F}$	(1)	$0.4 \cdot 0.6 \cdot 0.4 = 0.6 \cdot (0.4)^2$
$\bar{F}\bar{F}F$	(1)	$0.4 \cdot 0.4 \cdot 0.6 = 0.6 \cdot (0.4)^2$
$\bar{F}\bar{F}\bar{F}$	(0)	$0.4 \cdot 0.4 \cdot 0.4 = (0.4)^3$

The numbers in parentheses after the outcomes classify the outcomes as values of X, the number favoring the policies. There are three cases of $X = 2$, each with the same probability, $(0.6)^2 \times 0.4$; three cases of $X = 1$, each with the same probability; and one case each of $X = 0$ and $X = 3$.

Figure 6.4

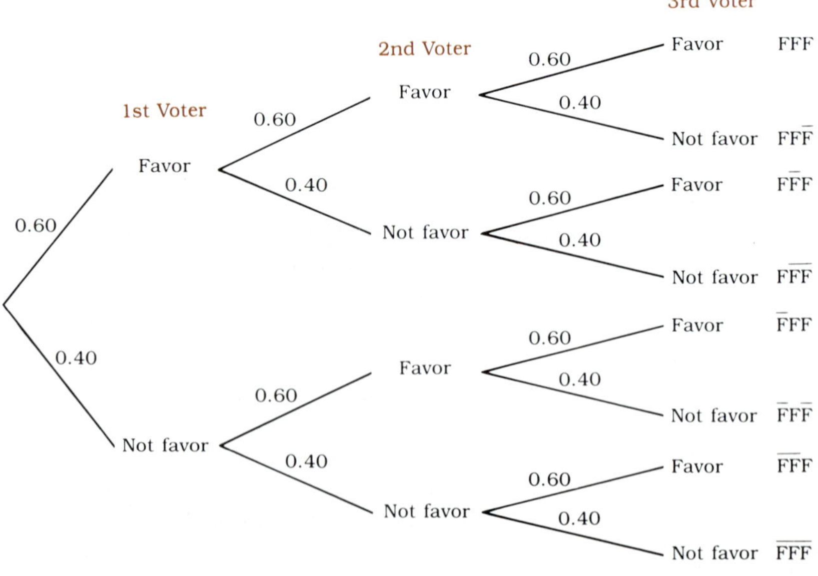

X	$P(X)$
0	$(0.4)^3 = 0.064$
1	$3 \cdot 0.6 \cdot (0.4)^2 = 0.288$
2	$3 \cdot (0.6)^2 \cdot 0.4 = 0.432$
3	$(0.6)^3 = 0.216$
	1.000

□

If we examine the properties of the experiment from which the preceding distribution came, we find that the responses of the voters were separated into only two (*bi*) categories: *favor* or *not favor.* We had three opportunities, called **trials,** to get a response. The random selection assured us of independence and thus told us that the probability that a voter favored the policy remained constant at 0.6. Similarly, the probability that a voter did not favor the policies remained constant at 0.4. Our work in Chapter 5 tells us then to multiply these appropriate probabilities along the limbs of the tree diagram.

binomial experiment ■ An experiment that has the following properties:

1. There are n independent, identically repeated trials, each of which has two outcomes. These outcomes are usually called *success* and *failure.*
2. X = the number of successes; X can take on values from 0 to n.
3. P(success on any one trial) = p, which remains constant for the life of the experiment. P(failure on any one trial) = q.

$$p + q = 1.$$

The binomial probability distribution becomes readily apparent in generalized form when we recall that a number with a zero exponent (except 0 itself) equals 1; for example $(0.6)^0 = 1$. Let us rewrite our distribution.

X	$P(X)$
0	$1 \cdot (0.6)^0 \cdot (0.4)^3$
1	$3 \cdot (0.6)^1 \cdot (0.4)^2$
2	$3 \cdot (0.6)^2 \cdot (0.4)^1$
3	$1 \cdot (0.6)^3 \cdot (0.4)^0$

Each probability is composed of three parts. The first is a whole number or *coefficient*, which tells us the number of limbs of the tree diagram contained in the event. The second is a power of 0.6, in this case p; and the third is a power of 0.4, the value of q. Note that the exponent on 0.6 matches the value of X, and that the exponent of 0.4 is $3 - X$, or, since $n = 3$, $n - X$. We have:

$$P(X) = \text{Number of ways } X \text{ successes can occur} \cdot p^X \cdot q^{n-X}.$$

It is unnecessarily tedious to draw a tree diagram for every binomial experiment in order to determine the coefficient; the binomial theorem from algebra can provide us with the numbers we need. The binomial coefficients can be found by using

$$\binom{n}{X} = \frac{n!}{(n-X)!X!}.$$

(For those who are unfamiliar with factorial (!) notation, an explanation and a set of practice exercises are contained in Appendix C.)

Thus

$$\boxed{P(X) = \binom{n}{X} p^X q^{n-X}.}$$

EXAMPLE 6.5 A new drug was found to be effective for 30% of the patients tested. If this drug is given to four randomly selected patients at Big City Hospital, what is the probability that it will be effective for two of them?

Solution First, we must identify the parts of the binomial experiment.

Success = Effective.

Failure = Not effective.

$p = 0.3.$

$q = 0.7.$

$n = 4.$

X = number of patients for whom the drug is effective.

We are asked to find $P(2)$. The preceding equation tells us that

$$\begin{aligned} P(2) &= \binom{4}{2} (0.3)^2 (0.7)^2 \\ &= \frac{4!}{2!2!} (0.3)^2 (0.7)^2 \\ &= 6 \cdot 0.09 \cdot 0.49 = 0.2646. \end{aligned}$$

Let us complete the probability distribution for $X = 0, 1, 3,$ and 4.

X	$P(X)$
0	$1 \cdot (0.3)^0 \cdot (0.7)^4 = 0.2401$
1	$4 \cdot (0.3)^1 \cdot (0.7)^3 = 0.4116$
2	$6 \cdot (0.3)^2 \cdot (0.7)^2 = 0.2646$
3	$4 \cdot (0.3)^3 \cdot (0.7)^1 = 0.0756$
4	$1 \cdot (0.3)^4 \cdot (0.7)^0 = 0.0081$
	1.0000

We can graph this distribution (Fig. 6.5) and also find the mean and the standard deviation:

X	$P(X)$	$X \cdot P(X)$	$X^2 \cdot P(X)$
0	0.2401	0	0
1	0.4116	0.4116	0.4116
2	0.2646	0.5292	1.0584
3	0.0756	0.2268	0.6804
4	0.0081	0.0324	0.1296
		1.2000	2.2800

$$\mu = \Sigma X \cdot P(X) = 1.2;$$

$$\sigma = \sqrt{\Sigma[X^2 P(X)] - \{\Sigma[X \cdot P(X)]\}^2}$$

$$= \sqrt{2.28 - 1.2^2}$$

$$= 0.9165. \quad \square$$

Figure 6.5

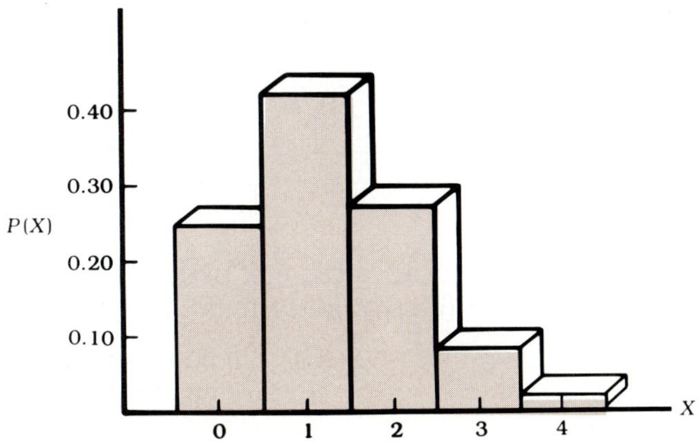

When n is large, calculations using the binomial formula are tedious, but we have an alternative—that of using tables. Table D.1 in Appendix D gives the binomial probabilities for $n \leq 25$ for various values of p. To use the table, first locate the value of n in the left column; then find the value of X as given in the problem; follow this row across until you come to the column for the proper value of p and read the resulting number, which is the desired probability. The entries 0.0000 mean that the probability is small and at four decimal places rounds to 0. If $n = 20$ and $X = 4$ successes with $p = 0.10$, $P(4) = 0.0898$ to four decimal places. Let us revisit Examples 6.4 and 6.5.

EXAMPLE 6.4 (cont.) A recently issued poll says that 60% of all voters favor the president's foreign policy and 40% do not. If we now randomly choose 20 voters and ask their opinions, find the probability that 14 will favor the president's policies.

Solution Now, $n = 20$, $X = 14$, and $p = 0.60$. Using Table D.1, we find that $P(14) = 0.1244$. □

EXAMPLE 6.5 (cont.) If a new drug was found to be effective for 30% of the patients tested, what is the probability that it will be effective for 15 patients in a randomly selected group of 25?

Solution With $n = 25$, $X = 15$, and $p = 0.30$, we use Table D.1 to find that $P(15) = 0.0013$. □

EXAMPLE 6.6 The article in News & Views 6.1 provides an example of the use of the binomial probability formula. If 40% of the women of a given time period embrace a fashion fad, what is the probability that in any randomly selected group of 15, four or less are participating in the fad?

Solution This problem asks us to find $P(X \leq 4)$ for $p = 0.40$ and $n = 15$.

$$\begin{aligned} P(X \leq 4) &= P(X = 0 \text{ or } X = 1 \text{ or } X = 2 \text{ or } \\ &\qquad X = 3 \text{ or } X = 4) \\ &= P(X = 0) + P(X = 1) + P(X = 2) \\ &\qquad + P(X = 3) + P(X = 4). \end{aligned}$$

The advantage of using Table D.1 should be apparent; all we need to do is add the proper probabilities from the table:

$$\begin{aligned} P(X \leq 4) &= 0.0005 + 0.0047 + 0.0219 + 0.0634 \\ &\qquad + 0.1268 = 0.2173. \end{aligned}$$ □

NEWS & VIEWS 6.1

Sociobiology's vaudeville team

E.O. Wilson and a young colleague survey genes and culture

One of the noisiest academic squabbles of the 1970s swirled about the teachings of sociobiology and its best-known preacher, Harvard Entomologist Edward O. Wilson. According to sociobiologists, some human behavior is under the sway of genetics: it evolves, the way anatomy and body chemistry do, to increase the chances of survival of the species. As if that idea were not controversial enough, Wilson introduced it to the public in a particularly unfortunate way: his 1975 book, *Sociobiology, the New Synthesis,* had 26 tightly reasoned chapters focusing on animals and one wildly speculative one on humans—the sort of sectión that publishers love to include and most academics have the sense to omit.

That chapter casually lumped together a number of contentious suggestions: that the traditional sexual division of labor (and by implication, male dominance) is a genetic fact; that humans have a limited amount of time left to understand their social evolution and control it; that it might be time for ethics to be "removed temporarily from the hands of the philosophers and biologicized."

Wilson's next book, *On Human Nature,* a sensible 1978 rewrite of the inflammatory generalizations of *Sociobiology's* last chapter, won a Pulitzer Prize but violated the first law of campus advancement: no book written in plain English has ever forwarded an academic career. Now Wilson has corrected that fault. *Genes, Mind and Culture,* written with a young Canadian theoretical physicist named Charles Lumsden and to be published in May, is a heap of jargon and theoretical math that runs scant risk of being widely understood. For instance, at one point the authors somberly produce a formula

$$P(n_1,t) = \binom{N}{n_1} p(t)^{n_1}(1 - p(t))^{N-n_1}$$

purporting to show how women's fashions change.* . . .

* The formula for the probability (P) that at any time (t) in history a certain number of women (n_1) are using a particular style.

Source: *Time,* January 26, 1981, p. 68.

For the binomial distribution, μ and σ may be found by using simpler formulas. After some algebraic manipulations:

$$\mu = np.$$
$$\sigma = \sqrt{npq}.$$

Using the information from Example 6.6, we find that

$$\mu = 4 \cdot 0.3 = 1.2;$$
$$\sigma = \sqrt{4 \cdot 0.3 \cdot 0.7} = 0.9165.$$

EXAMPLE 6.7 News & Views 6.2 tells us that Census Bureau results show that the proportion of youths aged 18–24 who voted in 1980 was 40%. If five randomly selected people who were in the

NEWS & VIEWS 6.2

Younger Americans marry later, vote less

About 4 Americans in 10 is under age 25, and as young adults they tend to marry later, have fewer children and are less likely to vote than their parents, the Census Bureau said yesterday.

The study, "Characteristics of American Children and Youth: 1980," issued yesterday, reports that there are about 92 million Americans under 25.

This group grew sharply with the "baby boom" of the 1950s and 1960s and is now producing a smaller boom by having its own babies. But overall, the report notes, the proportion of children in the population dropped from 30 percent in 1960 to 21 percent as of the 1980 census.

Thanks to this so-called baby boom "echo" effect, however, the number of children under 5 is expected to increase through 1990, before resuming its decline.

The report defines children as persons under 14 years of age.

A growth in the number of young women who work has helped produce a sharp jump in the percentage of children enrolled in nursery school, according to the report.

This enrollment among children aged 3 to 5 jumped from 3.7 million in 1966 to 4.9 million in 1980, the bureau said, "despite the decline in the total number of children in this age group."

"As more young mothers seek employment, they subsequently need more child care facilities for the daytime care of their young children. The enrollment of children in pre-primary facilities provides the mothers with the necessary child care services to enable them to work," according to the report by Jerry T. Jennings of the bureau's population division.

Black children were enrolled primarily in public facilities, the report noted, while white ones mostly were in private programs.

The study noted a sharp increase in the number of young people delaying marriage. The percentage of women aged 20 to 24 who hadn't married rose from 28 percent in 1960 to 50 percent in 1980. For men the corresponding increase was from 53 percent to 69 percent.

Other major findings of the study included:

- Approximately two-thirds of people aged 25 and under lived in metropolitan areas, predominantly in the suburbs.
- Young wives aged 18 to 24 expect to have a total of 2.2 children, on average. This is down from 2.7 children anticipated by wives of the same ages in 1967.
- The infant mortality rate in the United States in 1978 was 13.8 infant deaths per 1,000 live births, about half the rate recorded in 1960.

Large differences remained between the races, though, with a rate of 12.0 for whites and 21.1 for blacks.

- The proportion of youths aged 18 to 24 who voted in the 1980 presidential election was 40 percent, down from 50 percent in 1972.
- Out-of-wedlock births have tripled since 1950, although they have tapered off in recent years. The survey noted that the proportion is higher for black women, who tend to delay marriage to a greater extent than whites.
- The leading cause of death for children and youths of all ages was accidents, with motor vehicle accidents the most common.
- In 1979, about 16 percent of children under 18 lived in families with income below the poverty level, approximately the same percentage as in 1966.
- People aged 12 to 24 were more likely to become victims of both violent crimes and crimes of theft than older persons. This age group also constituted a substantial proportion of jail inmates, with 49 percent of those locked up being between ages 14 and 24.

Source: *Baltimore Sun,* March 5, 1982.

18–24 age bracket in 1980 are questioned, what is the probability that three of them voted? Complete the binomial probability distribution, draw its histogram, and find its μ and σ.

Solution We identify the components of the problem:

$$X = \text{number of students voting.}$$

$$n = 5.$$

$$p = 0.4.$$

$$q = 0.6.$$

We are to find $P(3)$.

$$P(3) = \frac{5!}{3!2!}(0.4)^3(0.6)^2$$
$$= 10 \cdot 0.064 \cdot 0.36 = 0.2304.$$

Let us complete the distribution.

X	$P(X)$
0	$1 \cdot (0.4)^0 \cdot (0.6)^5 = 0.0778$
1	$5 \cdot (0.4)^1 \cdot (0.6)^4 = 0.2592$
2	$10 \cdot (0.4)^2 \cdot (0.6)^3 = 0.3456$
3	$10 \cdot (0.4)^3 \cdot (0.6)^2 = 0.2304$
4	$5 \cdot (0.4)^4 \cdot (0.6)^1 = 0.0768$
5	$1 \cdot (0.4)^5 \cdot (0.6)^0 = 0.0102$
	1.0000

The histogram for this distribution is shown in Fig. 6.6 and

$$\mu = 5 \cdot 0.4 = 2;$$

$$\sigma = \sqrt{5 \cdot 0.4 \cdot 0.6} = 1.095. \quad \square$$

Figure 6.6

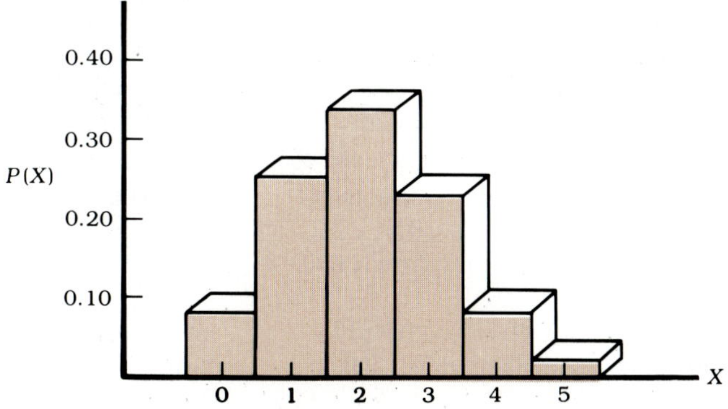

The binomial distribution usually comes into play whenever responses or sample results can be placed in two categories, such as:

Sex of a person.

Manufactured objects that are or are not defective.

Treatments or drugs that are or not effective.

Voters choosing between two candidates or two political parties.

Answering questions correctly or incorrectly.

Athletes who hit or do not hit, catch or do not catch, pass or do not pass.

Experiments are binomial if the data are recorded in only two categories; for example, tossing two dice n times and getting either doubles or not doubles. The possibilities are endless.

EXERCISES/Section 6.2

1. A large midwestern university discovered that 25% of its entering freshmen left school after two years of below-average work.
   a) If four randomly chosen freshmen are followed through their first two years of school, what is the probability that three of them will drop out?
   b) Complete the probability distribution.
   c) Graph the histogram of this distribution.
   d) Find μ and σ.
2. A large state university determined that 42% of its entering freshmen requires at least one remedial course in English or math.
   a) If six students are randomly chosen from this year's class, find the probability that one-half of them will need remedial work.
   b) Complete the probability distribution.
   c) Graph this distribution as a histogram.
   d) Find μ and σ.
3. The probability p that a driver will be involved in an accident downtown today is 0.10. A random sample of 15 drivers going downtown today is drawn.
   a) Find the probability that one driver will be involved in an accident today using Table D.1.
   b) What is μ; that is, how many drivers can be expected to be involved in accidents?
   c) What is σ?
4. The probability that a man aged 20–24 is married is 0.30. A random sample of 14 men in this age group is chosen.
   a) Find the probability that nine men in this sample are married.
   b) What is μ, the number of men in this group who can be expected to be married?
   c) What is σ?
5. A test contains ten multiple choice questions, each with five choices, but only one of which is correct. If a student comes unprepared and must guess to answer all questions, what is the probability that
   a) he will answer all questions incorrectly?
   b) he will answer all questions correctly?
   c) How many questions can he expect to answer correctly?
   d) What will be the class average if the entire class guesses?
6. A manufacturer guarantees a product for four years but finds that 20% of his merchandise is being replaced during this warranty period—a rather high figure if the profit margin is only

12%. In a random sample of 15 items,

a) what is the probability that none will fail within the warranty period? (Use Table D.1.)

b) what is the probability that at least one will fail within the warranty period?

c) how many can the manufacturer expect to replace in the warranty period; that is, what is μ?

7. About 70% of a store's customers respond to advertised sales. The other 30% shop when they have time or because they need an item, regardless of promotional advertising.

a) Of 14 randomly selected customers, what is the probability that six have seen the store's most recent promotional campaign? (Use Table D.1.)

b) What is the probability that at least ten have seen the store's campaign?

c) Find μ, the average number of shoppers responding to the sale ads, and σ.

8. About 60% of the population takes a vacation trip each year.

a) Of 25 randomly selected employees, what is the probability that eight will take a trip this year?

b) Find the probability that 5–9 employees (inclusive) will take a trip this year.

c) Find μ and σ.

9. About 75% of the population says they never read books.

a) What is the probability that, in a group of four randomly chosen people, all read books?

b) Complete the binomial distribution.

c) Draw its histogram.

d) Find μ, the average number who can be expected to read books, and σ.

10. About 25% of all dwellings have 5 rooms.

a) What is the probability that out of five randomly selected houses or apartments, four will have 5 rooms?

b) Complete the binomial distribution.

c) Draw its histogram.

d) Find μ and σ.

6.3 Normal Distribution

So far we have studied discrete probability distributions. We now examine the most important probability distribution that deals with continuous variables: the normal distribution. This is the familiar bell-shaped curve (Fig. 6.7). There are actually many curves that follow this pattern, one for every possible combination of means and standard deviations. The mean determines its center and, as we have seen before with Chebyshev's theorem, the standard deviation describes its spread, as shown in Fig. 6.8.

Each normal distribution in Figure 6.8 has the same mean but a different standard deviation, with $\sigma_1 > \sigma_2 > \sigma_3$. If we know the mean and the standard deviation of a normal distribution, we can determine everything else we need to know about it, such as its area. As we have already seen in this chapter, the area under the curve provides information about probability. As before, the total area under the curve equals 1, but now areas are not so obvious or easily calculated as they were with discrete variables. An added feature of the distribution is that it is symmetric with

Figure 6.7

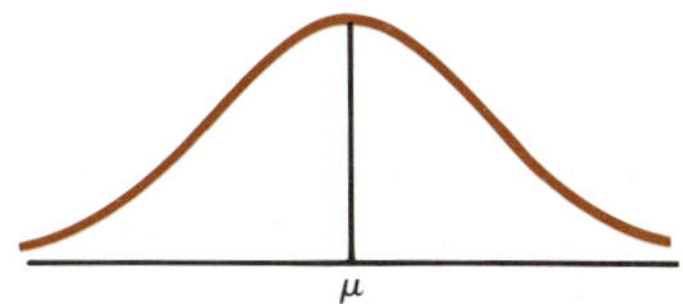

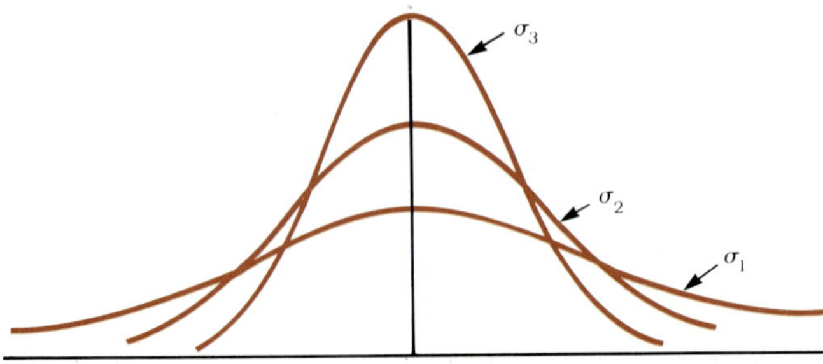

Figure 6.8

regard to the mean. We also need to recall that when dealing with continuous variables, as we did when setting up grouped frequency distributions, measured data fall within intervals that reflect the imprecision of the measurement. Consequently, we need to discuss probabilities of *intervals*, rather than of specific numbers, and determine them by finding the corresponding areas under the normal curve, such as that for interval a–b in Fig. 6.9.

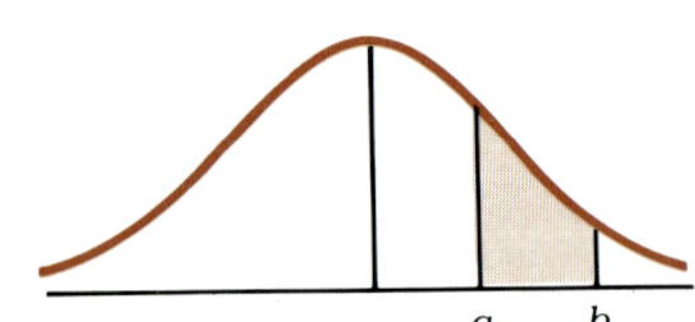

Figure 6.9

Since the areas are irregular in shape, we will determine them by using a table. It appears that we need a different table for each possible combination of μ and σ, but, fortunately, this is not the case. In Chapter 4 we discussed the procedure for changing a scale in units of measurement from one involving a specific mean and standard deviation to one that is unit-free and comparable to any other scale. This required the use of z scores, and we will now use them again. In fact, z scores are frequently called **standard normal scores** because they serve to standardize normally distributed variables with $\mu = 0$ and $\sigma = 1$. This allows us to use only one table of areas in order to determine probabilities: Table D.2.

When using a table, you should note any accompanying explanation. In this case there is a diagram, which indicates that the table gives areas under the curve for intervals starting at the mean, 0, and going to the right to z. The z scores are listed down the left side of the table (to one decimal place) and across the top (to the second decimal place). The corresponding area appears in the body of the table.

Figure 6.10

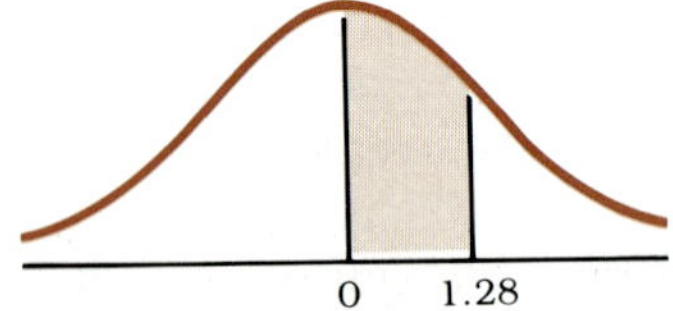

EXAMPLE 6.8 Find the area under the standard normal curve between 0 and $z = 1.28$ (Fig. 6.10).

Solution In Table D.2, we find 1.2 down the left side of the table and follow that row to the 0.08 column to find 0.3997. This is the desired area. □

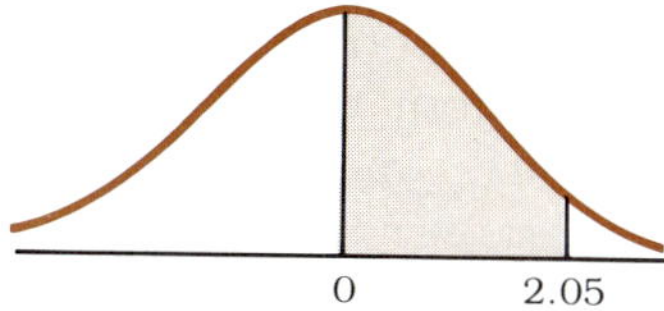

Figure 6.11

EXAMPLE 6.9 Find $P(0 < z < 2.05)$.

Solution This problem asks us to find a probability, so we need to determine the area under the curve for the interval given in Fig. 6.11. Finding 2.0 on the left and going over to the 0.05 column, we find 0.4798. Thus the probability of choosing someone at random and finding their z score on some normally distributed characteristic to be between 0 and 2.05 is 0.4798. □

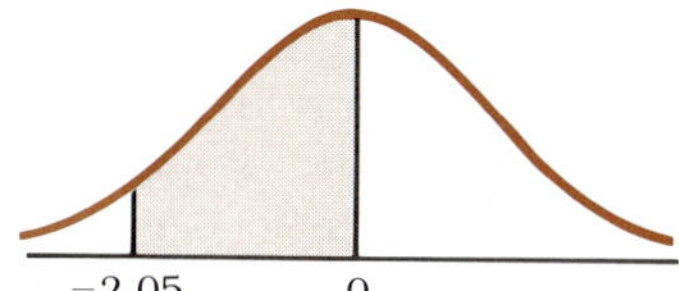

Figure 6.12

EXAMPLE 6.10 Find $P(-2.05 < z < 0)$.

Solution Because of symmetry in Fig. 6.12, the shape under the curve from 0 to +2.05 is the same as that from 0 to −2.05. Therefore $P(-2.05 < z < 0) = 0.4798$, as before. Remember that *probabilities will never be negative;* z scores may be negative (when they fall below the mean). □

However, we will need more probabilities than those from areas bounded by the mean. We can expand our use of the table, utilizing the symmetry properties of the normal distribution and the fact that the area under the curve equals 1. Be sure to draw a sketch of the desired area. This will help you see how to utilize numbers from Table D.2 to get the answers you want.

Figure 6.13

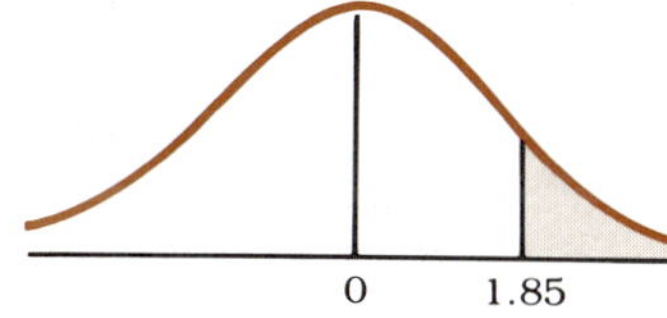

EXAMPLE 6.11 Find the following probabilities:

a) $P(z > 1.85)$
b) $P(z < -2.33)$
c) $P(z < 1.07)$
d) $P(-1.35 < z < 2.17)$
e) $P(0.95 < z < 1.38)$

Figure 6.14

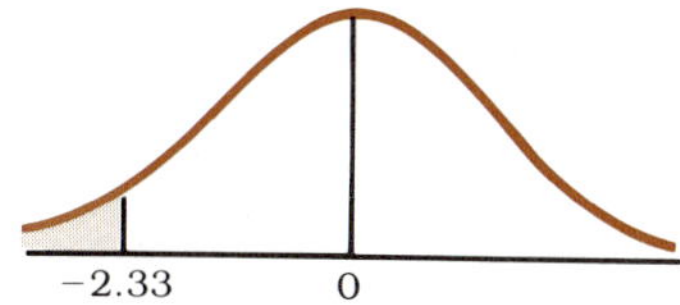

Figure 6.15

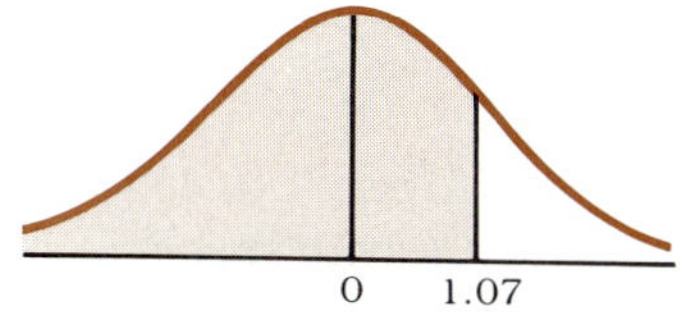

Solution

a) In Fig. 6.13, we are looking for the area to the right of 1.85, frequently called the *tail* of the distribution. In Table D.2 we find the area from 0 to 1.85 to be 0.4678. Remember that the total area under the curve is 1, so the area to the right of 0 is 0.5. We do not need the section from 0 to 1.85, so the desired area is $0.5 - 0.4678 = 0.0322$.
b) Similarly, in Fig. 6.14, the area from 0 to −2.33 is 0.4901 from the table and the area in the left tail is $0.5 - 0.4901 = 0.0099$.
c) In Fig. 6.15, the area to the left of 0 is 0.5. The table tells us that the area from 0 to 1.07 is 0.3577. Thus the total shaded area is $0.5 + 0.3577 = 0.8577$.

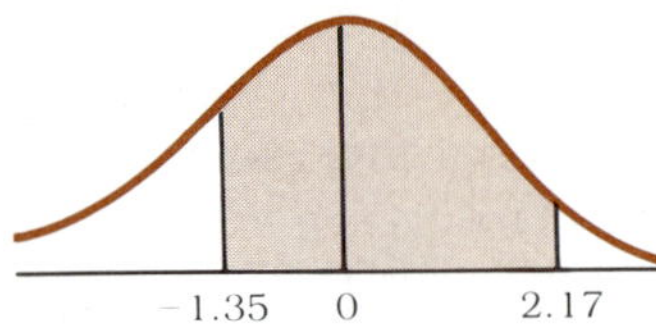

Figure 6.16

The solution in part (c) would answer the question: What is the probability that an individual chosen at random has, for some normally distributed characteristic, a z score that is less than 1.07? Solutions of this nature are used to answer questions not only about randomly chosen *individuals* but also about *populations* as a whole. The same solution would also answer the question: What proportion of the normally distributed population has a z score less than 1.07?

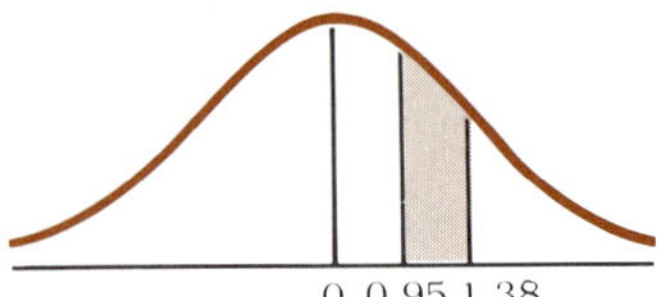

Figure 6.17

d) Using Fig. 6.16, the table gives us the areas from 0 to 2.17 (0.4850) and from 0 to −1.35 (0.4115). The desired area is the sum, 0.4115 + 0.4850 = 0.8965.

e) Using Fig. 6.17, the table gives us 0.3289 as the area between 0 and 0.95.

In Fig. 6.18, the area from 0 to 1.38 is 0.4162, but we only want to go from 0.95 to 1.38; we subtract, 0.4162 − 0.3289 = 0.0873. □

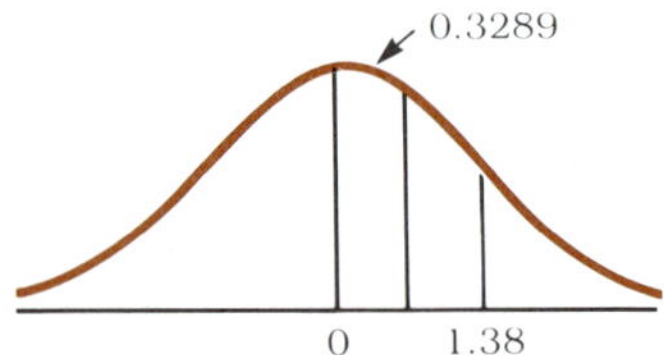

Figure 6.18

EXAMPLE 6.12 The reaction times, in seconds, for drivers in a certain age group who have consumed 2 oz of alcohol are normally distributed with a mean of 1.2 s and a standard deviation of 0.12 s.

a) What is the probability that a randomly chosen driver in this category will have a reaction time of 1.1–1.35 s?
b) What percent of all such drivers will have reaction times slower than 1.4 s? Remember that reaction times slower than 1.4 s means that $X > 1.4$.
c) A driver with a reaction time of 0.9 s will be faster than what percent of his peers?
d) What reaction time represents P_{60}?

Figure 6.19

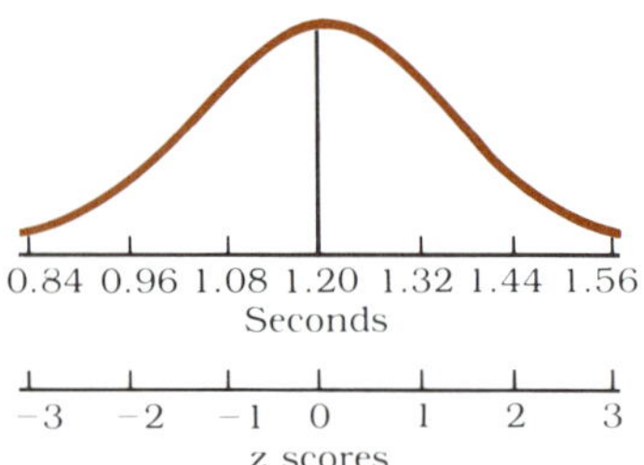

Solution This problem is not expressed in terms of z scores, and we must convert. Conversion is easy to visualize if we draw two scales, one for seconds and one for z scores (Fig. 6.19).

a) We need the area between 1.1 and 1.35 seconds in Fig. 6.20. We find the z scores of both endpoints.

$$z = \frac{X - \mu}{\sigma} = \frac{1.35 - 1.2}{0.12} = 1.25.$$

$$z = \frac{1.1 - 1.2}{0.12} = -0.83.$$

Figure 6.20

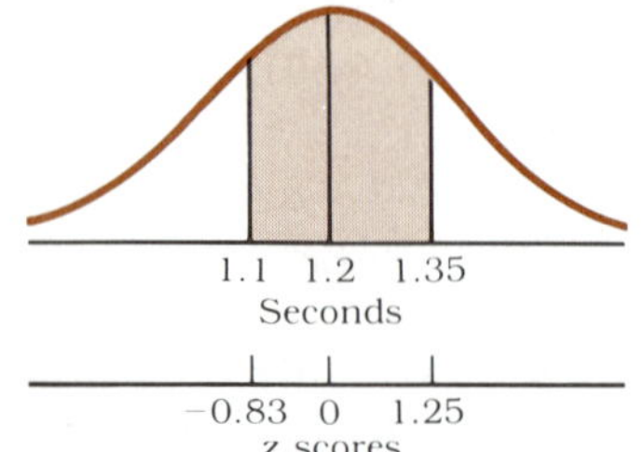

The area from −0.83 to 0 is 0.2967 and from 0 to 1.25 is 0.3944 (Table D.2). Summing to get the answer produces

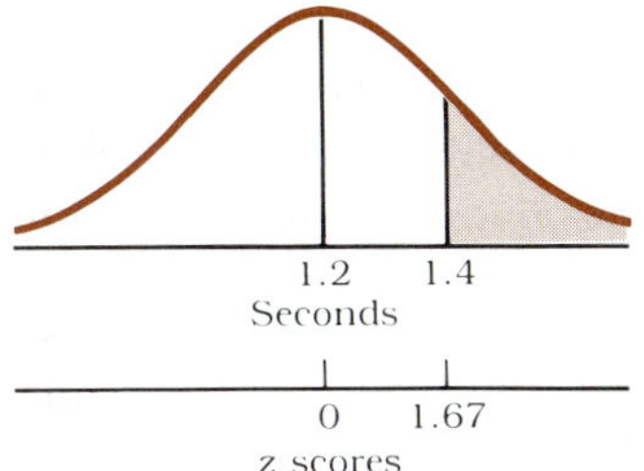

Figure 6.21

$0.2967 + 0.3944 = 0.6911$, the probability of choosing someone at random with a reaction time of 1.1–1.35 s.

b) Using Fig. 6.21 to find the z score of 1.4 s,

$$z = \frac{1.4 - 1.2}{0.12} = 1.67.$$

The table gives the area from 0 to 1.67 as 0.4525. Thus the area in the tail is $0.5 - 0.4525 = 0.0475$. This says that 4.75% of drivers in this age group and at this level of intoxication will have a reaction time slower than 1.4 s.

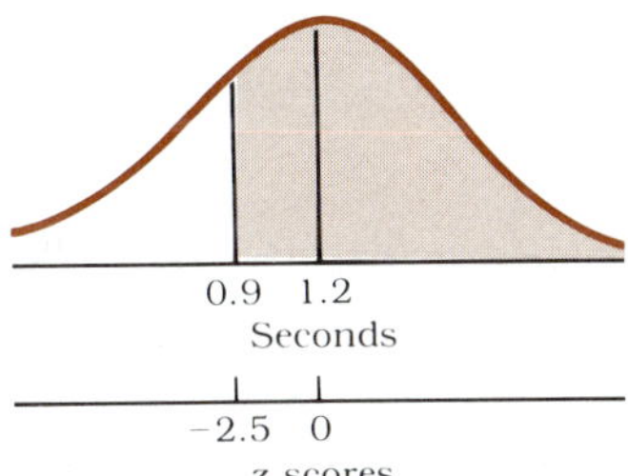

Figure 6.22

c) We are asked to find the area above 0.9 s in Fig. 6.22. The necessary z score is

$$z = \frac{0.9 - 1.2}{0.12} = -2.5.$$

The area from -2.5 to 0 is 0.4938. This is added to 0.5 to get $0.4938 + 0.5 = 0.9938$. Thus more than 99% of the drivers, as described, have reaction times slower than 0.9 s.

d) Now we are told something about percent of the given population. The sixtieth percentile (P_{60}) breaks the population into two groups: 60% of the total population (or area) will have numbers smaller than some dividing value, and 40% will have larger values. (Fig. 6.23).

Figure 6.23

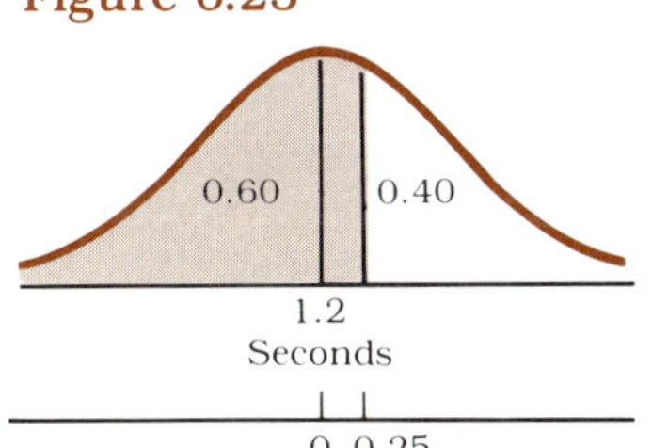

Since the only place we can find areas is in Table D.2, we need an area bounded by the mean on one side in order to read the table. In Fig. 6.23, we see that the area from 1.2 to the dividing line is 0.10 ($0.60 - 0.50$). The number in the body of the table that comes closest to 0.10 is 0.0987, which corresponds to a z score of 0.25. We need to convert this z score into a number on the seconds scale. We do this by going 0.25 standard deviations above the mean:

$$\begin{aligned} X &= \mu + z\sigma \\ &= 1.2 + (0.25)(0.12) \\ &= 1.23 \text{ s.} \end{aligned}$$

We have found that 60% of the given population has reaction times faster and 40% has reaction times slower than 1.23 s. □

The last part of this problem required more steps for its solution. Let us consider another example of this type before doing the exercises.

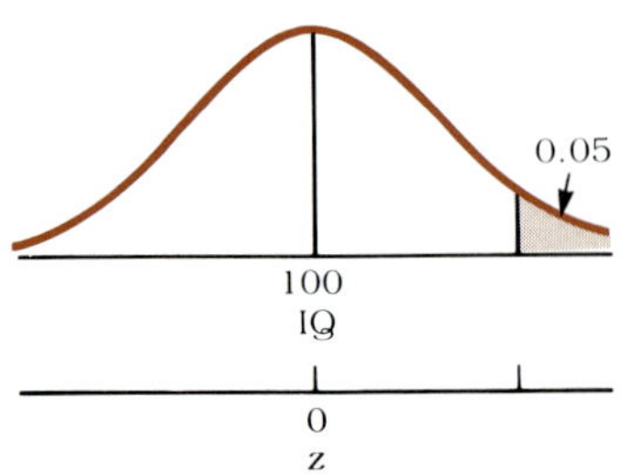

Figure 6.24

EXAMPLE 6.13 MENSA is an organization that accepts for membership only those whose IQ is in the top 5% of IQ's for the population. If IQ's are normally distributed with a mean of 100 and a standard deviation of 15, at what IQ should MENSA set its membership qualification?

Solution The top 5% will be the area in the right tail of Fig. 6.24. We need an area bounded by the mean, and we know the total area above 100 is 0.5. Thus the area we use in Table D.2 is 0.45. There are two numbers in the table equally distant from 0.45: 0.4495 and 0.4505. Thus

z→ ↓	0.04	0.05
1.6	0.4495	0.4505

and 1.65 is used. We need to find the IQ associated with a z score of 1.65.

$$\mu + z\sigma = 100 + (1.65)(15) = 124.75 \cong 125.$$

Thus MENSA would consider only those whose IQ is 125 or higher. □

EXERCISES/Section 6.3

1. Find:
   a) $P(0 < z < 1.75)$
   b) $P(0 < z < 2.44)$
   c) $P(0 < z < 0.83)$
   d) $P(-0.73 < z < 0)$
   e) $P(-1.98 < z < 0)$

2. Find:
   a) $P(z > 2.31)$
   b) $P(z < -0.75)$
   c) $P(z > 1.73)$
   d) $P(z < -0.42)$
   e) $P(z < -1.45)$

3. Determine:
   a) $P(-2.12 < z < 0.39)$
   b) $P(-2.45 < z < 1.36)$
   c) $P(-1.67 < z < 2.64)$
   d) $P(-0.75 < z < 0.25)$
   e) $P(-1.96 < z < 1.96)$

4. Determine:
   a) $P(z > -1.05)$
   b) $P(z < 1.25)$
   c) $P(z < 1.69)$
   d) $P(z > -3.06)$
   e) $P(z > -2.84)$

5. Determine:
   a) $P(1.03 < z < 2.98)$
   b) $P(-1.38 < z < -0.83)$
   c) $P(0.89 < z < 2.98)$
   d) $P(-3.01 < z < -1.35)$
   e) $P(-2.73 < z < -0.65)$

6. Find:
   a) $P(-0.64 < z < 1.28)$
   b) $P(z < 2.33)$
   c) $P(z > 2.58)$
   d) $P(1.42 < z < 2.84)$
   e) $P(-0.85 < z)$

7. Rods produced by G&R Company have measurements that are normally distributed with a mean of 66 cm and a standard deviation of 2 cm. If X = length of a randomly selected rod,

what is

a) $P(X < 66)$
b) $P(66 < X < 70)$
c) $P(60 < X < 65)$
d) $P(70 < X < 71)$
e) $P(X > 63)$
f) $P(X > 72)$

8. The amount of time, in hours, spent by small children in front of a television set is normally distributed with $\mu = 7$ hrs and $\sigma = 2.75$ hrs. If X = time spent watching television by a randomly selected child, find:

a) $P(X < 5.5)$
b) $P(X > 6.3)$
c) $P(4 < X < 10)$
d) $P(7.5 < X < 11)$
e) $P(X < 2)$
f) $P(X > 13)$

9. A measure of facility utilization is an activity score that indicates turnover and efficiency in handling. If activity scores are normally distributed with $\mu = 240$, $\sigma = 30$, and X = activity score of a randomly selected warehouse, find:

a) $P(X < 275)$
b) $P(210 < X < 290)$
c) $P(X > 300)$
d) $P(X < 200)$
e) $P(265 < X < 310)$
f) $P(250 < X < 260)$

10. Professor Jones has given the same final exam for the past ten years. She has found that grades are normally distributed with a mean of 70 and a standard deviation of 12. What is the probability that a student randomly selected from current enrollment will receive a grade of

a) 90 or more?
b) 60 or below?
c) 75–85?
d) What proportion of this year's enrollment can be expected to receive a grade of 65–87?
e) What should be the cutoff for an A if the top 10% of the class is to get an A?

11. If College Board (CEEB) scores are normally distributed with a mean of 700 and a standard deviation of 100,

a) what is the probability that a randomly selected college-bound senior has a score of 600 or above?
b) what percent of students taking the exam have a score of 750 or above?
c) what percent have a score below 450?
d) what percent have a score of 400–675?
e) what score represents the thirtieth percentile, P_{30}?
f) what score represents P_{75}?

12. Batting averages seem to be normally distributed with a mean of 0.270 and a standard deviation of 0.050. What is the probability that a randomly selected ball player has a batting average of

a) 0.250–0.350?
b) less than 0.200?
c) 0.290–0.390?
d) What percent of all players have batting averages over 0.400?
e) Find P_{45}.
f) Find P_{90}.

13. A certain professor has discovered that, over the years, the average grade on a 150-point comprehensive exam is 110 with a standard deviation of 12. He has decided that only the top 5% get an A on the exam; the next 15% a B; the next 40% a C; the next 30% a D; and the bottom 10% an F. What grade must a student get to receive:

a) A? b) B? c) C? d) D?

14. A botanist testing a new type of fertilizer that stimulates stem growth determined that after four weeks his seedlings had a mean height of 10.1 cm with $\sigma = 3.9$ cm. What proportion of his seedlings were

a) taller than 15.2 cm?
b) 8.7–12.2 cm tall?
c) shorter than 5 cm?
d) Find P_{25}. e) Find P_{70}?

15. The weekly production rate of disabled people hired by ABC, Ltd. is normally distributed with $\mu = 23.3$ items and $\sigma = 2.7$ items. What is the probability that the newly hired worker will produce

a) more than 27 items per week?
b) 19–26 items?
c) at least 18 items?
d) What is the cutoff for the top 20%, that is, the fastest 20% of the workers?
e) What is the maximum rate for the slowest 1/3 of the workers?

6.4 Normal Approximation to the Binomial Distribution

When we first examined the binomial probability distribution, we looked at examples where n was small and the use of

$$P(X) = \binom{n}{X} p^X q^{n-X}$$

was relatively easy (especially with access to a table of probabilities). Now let us examine a case where n is slightly larger, say 20, and p and q are each 0.5.

EXAMPLE 6.14 In any randomly selected group of 20 people, what is the probability that at least 14 of them will be women?

Solution We have a binomial problem: Success means finding a woman, $n = 20$, $p = 0.5$, and $q = 0.5$. Using Table D.1, we can answer the question by adding the probabilities associated with $X \geq 14$.

$$\begin{aligned} P(X \geq 14) &= P(14) + P(15) + P(16) + P(17) + P(18) \\ &\quad + P(19) + P(20) \\ &= 0.037 + 0.015 + 0.005 + 0.001 + 0 + 0 + 0 \\ &= 0.058. \end{aligned}$$

Figure 6.25 shows the histogram of the distribution. □

This method is fairly tedious. Imagine the effort required if $n = 100$, with no table available. However, a quick glance at the histogram provides a clue to a much easier procedure. The histogram is symmetric and is easily approximated by overlaying on it a normal distribution, also shown in Fig. 6.25. Rather than adding up the areas of the histogram bars, we can find the corresponding area under the normal curve, which approximates the binomial probability well enough. The small parts of the rectangles outside the curve are offset by areas under the curve that are not included in any rectangles. The effort saved in the calculations for the normal table versus those of the binomial probability equation is worth the trade-off of an approximation for an exact value.

There is much formal, higher mathematical justification for using the normal distribution to approximate a binomial distribution; the procedure is not as intuitive as it may at first appear to be. Certain requirements must be met by the form of the prob-

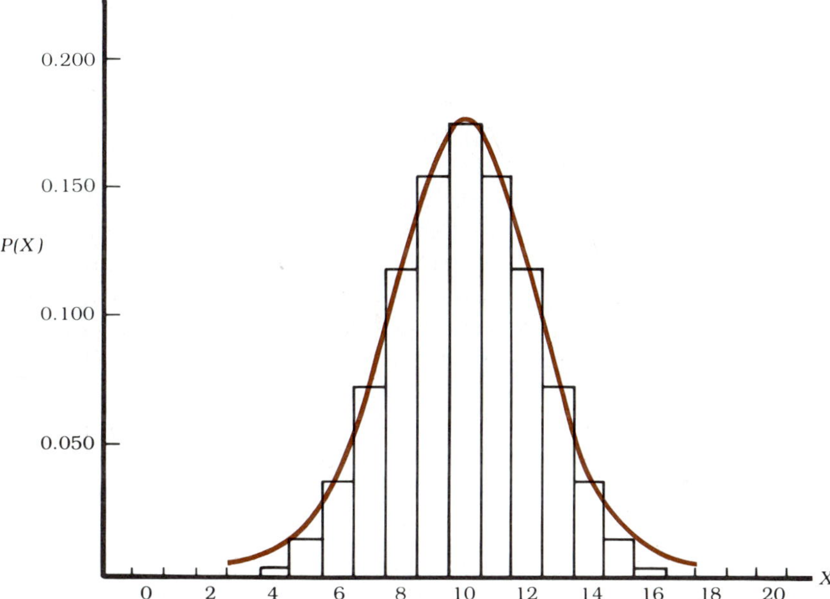

Figure 6.25

lem itself, and certain extra steps must be followed before the normal table is used.

First, statistical theory requires that both $np \geq 5$ and $nq \geq 5$ in order for the approximation to be satisfactory. Intuitively, this guarantees that if p and q stray too much from 0.5, then n must increase. As the probability of an individual success decreases (an event becomes rarer), then n must increase to compensate. If $p = 0.01$, then n must be at least 500. If we were to look for a success that has only 1 chance in 100 of showing up, we would not choose a small sample. We must also put this condition $nq \geq 5$ on q because success and failure are complementary. If failure is relatively rare, a large sample is also required. For extremely rare situations, another distribution, the *Poisson distribution*, may be considered but we will not do so here.

Second, we need to find μ and σ in order to calculate z scores. Recall that

$$\mu = np \quad \text{and} \quad \sigma = \sqrt{npq}.$$

Third, we must determine the **continuity corrections,** especially if n is small. The normal distribution only *approximates* the binomial distribution because the normal is a continuous distribution, whereas the binomial deals with discrete variables. In order to use a normal table, the binomial problem must be converted into one involving a continuous variable. We did this automatically when we drew a histogram for a binomial distribution earlier in this chapter (see Example 6.5, Fig. 6.5). We drew the histogram to cover all values on the horizontal axis from −0.5 to 4.5. The bars are centered on each whole number and go from −0.5–0.5 for 0, 1.5–2.5 for 2, and so on. The graph displays intervals that were expanded to halfway between consecutive integers. This is similar to the procedure we followed in creating class boundaries in Chapter 3, and we now carry it over into calculations. In order to solve Example 6.14 using a normal approximation, we must think of $X = 14$ as meaning the interval from 13.5 to 14.5, and $P(X \geq 14)$ in binomial terms becomes $P(X \geq 13.5)$ for the normal approximation. Let us solve Example 14 again—this time using a normal approximation (Fig. 6.26).

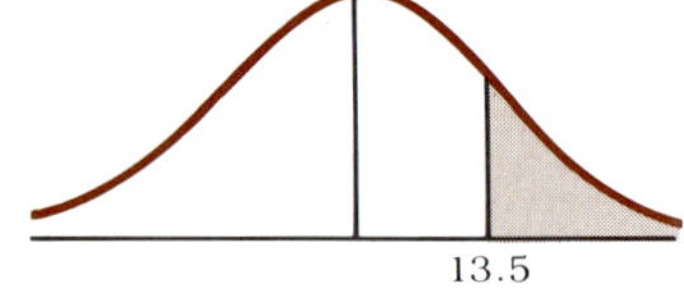

Figure 6.26

$$\mu = np = 20 \cdot 0.5 = 10;$$

$$\sigma = \sqrt{20 \cdot 0.5 \cdot 0.5} = 2.24;$$

$$z = \frac{13.5 - 10}{2.24} = 1.56.$$

The area to the right of $z = 1.56$ is $0.5 - 0.4406 = 0.0594$. This compares favorably with the 0.058 obtained by using the formula.

Note how a slight change in wording could change this problem: If instead of *at least 14*, it had read *greater than 14*, where would the shading have begun on the normal distribution? *Greater than 14* means *15 or more*. The interval for 15 begins at 14.5, so we would have wanted the area above 14.5; we would not have included 14.

Let us review the four-step procedure for using a normal approximation in a binomial problem.

RULES

1. Check to see that both $np \geq 5$ and $nq \geq 5$.
2. Calculate $\mu = np$ and $\sigma = \sqrt{npq}$ to use to find the z scores.
3. Determine the necessary continuity correction intervals.
4. Proceed as if you were working a normal distribution problem, as discussed in Section 6.3.

EXAMPLE 6.15 A large supermarket stocks its own brand of coffee and several major national brands. About 30% of customers prefer the house brand. Of the next 200 coffee buyers, what is the probability that (a) at most 50 will buy the house brand? (b) 65–75, inclusive, will buy the house brand?

Solution Let X = the number of buyers of the house brand coffee.

$$n = 200; \qquad p = 0.30; \qquad q = 0.70.$$

Step 1.

$$np = 200 \cdot 0.3 = 60 > 5;$$

$$nq = 200 \cdot 0.7 = 140 > 5.$$

Step 2.

$$\mu = np = 60;$$

$$\sigma = \sqrt{npq} = \sqrt{200 \cdot 0.3 \cdot 0.7} = 6.48.$$

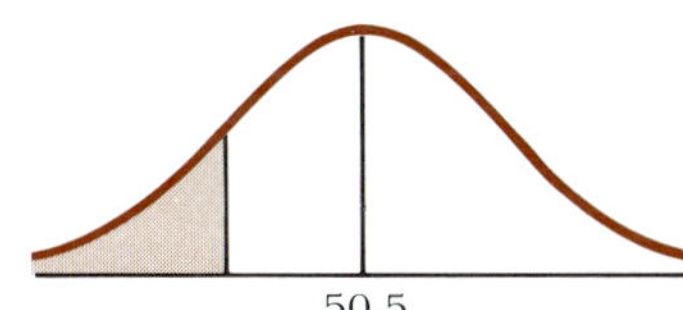

Figure 6.27

For part (a):

Step 3. At most 50 includes 50, so we want to be sure to create an interval to 50.5 (Fig. 6.27).

Step 4.

$$z = \frac{50.5 - 60}{6.48} = -1.47.$$

The area to the left of $z = -1.47$ is $0.5 - 0.4292 = 0.0708$.

For part (b):

Step 3. 65–75, inclusive, will stretch from 64.5–75.5. (Fig. 6.28).

Figure 6.28

Step 4.

$$z = \frac{75.5 - 60}{6.48} = 2.39.$$

$$z = \frac{64.5 - 60}{6.48} = 0.69.$$

The area between $z = 0.69$ and $z = 2.39$ is $0.4916 - 0.2549 = 0.2367$. □

To show the necessity of the continuity correction, let us do one more example.

EXAMPLE 6.16 Voter registrations indicate that, in Congressman Jones's district, 60% are Democrats and 40% are Republicans. In a random sample of 14 registered voters, what is the probability that at least 12 are Democrats?

Solution

$$n = 14; \qquad p = 0.6; \qquad q = 0.4.$$

Step 1.

$$np = 14 \cdot 0.6 = 8.4 > 5;$$

$$nq = 14 \cdot 0.4 = 5.6 > 5.$$

Step 2.

$$\mu = 8.4;$$

$$\sigma = \sqrt{14 \cdot 0.6 \cdot 0.4} = 1.83.$$

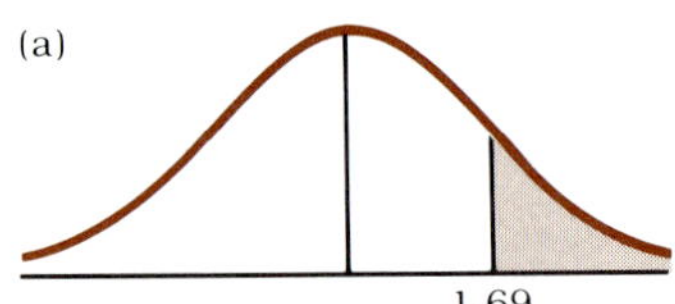

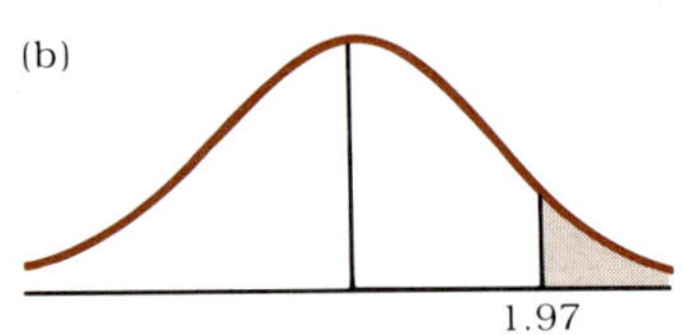

Figure 6.29

When correcting for continuity,

Step 3. At least 12 requires an interval from 11.5 (Fig. 6.29a).

Step 4.

$$z = \frac{11.5 - 8.4}{1.83} = 1.69.$$

The area above 1.69 is 0.5 − 0.4545 = 0.0455.

If the continuity correction is not made,

Step 3. At least 12 requires an interval from 12 (Fig. 6.29b).

Step 4.

$$z = \frac{12 - 8.4}{1.83} = 1.97.$$

The area above 1.97 is 0.5 − 0.4756 = 0.0244. □

Table D.1 gives, to three decimal places, $P(X \geq 12) = 0.040$. Thus the probability found by using the continuity correction is a better approximation of the real value.

EXERCISES/Section 6.4

1. Find the normal approximation to the binomial probability $P(X = 9)$ when $n = 20$ and $p = 0.7$.

2. If $n = 100$ and $p = 0.45$, use a normal approximation to find the binomial probability $P(X \leq 55)$.

3. In a binomial experiment, if $n = 50$ and $p = 0.5$, use a normal approximation to find:
   a) $P(X = 26)$
   b) $P(X \geq 30)$
   c) $P(20 \leq X \leq 30$, inclusive)

4. Find the normal approximation to the binomial probability $P(X \geq 30)$, if $n = 40$ and $p = 0.8$.

5. If $n = 30$ and $p = 0.65$, use a normal approximation to find the binomial probability $P(X \leq 18)$.

6. Find the probability that a random sample of 200 births contains exactly 100 girl babies.

7. In the United States about 45% of women over the age of 15 own diamond jewelry. What is the probability that, in a random sample of 25

women, 9–15 (inclusive) women will own diamonds?

8. The article in News & Views 6.2 reported that 49% of jail inmates are aged 12–24. In a random sample of 100 inmates, what is the probability that at least 40 of them are 12–24 years of age?

9. In the United States 25% of men who shave use an electric razor. In a group of 60 men, what is the probability that fewer than six use an electric razor?

10. Seventy percent of women use lipstick. What is the probability that, in a randomly selected group of 40 women, no more than 30 will be wearing lipstick?

11. Thirty-seven percent of the population smokes. A restaurant manager is planning a separate smoking area for customers. What is the probability that 45 or more customers in 100 will be smokers?

12. Thirty-two percent of the population plans to save or invest extra money. What is the probability that a random sample of 50 people contains more than 20 savers?

13. About 95% of workers are generally satisfied with their jobs. What is the probability that in a sample of 200 workers there will be more than 18 malcontents?

14. If 38% of the U.S. population can speak a second language, what is the probability that, in a group of 40 people, more than 20 can speak more than one language?

15. Sixty-nine percent of the population believes in some form of life after death. What is the probability that a random sample of 300 people contains fewer than 75 who do not believe in an afterlife?

Study Notes

KEY TERMS

binomial experiment ■ An experiment that has the following properties:

1. There are n independent, identically repeated trials, each of which has two outcomes. These outcomes are usually called *success* and *failure*.
2. X = the number of successes; X can take on values from 0 to n.
3. P(success on any one trial) = p, which remains constant for the life of the experiment.
4. P(failure on any one trial) = q.

$$p + q = 1.$$

binomial probability ■ A probability associated with a binomial experiment. It may be determined using the equation

$$P(X) = \binom{n}{X} p^X q^{n-X}$$

probability distribution ■ A set of all possible values of a variable together with their associated probabilities.

random variable ■ A variable having values that are determined by a random process.

standard normal score ■ z score.

OUTLINE

I. Probability distributions
   A. Use random variables.
      1. Discrete variables: X = number of. . . .
      2. Continuous variables.
   B. Draw histograms with bars centered over discrete values.
      1. Areas under the graph are probabilities.
      2. Σ Areas = 1.

C. $\mu = \Sigma[X \cdot P(X)]$

D. $\sigma = \sqrt{\Sigma[(X - \mu)^2 \cdot P(X)]}$

or

$\sigma = \sqrt{\Sigma[X^2 \cdot P(X)] - \{\Sigma[X \cdot P(X)]\}^2}$

II. Binomial distribution

A. Binomial experiment has three required properties:

1. n independent trials, each with two outcomes, success and failure.
2. X = the number of successes.
3. $P(\text{success}) = p$.
   $P(\text{failure}) = q$.
   $p + q = 1$.

B. $P(X) = \binom{n}{X} p^X q^{n-X}$

C. $\mu = np$.

D. $\sigma = \sqrt{npq}$.

III. Normal distribution

A. $P(a < X < b)$ = area between a and b under the normal curve.

B. z scores (standard normal scores) are necessary for finding areas.

C. When finding probabilities, draw a diagram of the desired area to determine what arithmetic is necessary to supplement numbers from the normal table.

IV. Normal approximation to the binomial distribution

A. np must be greater than or equal to 5, and nq must be greater than or equal to 5.

B. Calculate $\mu = np$ and $\sigma = \sqrt{npq}$

C. Make continuity corrections by making each value of X include an interval from $X - 0.5$ to $X + 0.5$.

D. Solve problems using the normal table.

REVIEW PROBLEMS

1. A coin purse contains 8 pennies, 6 nickels, 3 dimes, and 3 quarters. Let X = the value (in cents) of the coin chosen.

a) What are the possible values of X?
b) Determine $P(X)$ for each value of X.
c) Calculate μ and σ.

2. The face cards (including aces) are removed from a deck of cards. The remaining 36 cards have numbers, 2–10. Two cards are removed from this deck with replacement and X = the sum of the numbers on the two cards.

a) What are the possible values of X?
b) Determine $P(X)$ for each value of X.
c) Calculate μ and σ.

3. A bowl contains five markers numbered 1–5. A marker is withdrawn, its number is recorded, and it is replaced; then a second marker is withdrawn and its number is recorded. Let X = sum of the numbers on the markers.

a) What are the possible values of X?
b) Determine $P(X)$ for each value of X.
c) Calculate μ and σ.

4. The following table describes a probability distribution.

X:	5	6	7	8	9
$P(X)$:	0.2	0.4	0.1	0.15	0.15

Determine μ and σ.

5. The following table describes a probability distribution.

X:	0	2	4	6	8
$P(X)$:	0.25	0.1	0.2	0.4	0.05

Determine μ and σ.

6. Last year a nearby state enacted strict auto emission requirements for cars registered in the state. Despite this, its air quality monitoring board has disclosed that 30% of the registered cars fail to meet state standards. Twelve cars from that state pass through a toll booth in the state. What is the probability that

a) three of them fail to meet state requirements?
b) at most two of them fail to meet state requirements?

c) at least four of them fail to meet state standards?
d) 1–3 (inclusive) cars fail to meet state standards?

7. One of the players on a local baseball team is batting 275. [P(hit) = 0.275.] Each player usually bats six times during a game. Assuming that this player does bat 6 times during his next game, determine the probability that he gets a hit
a) twice.
b) at least 4 times.
c) at most once.
d) 3–5 times, inclusive.

8. In the last year, Acme Motors recalled 300,000 of the 1,500,000 cars it manufactured and sold. Pete Leary spotted 8 Acme Motors cars on his evening walk. Determine the probability that, of the 8 seen, the number recalled was
a) 3.
b) at least 5.
c) at most 3.
d) 2–6, inclusive.

9. A package of ten begonia seeds carries the notice: 80% germination rate. A gardener plants all ten of the seeds. What is the probability that the number which will germinate is
a) 7.
b) at most 5.
c) at least 8.
d) 3–9, inclusive.

10. Determine:
a) $P(0 < z < 1.78)$
b) $P(-0.56 < z < 0)$
c) $P(0.46 < z < 1.99)$
d) $P(-2.55 < z < -0.30)$
e) $P(-0.83 < z < 1.49)$
f) $P(0 < z < 0.09)$
g) $P(-2.04 < z < 0)$
h) $P(2.02 < z < 3.09)$
i) $P(-1.22 < z < -0.09)$
j) $P(-3.01 < z < 2.38)$

11. For a normal population having a mean of 180 and a standard deviation of 20, determine:
a) $P(180 < X < 208)$
b) $P(156 < X < 180)$
c) $P(192 < X < 212)$
d) $P(145 < X < 176)$
e) $P(135 < X < 184)$
f) $P(180 < X < 195)$
g) $P(171 < X < 180)$
h) $P(227 < X < 240)$
i) $P(128 < X < 156)$
j) $P(152 < X < 194)$

12. A stereo dealer observed that last year the average purchase was for $525 with a standard deviation of $125. Assuming that the purchases made last year were normally distributed, what percent of those purchases were
a) greater than $700?
b) less than $400?
c) greater than $300?
d) less than $750?
e) $350–900, inclusive?
f) Above what price was the most expensive 10% of those sold?

13. A statistics professor disclosed to one of her classes that the average grade on the last quiz was 72 with a standard deviation of 20. Assuming that the distribution was normal, what percent of her students had grades
a) below 60?
b) above 90?
c) 75–100, inclusive?
d) If the top 5% of the class were to get A's, what would be the cutoff grade for an A?

14. A survey disclosed that family income in an area of town averaged $40,000 per year with a standard deviation of $8,000. What percent of the families in that area have an annual salary of:
a) $50,000 or more?
b) $25,000 or less?
c) $55,000 or less?
d) $30,000 or more?
e) $35,000–55,000, inclusive?
f) If the poverty line is determined by selecting the 15% of the population with the lowest incomes, below what income is a family considered to be living in poverty?

15. A small manufacturing operation has 125 employees on its payroll. The payroll officer has determined that each year about 15% of the total number of employees need to take a week or more sick leave. Assuming that next year the number of employees remains the same, use the normal approximation to the binomial distribution to approximate the probability that the number of employees taking a week or more sick leave is:
a) 10 or less.
b) 20 or more.

c) 25 or less.
d) 15 or more.
e) 8–18, inclusive.

16. Forty percent of the registered voters in a nearby community are Democrats. A polling firm sets out to interview 30 of them. Use the normal approximation to the binomial distribution to determine the probability that the number of Democrats surveyed is:
a) at most 18.
b) at least 15.
c) 15–19, inclusive.
d) at least 10.
e) at most 8.

17. A recent study indicates that about 25% of the U.S. population has high blood pressure and that about one-half of them do not realize it; thus about 12.5% are unknowing victims. A local GP expects to see 50 patients today. Assuming that her patients are representative of the entire population, and using the normal approximation to the binomial distribution, what is the probability that the number of unsuspecting victims of high blood pressure is:
a) at least 10.
b) at most 5.
c) 5–9, inclusive.

18. Forty-five percent of the students at a local community college are enrolled in the general studies curriculum. A random sample of 40 students is selected. Use the normal approximation to the binomial distribution to estimate the probability that the number of students in this sample who are enrolled in the general studies curriculum is:
a) 15–20, inclusive.
b) no more than 10.
c) at least 25.
d) 19.

Sampling distributions

7

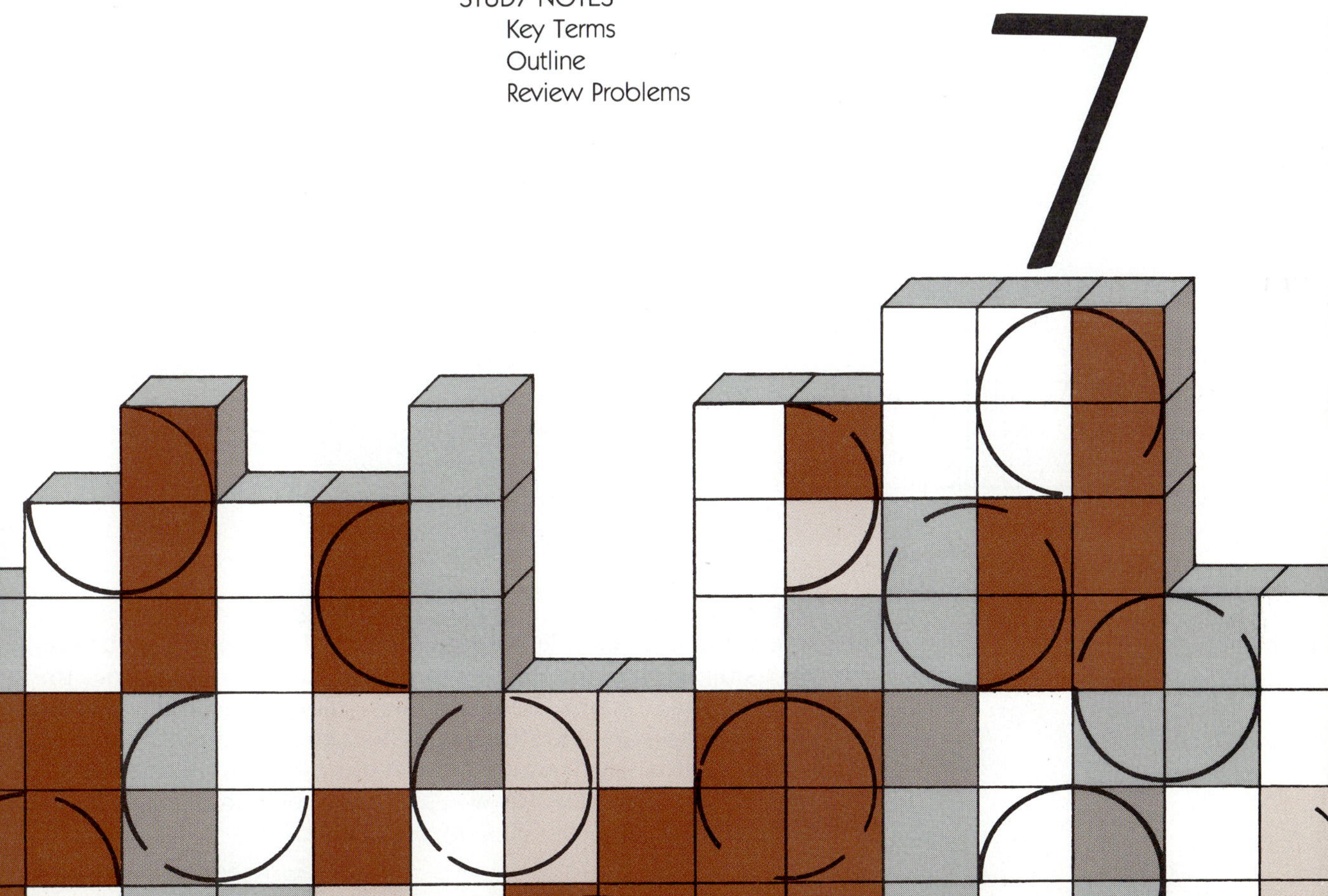

7.1 Central Limit Theorem

Let us consider a college class of 25 students, who are to determine the mean height of adult males on campus, as a class project. They decide that each will take a random sample and calculate the sample mean. Are all 25 sample means likely to be identical? Probably not. There will be some variability among means from one sample to another. Yet the class is to ascertain the mean height for the population from these samples. It should be evident from this that we need one more tool before we can begin to draw conclusions about a population: We need to account for sample variability; to examine how sample results vary and how to take this variability into account.

To do this, we will look at **sampling distributions.** Remember that a distribution is a list and now will be a list of sample results.

> ***sampling distribution*** ■ The list of all possible values of a sample statistic obtained by taking samples of size n. All samples must be the same size and the sample statistic may be any descriptive measure.

Let us look at a sampling distribution by considering a population with numbers that are easy to handle. Consider the population consisting of values of X: 1, 2, 3, 4, 5. Each occurs once. It is as if we had a hat containing five slips of paper, each with a different number. The histogram of the population is a uniform distribution, as shown in Fig. 7.1. It has a mean $\mu = 3$ and a standard deviation

Figure 7.1

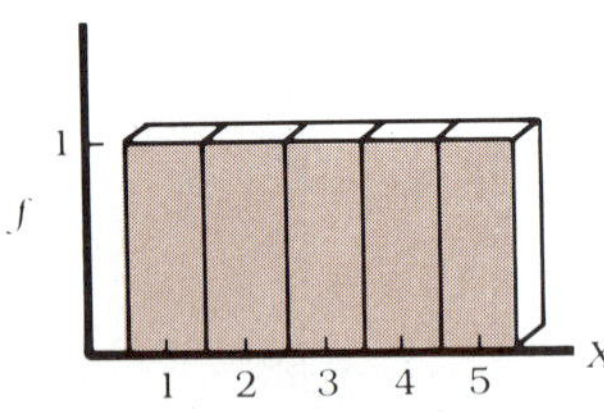

$$\sigma = \sqrt{\frac{\Sigma(X - \mu)^2}{N}}$$

$$= \sqrt{\frac{(1-3)^2 + (2-3)^2 + (3-3)^2 + (4-3)^2 + (5-3)^2}{5}}$$

$$= \sqrt{\frac{4 + 1 + 0 + 1 + 4}{5}} = \sqrt{2}.$$

Now we will sample this population. To guarantee that every possible value of a sample statistic is included, we must obtain every possible sample. We choose samples of size $2 = n$ in order to simplify the calculations. We want these samples to have useful probability properties like independence, so we sample with replacement, that is, we draw a slip from our imaginary hat, record its number, replace it, and draw again. Since our population is so small, we can list all possible samples (Table 7.1).

TABLE 7.1

(1, 1)	(2, 1)	(3, 1)	(4, 1)	(5, 1)
(1, 2)	(2, 2)	(3, 2)	(4, 2)	(5, 2)
(1, 3)	(2, 3)	(3, 3)	(4, 3)	(5, 3)
(1, 4)	(2, 4)	(3, 4)	(4, 4)	(5, 4)
(1, 5)	(2, 5)	(3, 5)	(4, 5)	(5, 5)

TABLE 7.2

1	1.5	2	2.5	3
1.5	2	2.5	3	3.5
2	2.5	3	3.5	4
2.5	3	3.5	4	4.5
3	3.5	4	4.5	5

To obtain a sampling distribution, we can choose any statistic—mean $\bar{X}$, standard deviation s, range, mode, median, or others—and calculate a value for each sample. The most informative one for our purposes is the sample mean $\bar{X}$. Let us calculate the sample mean of each sample in Table 7.1 and list them in corresponding positions (Table 7.2).

Table 7.2 thus is a sampling distribution. It is a list of all possible sample means from samples of size 2 taken from our original population of individual slips of paper. As such, Table 7.2 provides us with a new population: a population of *sample* means. Let us examine this new population by graphing it (Fig. 7.2). The means increase in increments of 0.5 from 1 to 5. Notice that the horizontal axis is now labeled $\bar{X}$; it contains sample means, *not* values of X for individuals.

What is the mean of this new population, the mean of the sample means, $\mu_{\bar{X}}$? Since the mean is the balance point, no calculations are required to see that

$$\mu_{\bar{X}} = 3.$$

Figure 7.2

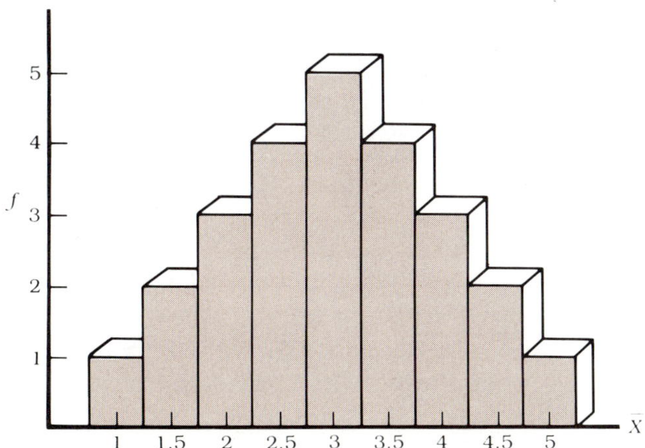

The standard deviation of the sample means, $\sigma_{\bar{X}}$ is

$$\sigma_{\bar{X}} = \sqrt{\frac{\Sigma(\bar{X} - \mu)^2}{N}} = \sqrt{\frac{\Sigma(\bar{X} - 3)^2}{25}} = 1.$$

You can work this calculation out for yourself substituting in each of the values in Table 7.2.

We are now ready to summarize and generalize about the sampling distribution. We have seen three characteristics:

1. Figure 7.2 is symmetric; and, with a little imagination, we can smooth it out to resemble a normal distribution.
2. The mean of the sampling distribution $\mu_{\bar{X}}$ is identical to the mean μ of the original population.
3. The standard deviation of the sampling distribution $\sigma_{\bar{X}}$ is obtainable from the standard deviation σ of the original population.

This last point is not obvious. To convert $\sigma = \sqrt{2}$ into $1 = \sigma_{\bar{X}}$, we need to divide $\sqrt{2}$ by $\sqrt{2}$. The one number we have not used in summarizing is n, in this case 2. Thus if we divide $\sigma = \sqrt{2}$ by $\sqrt{n}$, we get $1 = \sigma_{\bar{X}}$.

$$\sigma_{\bar{X}} = \frac{\sigma}{\sqrt{n}}.$$

We have just worked through an illustration of the **central limit theorem.**

> ***central limit theorem*** ■ Given any population with mean μ and standard deviation σ, if all possible random samples of size n are taken, the resulting sampling distribution of sample means $\bar{X}$'s has the following properties:
>
> 1. The sampling distribution will be normally distributed if the original population is normally distributed. If the original population is not normally distributed, the sampling distribution will be approximately normally distributed for samples of size larger than 30.
> 2. $\mu_{\bar{X}} = \mu$.
> 3. $\sigma_{\bar{X}} = \sigma/\sqrt{n}$

In Section 7.2, we will examine what to do in the case of a population when small samples ($n < 30$) are taken.

The standard deviation $\sigma_{\bar{X}}$ is called the **standard error of the mean.**

> ***standard error of the mean*** ■ The standard deviation of the sampling distribution of sample means of samples of size n.
>
> $$\sigma_{\bar{X}} = \frac{\sigma}{\sqrt{n}}.$$

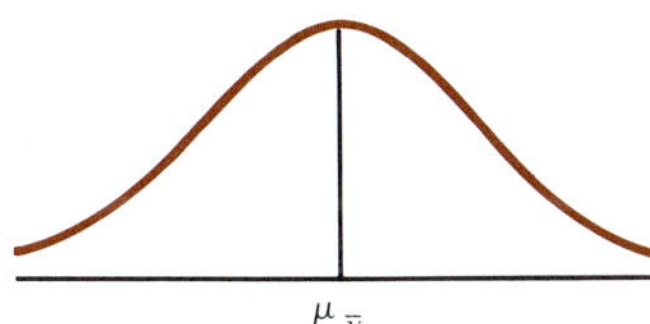

Figure 7.3

The central limit theorem is extremely important when we use sample means in order to make judgments about the populations from which they came. It tells us that we do not have to take many samples to see what kinds of patterns exist. We do not need to know anything about the shape of the original population. As long as we deal with a sample mean $\bar{X}$, we can use our knowledge of the normal distribution to help us make probabilistic statements.

The central limit theorem provides us with the ability to estimate an unknown population mean. We know that the center of the sampling distribution in Fig. 7.3 represents the unknown mean of the original population. If we can get one sample mean $\bar{X}$ to be very close to $\mu_{\bar{X}}$, then we will have a good estimate of μ. How can we get the sample result $\bar{X}$ to be close to $\mu_{\bar{X}}$? It could fall anywhere along the scale for $\bar{X}$ in Fig. 7.3. But the probability (area under the curve) of obtaining an $\bar{X}$ near the center is great. Can we control the amount of the distribution near the center? In other words do we have any control over the spread of the distribution?

(a)

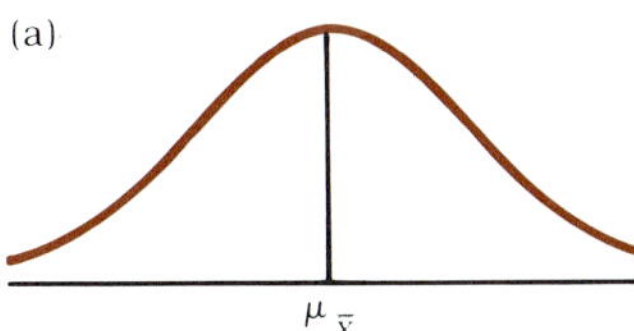

(b)

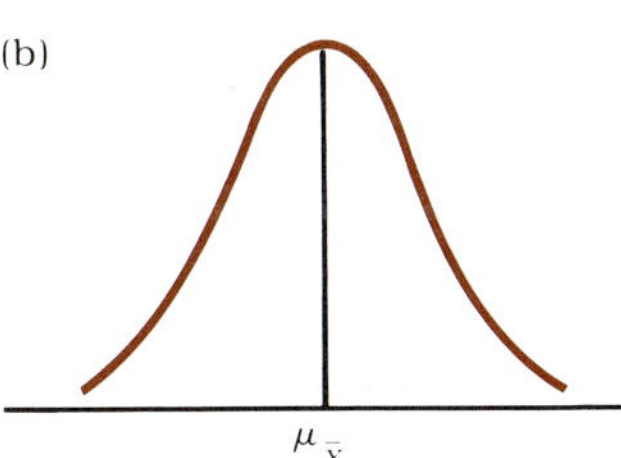

Figure 7.4

Is it possible to have a smaller spread, as in Fig. 7.4(b), so that a greater portion of the distribution is clustered around $\mu_{\bar{X}}$? Since spread is measured by standard deviation, we need to examine $\sigma_{\bar{X}}$. Do we have any control over $\sigma_{\bar{X}} = \sigma/\sqrt{n}$? Yes, by selecting n. Since n is in the denominator, as n gets larger, $\sigma_{\bar{X}}$ gets smaller, and a greater portion of the distribution clusters around $\mu_{\bar{X}}$. The probability of obtaining a sample mean $\bar{X}$ close to $\mu_{\bar{X}}$ and thus close to μ increases, as shown in Fig. 7.5. Controlling the sample size n allows us to control the degree of accuracy in predicting an unknown population mean, and vice versa. We can decide in advance how close we want to be in estimating a mean and then calculate the proper n to accommodate this requirement. All of this is based on probabilities as given by the normal distribution. This discussion leads to the technical work we will do in Chapter 8 to lay the groundwork for estimation procedures. The central limit theorem will be the basis for that work.

Figure 7.5

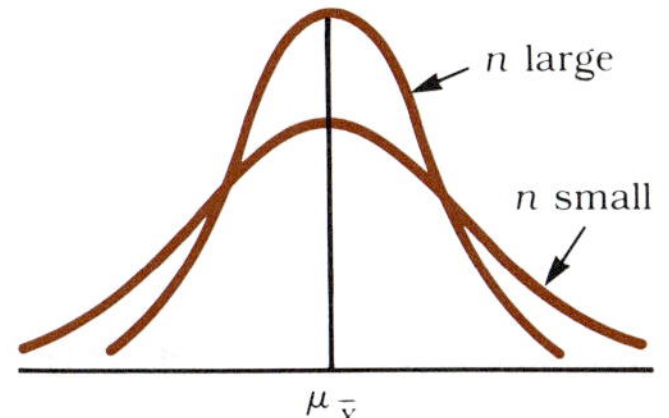

However, before we get that far, let us look at some examples and practice the arithmetic required by the central limit theorem.

EXAMPLE 7.1 Given a population with $\mu = 500$ and $\sigma = 40$, if a random sample of size 100 is taken, find:

a) $P(490 < X < 510)$
b) $P(490 < \bar{X} < 510)$

Solution Be careful with the notation. $P(490 < X < 510)$ asks for the probability that an *individual* value of X falls between 490 and 510. $P(490 < \bar{X} < 510)$ asks for the probability that a *sample mean* has a value between 490 and 510. In order to do either part of the problem, we must know the shape of the distribution being used. For (a) we do not know the shape of the population of individual values and so cannot find $P(490 < X < 510)$. For (b) the central limit theorem tells us to assume that the distribution of $\bar{X}$'s is normal with mean $\mu_{\bar{X}} = 500$ (Fig. 7.6).

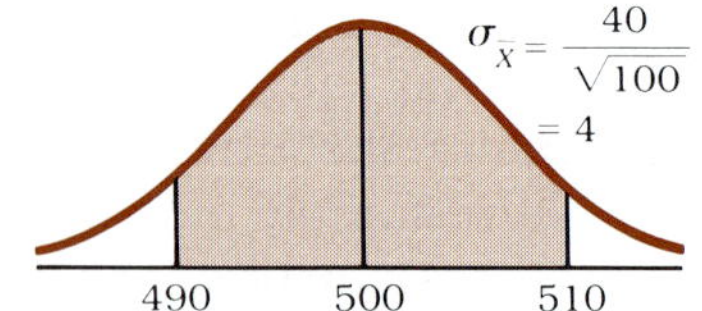

Figure 7.6

The z score of 490 is

$$z = \frac{490 - 500}{4} = \frac{-10}{4} = -2.5,$$

and for 510 is

$$z = \frac{510 - 500}{4} = 2.5.$$

The required probability is $2 \cdot 0.4938 = 0.9876$. □

Note that the z score still has the same format, but that we are now using sample means:

$$z = \frac{\text{Individual sample mean} - \text{Population mean}}{\text{Standard deviation}}$$

$$= \frac{\bar{X} - \mu_{\bar{X}}}{\sigma_{\bar{X}}}.$$

The central limit theorem requires that we substitute to obtain:

$$\boxed{z = \frac{\bar{X} - \mu}{\sigma/\sqrt{n}}.}$$

Whenever we are dealing with sample results, we have to remember to divide by $\sqrt{n}$.

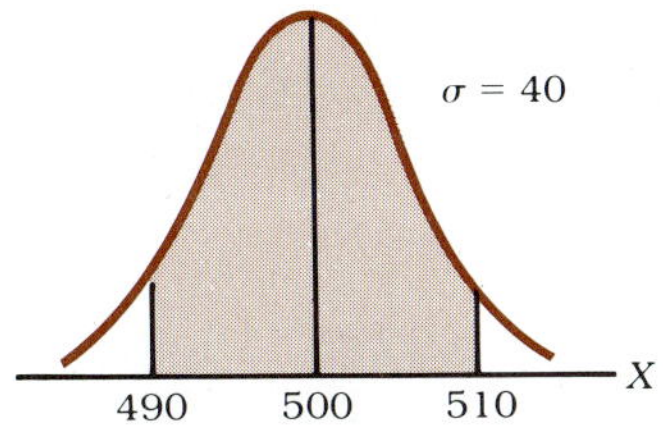

Figure 7.7

EXAMPLE 7.2 Rework Example 7.1, assuming that the population distribution is normal.

Solution Now we can solve part (a); part (b) does not change. Using Fig. 7.7, we find the z score of 490:

$$z = \frac{490 - 500}{40} = \frac{-10}{40} = -0.25,$$

and for 501, $z = 0.25$.

The required area is $2 \cdot 0.0987 = 0.1974$. Since we are dealing with an individual value of X, we do this problem just as we did the problems in Chapter 6, i.e., without dividing by $\sqrt{n}$. □

Now compare the results of (a) and (b), as shown in Fig. 7.8. Figure 7.8 illustrates the two distributions required by this problem. They have the same mean but drastically different spreads because of the effect of sample size on the standard deviation. A much larger portion of the curve for (b) is between 490 and 510, 0.9876, as opposed to 0.1974 for (a). The likelihood of obtaining a sample mean within 10 units of 500 is 0.9876, whereas the likelihood of obtaining a randomly chosen individual value within 10 units of 500 is only 0.1974. We have a higher probability of being close to the middle by using more information; by dealing with a sample of 100 instead of only one value. As complicated as the central limit theorem may seem, it provides us with a powerful tool for dealing with any population. It also verifies our commonsense reaction that more information is more helpful than less information.

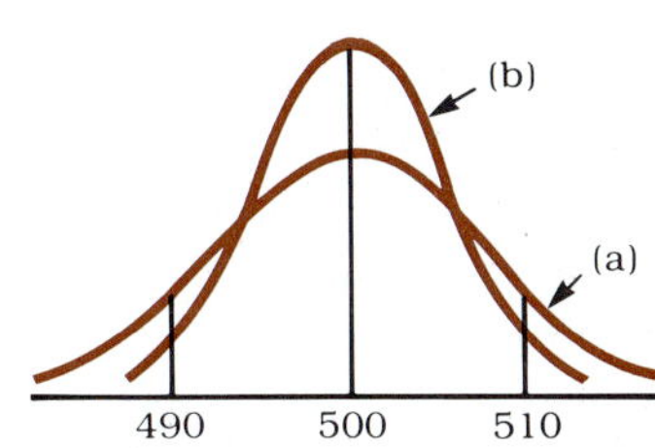

Figure 7.8

EXAMPLE 7.3 A manufacturer of dishwashers decided to check his warranty. If the life of dishwashers is normally distributed with $\mu = 6.5$ yrs and a standard deviation of 2.5 yrs, what is the probability that a random sample of 25 owners who sent in warranty cards to register their purchase would produce a mean life $\bar{X}$ of less than 6 yrs?

Solution Figure 7.9 tells us that

$$a = \frac{6 - 6.5}{2.5/\sqrt{25}} = \frac{-0.5}{0.5} = -1.$$

Figure 7.9

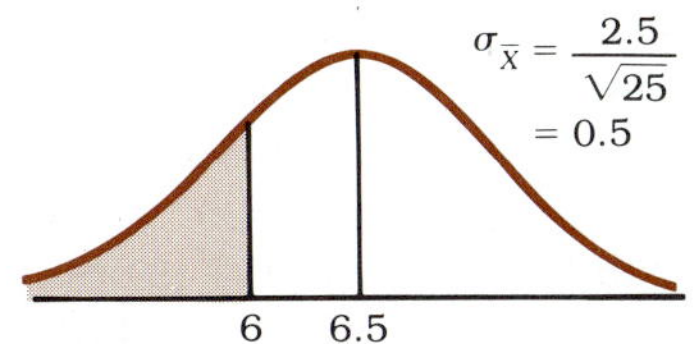

The area to the left of $z = -1$ is $0.5 - 0.3413 = 0.1587$. Nearly 16% of samples of size 25 will have sample means less than 6 years. If the warranty were for 6 years, it appears that about 16% of the dishwashers would need attention within the warranty period. □

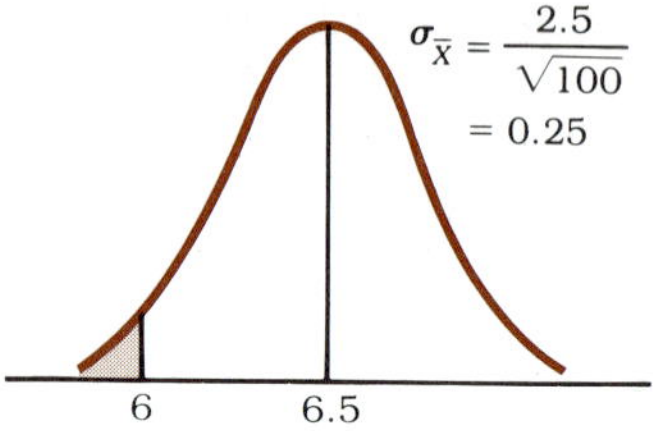

Figure 7.10

If we change the size of the sample to $n = 100$, what effect does this have on the answer? Using Fig. 7.10, we find that

$$z = \frac{6 - 6.5}{0.25} = -2.$$

The area in the left tail is now $0.5 - 0.4772 = 0.0228$. About 2% of samples of size 100 will have sample means less than 6 years. The work of estimating the likelihood of certain sample results ($\bar{X} < 6$) utilizes sample size. Now, what if the manufacturer contacts the 100 randomly chosen owners and obtains an $\bar{X} = 4.5$? His preliminary work indicates that this is very unlikely. What can the manufacturer conclude? Either the study actually produced an unlikely result or the assumptions about mean life, $\mu = 6.5$, are wrong. This is exactly the kind of thought process that we will pursue in Chapter 8.

EXERCISES/Section 7.1

1. If $n = 49$, $\mu = 10$, and $\sigma = 3.5$, find:
   a) the standard error of the mean.
   b) $P(9.2 < \bar{X} < 11.1)$.
2. A sample of size 100 is taken from a population with $\mu = 75$ and $\sigma = 18$. Find the probability that the sample mean will lie between 70–80.
3. a) A sample of size 64 is taken from a population with a mean of 121 and a standard deviation of 28. What is the probability that the sample mean will be greater than 125?
   b) What is the new standard error of the mean if $n = 200$?
   c) What is the new probability if $n = 400$?
4. a) A sample of size 36 is taken from a normally distributed population with $\mu = 50$ and $\sigma = 10.2$. What is the probability that the sample mean will be greater than 48?
   b) If the sample size is increased to 64, what is the new standard error of the mean?
   c) Find the probability that $\bar{X}$ will be greater than 48 with $n = 64$.
5. Statistics show that the average person receives 422 pieces of mail each year. Assume that the amount of mail received is normally distributed with $\sigma = 55$.
   a) What is the probability that a person chosen at random received 400–440 pieces of mail last year?
   b) If a sample of size 40 is chosen, what is the probability that its mean will lie between 400–440 pieces of mail?
6. Americans, on the average, eat 77 pounds of potatoes per year. Assume that potato consumption is normally distributed with $\sigma = 20$ lbs.
   a) What is the probability that a randomly chosen person will eat less than 60 pounds of potatoes next year?
   b) If a random sample of 100 people is chosen, what is the probability that the sample mean $\bar{X}$ will be less than 76 pounds?
7. The average number of people per household is 2.9. Assume that family size has a standard deviation of 1.2 people. What is the probability that a sample of 64 households will have a mean family size 2.5–3.3?
8. The total average earnings for manufacturing industries in the United States is \$4.81 per hour. Assume that the distribution of earnings has a standard deviation of \$1.25. What is the probability that a sample of 50 people would have a mean hourly wage rate less than \$4.30?

9. Using the information in Exercise 7, find e such that $P(2.9 - e < \bar{X} < 2.9 + e) = 0.95$; that is, construct an interval centered on 2.9 such that 95% of the distribution of sample means is contained in the interval.

10. Using the information in Exercise 8, find e such that $P(4.81 - e < \bar{X} < 4.81 + e) = 0.99$; that is, construct an interval centered on 4.81 such that 99% of the distribution of sample means is contained in that interval.

7.2 Student's *t* Distribution

The central limit theorem is very helpful if we have large samples. We do not even need to know σ; we can estimate it with the sample standard deviation s. However, if $n \leq 30$ and the population standard deviation is unknown—a very common situation—then the normal distribution is no longer an adequate approximation of the distribution of sample means. In such cases we must modify our approach. We use **Student's *t* distribution** instead of the normal distribution whenever $n \leq 30$ and only s is given. Student's t distribution is the distribution of t scores where

$$t = \frac{\bar{X} - \mu}{s/\sqrt{n}},$$

just as the normal distribution is the distribution of z scores.

Early in this century an employee of the Irish Guinness brewery, William Gosset, found himself faced with small samples in his quality control work. The normal distribution was not satisfactory, so he conducted research that led him to what is now called the t distribution. His company forbade publishing what it thought was a trade secret, and so he published his results anonymously as "A Student."

The t distribution is actually many distributions (one for each sample size) centered symmetrically about 0, as the normal distribution is. Their standard deviations are greater than 1, so the distributions are more spread out than the normal distribution, as shown in Fig. 7.11. However, as n approaches 30 the center of the distribution rises and the tails fall, until for any sample of size larger than 30, there is no useful difference between the normal distribution and a t distribution.

Table D.3 is the table we will use with t distributions. This table is written differently from Table D.2 because work in statis-

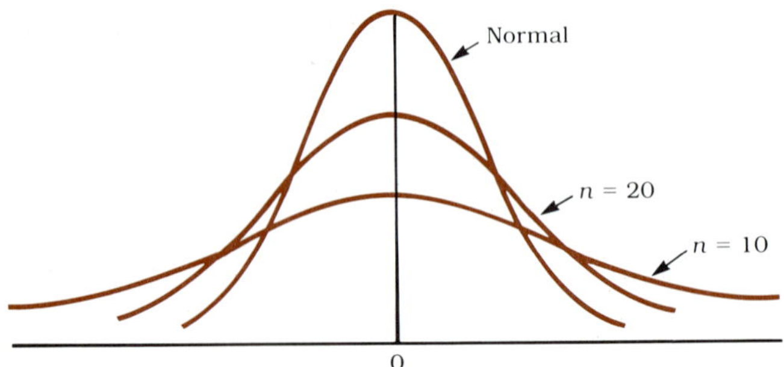

Figure 7.11

tical inference stresses the areas in the tails of the distribution rather than areas from the center outward. In using Table D.3, we must first find the area in a tail, locate it along the top of the table, and then find numbers in the body of the table to be used as t scores; we use t scores in the same way that we used z scores. The other piece of information we need has to do with degrees of freedom (df). The concept of *degrees of freedom* comes from the theoretical development of the t distribution. Intuitively, we can see the use of this idea in the following example.

EXAMPLE 7.4 If we want five numbers whose sum is 10, how many can be freely chosen?

Solution

$$\_\_\_\_ + \_\_\_\_ + \_\_\_\_ + \_\_\_\_ + \_\_\_\_ = 10.$$

We can choose any number for the first blank, say 6. Again, choose anything for the second, third, and fourth blanks, say 7, 9, 1.

$$6 + 7 + 9 + 1 + \_\_\_\_ = 10.$$

The fifth blank *must* now be -13 in order to satisfy the equation. We had five numbers to choose, but only four of them freely, because of the condition imposed by the equation. Thus, degrees of freedom df = 5 − 1 = 4. □

In general, for Student's t distribution.

$$\boxed{\text{df} = n - 1.}$$

Our use of the t distribution will be limited to the areas in the tails of the distribution; the t score is the point (or dividing line)

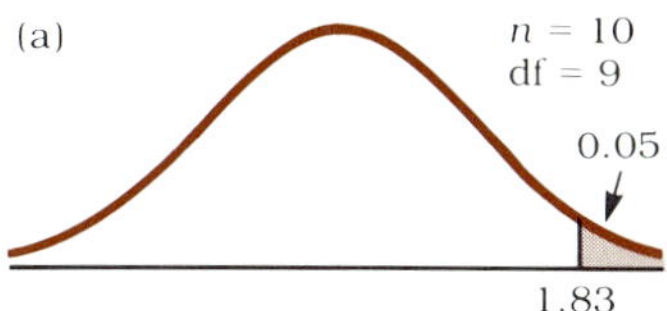

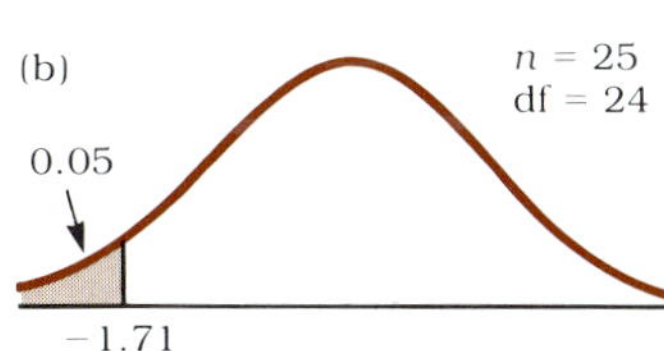

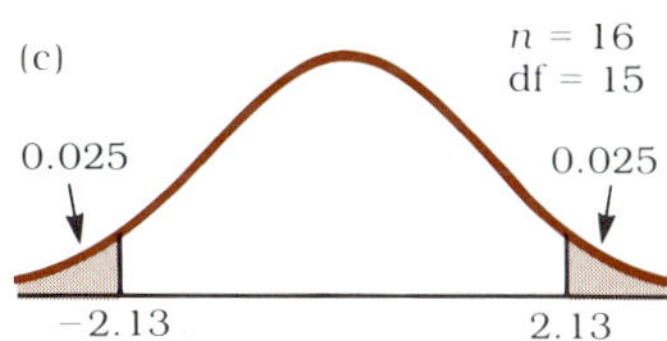

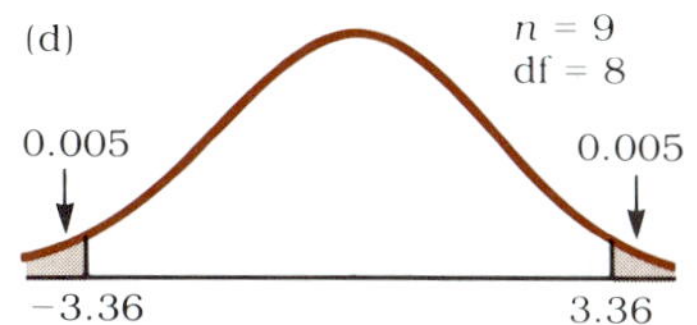

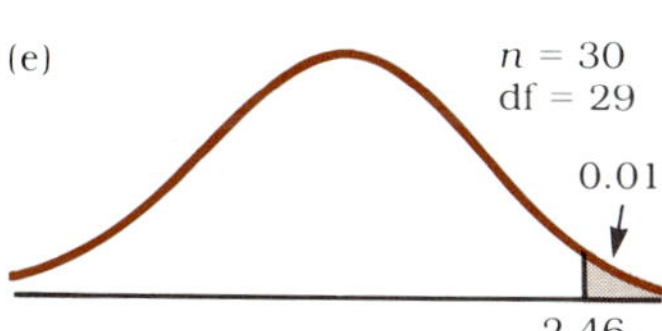

Figure 7.12

that identifies the area in the tail. Thus we need some practice in using Table D.3 as preparation for the t distribution work in Chapter 8.

EXAMPLE 7.5 Find the t score(s) for the areas identified in Fig. 7.12(a–e).

Solution

a) If $n = 10$, df $= 10 - 1 = 9$.
We go to the second t column to find 0.05. Reading across the 9 row, we get 1.83 as the t score that provides the dividing line in Fig. 7.12(a).

b) df $= 25 - 1 = 24$.
We use the 24 row and the 0.05 column to find 1.71. Now we must remember that we are below the mean of 0, so the t score is negative (-1.71) in Fig. 7.12(b).

c) df $= 16 - 1 = 15$.
The 15 row produces in the 0.025 column a t score of 2.13. The areas of both tails are given in Fig. 7.12(c), so we need two t scores, -2.13 and 2.13.

d) df $= 9 - 1 = 8$.
Again, we need two t scores, one positive and one negative. From the 8 row and the 0.005 column, we get the t scores shown in Fig. 7.12(d).

e) df $= 30 - 1 = 29$.
Using the 29 row and the 0.01 column, we get the t score of 2.46 in Fig. 7.12(e). □

EXAMPLE 7.6 A car dealer decides to keep records on the cars he sells of the average mileage when the cars are brought in for their first warranty checkup. The manufacturer calls for an inspection at 1000 miles. If a sample of 25 cars produced a standard deviation of 100 miles using $\mu = 1000$, find e such that $P(1000 - e < \bar{X} < 1000 + e) = 0.95$. In other words, we are to construct an interval around 1000 so that 95% of the sample means the dealer calculates will fall within this interval.

Solution Since we are dealing with a small sample ($n = 25$) and a sample standard deviation $s = 100$, we have a t distribution problem (df $= 25 - 1 = 24$). If 95% of the sample means are in the center of the distribution, that leaves 5% for both tails, or 2.5% for each tail (Fig. 7.13(a)).

Using Table D.3 we can find the t scores to provide these dividing lines. They are $+2.06$ and -2.06, as shown in Fig. 7.13(b). Now we need to convert these t scores into numbers of miles.

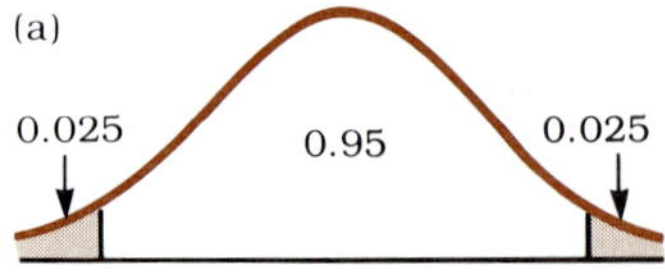

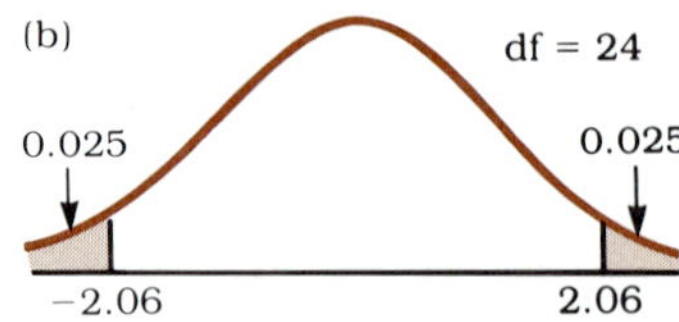

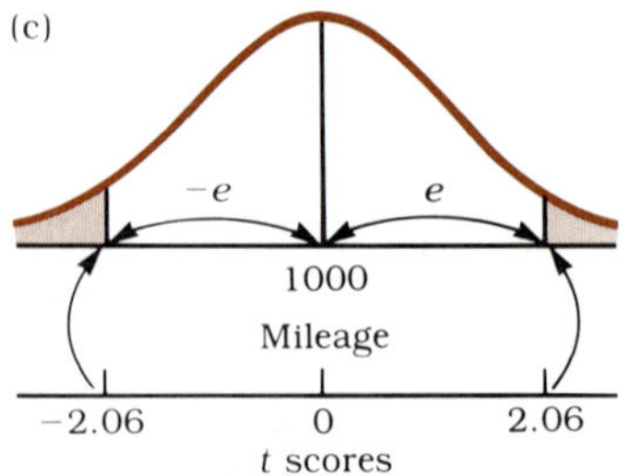

Figure 7.13

Figure 7.13(c) tells us that e represents 2.06 as a t score. But t scores measure units of standard deviation just as z scores do. We have to convert:

$$e = 2.06 \cdot \frac{s}{\sqrt{n}} = 2.06 \cdot \frac{100}{\sqrt{25}} = 41.2 \text{ miles.}$$

The interval stretches from $1000 - 41.2 = 958.8$ miles to $1000 + 41.2 = 1041.2$ miles, or

$$P(958.8 < \bar{X} < 1041.2) = 0.95.$$

Ninety-five percent of all sample means from samples of size 25 will lie within the interval. □

Remember, the use of z scores and t scores is very similar. Working with means, we use t scores when σ is unknown (s is known) and $n \leq 30$. Otherwise, we use z.

EXERCISES/Section 7.2

1. Find the t score(s) for each part of Fig. 7.14.
2. Find the t score(s) that divides off the tail(s) with the areas shown in each part of Fig. 7.15.
3. A supermarket chain had profits last year averaging \$50,000 per month per store. If a sample of 16 stores produced a standard deviation of \$8,000 for last month's sales, find e

Figure 7.14

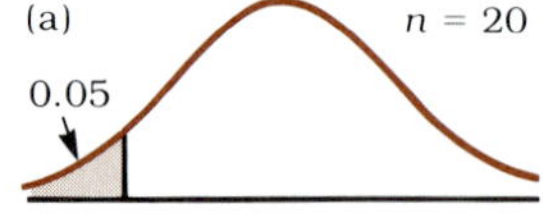

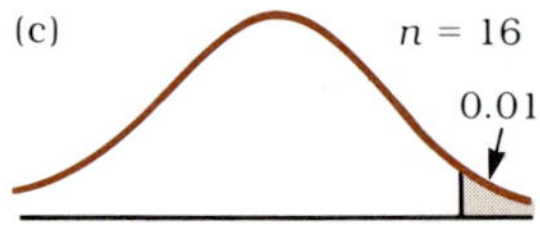

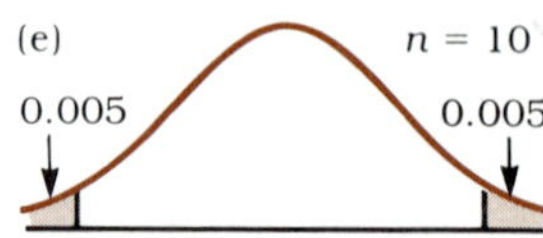

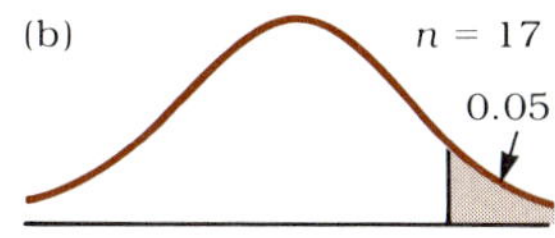

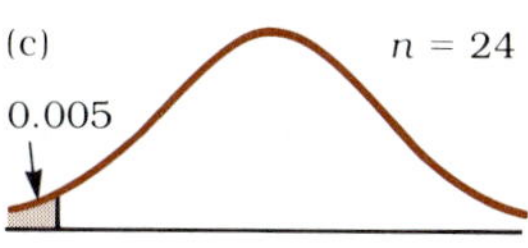

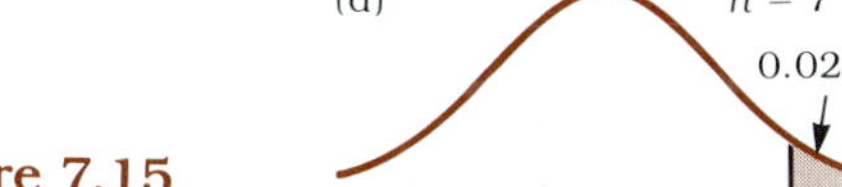

(e) $n = 23$

0.025 0.025

Figure 7.15

such that $P(50{,}000 - e < \bar{X} < 50{,}000 - e) = 0.99$.

4. The advertised weight on a jar of coffee is 16 oz. A sample of 20 jars resulted in $s = 2.4$ oz. Find e such that the probability is 0.95 that the sample mean is within e units of 16 oz: $P(16 - e < \bar{X} < 16 + e) = 0.95$.

5. The average shipping distance for ABC Trucking is 1600 miles. If a sample of 28 trucks produced a standard deviation of 430 miles, find e such that the probability is 0.95 that the sample mean is within e units of $\mu = 1600$: $P(1600 - e < \bar{X} < 1600 + e) = 0.95$.

Study Notes

KEY TERMS

sampling distribution ■ The list of all possible values of a sample statistic obtained by taking samples of size n. All samples must be the same size and the sample statistic may be any descriptive measure.

standard error of the mean ■ The standard deviation of the sampling distribution of sample means of samples of size n.

$$\sigma_{\bar{X}} = \frac{\sigma}{\sqrt{n}}.$$

OUTLINE

I. Central limit theorem: Given any population with mean μ and standard deviation σ, if all possible random samples of size n are taken, the resulting sampling distribution of sample means $\bar{X}$'s has the following properties.

A. The sampling distribution will be normally distributed if the original population is normally distributed. If the original population is not normally distributed, the sampling distribution will be approximately normally distributed for $n > 30$.

B. $\mu_{\bar{X}} = \mu$.

C. $\sigma_{\bar{X}} = \sigma/\sqrt{n}$.

II. A. $z = \dfrac{\bar{X} - \mu}{\sigma/\sqrt{n}}$ for sample means.

B. $z = \dfrac{X - \mu}{\sigma}$ for individual values.

III. Student's t distribution

A. $t = \dfrac{\bar{X} - \mu}{s/\sqrt{n}}$

B. $\text{df} = n - 1$.

REVIEW PROBLEMS

1. Consider the set of numbers: 0, 1, 2, 3.
   a) List all the possible samples of size 2 that can be extracted from the set. Assume that the first element is replaced before the second is drawn.
   b) Determine the mean of each sample.
   c) Set up the probability distribution for the set of means.
   d) Draw the histogram for the original set of numbers and for the distribution of the means.
   e) Determine $\mu_{\bar{X}}$ and $\sigma_{\bar{X}}$ for the distribution of the means.
2. The average career of professional athletes is estimated to be 5.5 yrs with a standard deviation of 1 yr. A random sample of 25 professional athletes is chosen. What is the probability that the average career for this sample is:
   a) 5 years or less?
   b) 5.25–6 years?
3. The automobile dealers association in a certain region reported that the average price paid for a new car recently was \$7900 with a standard deviation of \$1000. A marketing research firm plans to analyze this claim by interviewing 100 randomly chosen buyers of new cars in that region. What is the probability that:
   a) $\bar{X}$ will be less than \$7700?
   b) $\bar{X}$ will be more than \$8000?
   c) $\$7800 \leq \bar{X} \leq \$8100$?
4. The average playing time on an audio tape cassette is 45 min (both sides) with a standard deviation of 4 min. The manager of a record store chooses ten cassettes at random. What is the probability that the average playing time for this sample is:
   a) less than 43 min?
   b) more than 48 min?
   c) 47–48 min?
5. The average length of time needed to complete a certain math professor's quiz is 18 min with a standard deviation of 5 min. A random sample of five students from her class is chosen. What is the probability that the average completion time for this sample is:
   a) at least 15 min?
   b) no more than 20 min?
   c) 16–24 min?
6. An appliance dealer had profits last year averaging \$100 per "big ticket" item sold. If a sample of ten such sales had a standard deviation of \$25, find the value of e that satisfies $P(100 - e < \bar{X} < 100 + e) = 0.90$.
7. The advertised volume of a bottle of soda is 2 l. A sample of 20 bottles yielded $s = 0.05$ l. Find the value of e that satisfies $P(2 - e < \bar{X} < 2 + e) = 0.95$.
8. The average commuting distance for workers in a certain office is 27.5 km. A sample of five workers from that office had $s = 8.25$ km. Find the value of e that satisfies $P(27.5 - e < \bar{X} < 27.5 + e) = 0.99$.
9. The average price per couple for dinner in a local French restaurant is \$58. A sample of eight couples had $s = \$9$. Find the value of e such that $P(58 - e < \bar{X} < 58 + e) = 0.99$.
10. The average amount of time spent on a statistics assignment is 40 minutes. A sample of three students had $s = 15$ minutes. Find e such that $P(40 - e < \bar{X} < 40 + e) = 0.95$.

PART III

INFERENTIAL STATISTICS

Inference concerning the means of one population

8

8.1 Introduction to Statistical Inference

We have discussed methods of collecting, organizing, summarizing, and describing data. Now we are ready to begin consideration of statistical inference, the most important topic in a study of statistical methods. Statistical inference techniques enable us to make some judgments, draw some conclusions, and/or do some estimating and predicting from our data. These techniques answer three types of questions:

1. Is the claim about the value of a parameter correct?
2. What is the value of a parameter?
3. Is there a relationship between two (or more) variables and, if so, what is that relationship?

Let us discuss each of these questions individually to make sure that we understand exactly what is being asked. The first type of question implies several things. To begin with, someone has made a claim. Either you or someone else has a notion of what is going on with regard to a certain data characteristic. Our job is to determine whether that notion is correct. We may think so and try to buttress this point of view. Or, as is often the case, we may disagree and try to prove the other person wrong. Either way, we will answer this question by means of a **hypothesis test.**

The second type of question, about the value of a parameter, assumes that we begin with no preconceived notions and try to determine the properties of the data characteristic in question. This will be done by using **estimation** techniques.

The third type of question, about a possible relationship between data characteristics is handled by **correlation analysis** for continuous data and by means of **chi-square analysis** for attribute and discrete data. We will get to these last two techniques after spending some time on hypothesis testing and estimation.

After we understand the basic format of and reasoning behind a hypothesis test and estimation procedures, we will be concerned only with their variations. Because the basic procedures are fundamental to all kinds of hypothesis tests and estimates, we will spend considerable time and effort working with them. If we can visualize the pattern of the problems in this chapter and the next two chapters, we will have learned a great deal.

8.2 Introduction to Hypothesis Testing: The Null Hypothesis

Our society seems to regard height as an important characteristic, awarding status and prestige to those who are tall. Sports publicists exaggerate the heights of athletes, fashion models tend to be tall and lean, and many people perceive their boss to be taller than he or she actually is. Suppose that I were to tell you that the average height of adult males is 5′8″. You would probably think that I am wrong and have erred on the short side. How do you go about proving my claim wrong? All you have to do, as common sense should suggest even if you knew nothing about statistics, is to go out and measure some adult males; that is, gather data from an appropriate sample using techniques that we have discussed previously. These data can be summarized by techniques you know: Calculate the sample mean, $\bar{X}$; compare your sample result to my claim; and draw your own conclusion about whether I was right or wrong. We will assume that you have a correctly chosen random sample and have made no arithmetic mistakes.

So far, this sounds fairly simple. The problem arises when we start to examine the possible sample results for $\bar{X}$. You know from the discussion of the central limit theorem that, even if I am correct, you probably will not get an $\bar{X}$ exactly equal to 5′8″. Let us say that you get a sample mean $\bar{X}$ of 5′8½″. In our society and in statistics we assume innocence (or truth) until guilt (or falsity) is proven. Therefore you must decide whether this value of 5′8½″ is sufficiently far from my claim of 5′8″ to prove me wrong. Could the difference be accounted for merely by sample variability? You would probably not stick your neck out for 5′8½″. What about $\bar{X}$ = 5′9″? Would you now think that your result differed sufficiently from mine to say that I was wrong? If you got a sample result of 5′11″, would you say that I was wrong? Probably by now you are leaning toward a guilty (or claim is wrong) verdict. Certainly, $\bar{X}$ = 6′2″ (assuming that your sample was fairly chosen) would be a strong case against my claim.

Each of you probably has some cutoff point in your head to use in order to make a decision about my claim. If your sample results are on the 5′8″ side of your cutoff point, you will go along with my claim or at least state that you cannot prove me wrong. If your sample results are beyond your cutoff point, away from 5′8″, you will conclude that my claim should be rejected as wrong. Drawing such conclusions cannot be left to whim or intuition but must be the result of a well-reasoned argument that can be agreed upon by all. That is our task: to establish criteria whereby we can all draw the same conclusions given the same data. Statistical inference methods give us the tools to accomplish this task.

Let us begin by examining my contention about the average height of adult males. I presented a hypothesis or claim. This is called the **null hypothesis** and is represented by the symbol H_0.

> ***null hypothesis*** ■ A statement about a population parameter that is being tested by the use of sample results and a decision-making process.

A hypothesis test *always* focuses on the null hypothesis with the result being rejection or no rejection of this claim. Again, the truth of the claim is assumed until data and the decision process can indicate falseness of the claim. However, this method is not without risks: The claim may actually be true, but the data may lead to its rejection and result in an error. The possible consequences of a final decision are shown in Table 8.1.

Now, if 5′8″ is the true mean height for adult males and your data support this claim, you will accept it and will have made a correct decision. If, on the other hand, the claim of 5′8″ is not true and your data do not support it, you will reject the claim—again a correct decision. However, there are two other possibilities. If my claim of 5′8″ is correct but you happen to get a sample with $\bar{X}$ = 6′1″, you will probably reject my claim. If you do, you have made the mistake of rejecting something that is true. This is called a **Type I error.** On the other hand, if my claim is false, you also may make a mistake by accepting it based on your sample results. If so, you have made a **Type II error.** Sample results do vary, and it is possible for you to get an extreme $\bar{X}$ that leads you to an incorrect conclusion.

> ***Type I error*** ■ Rejecting a true null hypothesis.
>
> ***Type II error*** ■ Failing to reject a false null hypothesis.

In order to make these risks manageable or predictable, we attach probabilities to them. The probability of making a Type I

TABLE 8.1

	H_0 Is	
Decision	**True**	**False**
Fail to Reject H_0	Correct decision	Type II error
Reject H_0	Type I error	Correct decision

error is denoted by **alpha** (α).

$$P(\text{rejecting a true null hypothesis}) = P(\text{Type I error}) = \alpha.$$

The probability of making a Type II error is called **beta** (β).

$$P(\text{accepting a false null hypothesis}) = P(\text{Type II error}) = \beta.$$

Is it possible to make either alpha or beta zero? In other words, can we guarantee that we will never make one of these types of errors? We can avoid rejecting true hypotheses, that is, $\alpha = 0$, by never rejecting anything. What does this do to β? If we never reject anything, we accept everything and thus accept all false claims. Thus β will be large. Similarly, if we try to make $\beta = 0$ by never accepting false claims, we must never accept anything, reject everything, and so make α large. In short we do not want either α or β to be 0. The common practice is to choose a value for α, the most frequent values used being 0.01 and 0.05. This means that when we are examining a situation in which H_0 is true, 1% or 5% of the time we will make incorrect decisions by rejecting true claims. The value we choose for α can be determined by the severity of the consequences of making a mistake. We will focus on α because the goal of a hypothesis test is to test the truth of the claim in H_0 and to make a decision about rejection. A discussion of beta is beyond the elementary scope of this text.

EXERCISES/Section 8.2

1. Consider the following as the null hypothesis: *An accused murderer is innocent.* A guilty verdict ensures the death penalty.
   a) Does a guilty verdict mean accepting or rejecting the null hypothesis?
   b) What is a Type I error in this case?
   c) What is a Type II error in this case?
2. You are driving down a country road when a policeman with radar pulls you over. His hypothesis is: *You are speeding.*
   a) What are the four possible interpretations of Table 8.1 in this case?
   b) What is a Type I error?
   c) What is a Type II error?
   d) Which type of error is more serious?
   e) If the judge makes a mistake when you go to court, which type of error is in your favor?
3. ABC Fast Food Service has been cited by the health department for unsanitary conditions that could lead to food poisoning. You decide to grab a quick bite to eat under the assumption (or hypothesis) that ABC's food is wholesome.
   a) What are the four possible interpretations of Table 8.1 in this case?
   b) What is a Type I error?
   c) What is a Type II error?
   d) Which type of error is more serious?
   e) Which do you prefer:
      1) α large and β small? or
      2) α small and β large?

4. Some medical tests produce *false positives.* An example of such results is the patch test for tuberculosis. A positive patch test result suggests that the patient has tuberculosis. Thus H_0 says that the patient has TB.
   a) What are the four possible interpretations of Table 8.1 in this case?
   b) What is a Type I error?
   c) What is a Type II error?
   d) Which type of error is more serious?
   e) As a patient, which would you prefer:
      1) α large and β small? or
      2) α small and β large?

8.3 Introduction to Hypothesis Testing: The Alternative Hypothesis

Anyone seeking to test a claim is usually motivated by a desire to contradict the claim—to prove it wrong. If the null hypothesis is rejected, we need an alternative to H_0 that can be accepted; it is called the **alternative** or **motivated hypothesis,** H_A.

> ***alternative hypothesis*** ■ A statement that is to be accepted if the null hypothesis is rejected. It utilizes the same parameter as the null hypothesis but gives motivation for the rest of the test procedure. This is the statement we want to accept.

In our height example, the alternative hypothesis is that the mean height of adult males is larger than 5′8″.

We now have the first two parts of a hypothesis test, H_0 and H_A. We must next decide on a cutoff point, a place that we can use as a decision point to compare to our sample results and help us to draw a conclusion. Decision points obviously depend on the problem, so let us consider specific examples.

EXAMPLE 8.1 An auto manufacturer claims that its new model cars, on the average, will get at least 38 miles per gallon for highway driving. You have been driving one of last year's models and do not believe that the company could have made such an improvement in fuel efficiency. What is the null hypothesis? What is the alternative hypothesis?

Solution The null hypothesis names the parameter being discussed and gives a value to that parameter. Thus we have

$$H_0: \mu = 38.$$

Note that we have written H_0 with an *equals* sign, which will always be the case. Before we finish discussing hypothesis testing, we will calculate z scores and will need a value to substitute for μ. Since we assume H_0 is true, we substitute the value it gives for μ in the z calculation. Hence we need the equals sign.

For the alternative hypothesis we consider the motivation and what we want to contradict. We want to contradict *at least 38 miles per gallon*. We are examining the same parameter and the same number as with H_0, but now we connect them with a different symbol to indicate motivation. We contradict *at least* with *less than*.

$$H_A: \mu < 38.$$

Note that the *null hypothesis and the alternative hypothesis differ only by the symbol relating the parameter and the number.* □

EXAMPLE 8.2 Let us change the orientation. The Environmental Protection Agency says that your company's cars are getting, on the average, no more than 23 miles per gallon. You are a design engineer and are interested in making your cars competitive in fuel economy. You would like to prove that the EPA is wrong. What is the null hypothesis? What is the alternative hypothesis?

Solution Again H_0 will name the parameter (μ) and contain an equals sign.

$$H_0: \mu = 23.$$

The alternative hypothesis has μ and 23. We want to contradict *no more than*, which we do with *more than* or *greater than*.

$$H_A: \mu > 23. \quad \square$$

EXAMPLE 8.3 The claim is made that the average mileage of a line of cars is 29 miles per gallon. You wish to contradict this. What is the null hypothesis? What is the alternative hypothesis?

Solution H_0 will be as before.

$$H_0: \mu = 29.$$

For H_A we want to contradict *is* and will do so with *is not* or *does not equal*.

$$H_A: \mu \neq 29. \quad \square$$

Let us summarize what we found for the three null hypotheses and the three possible alternative hypotheses.

Example 8.1	Example 8.2	Example 8.3
H_0: $\mu = 38$ H_A: $\mu < 38$	H_0: $\mu = 23$ H_A: $\mu > 23$	H_0: $\mu = 29$ H_A: $\mu = 29$

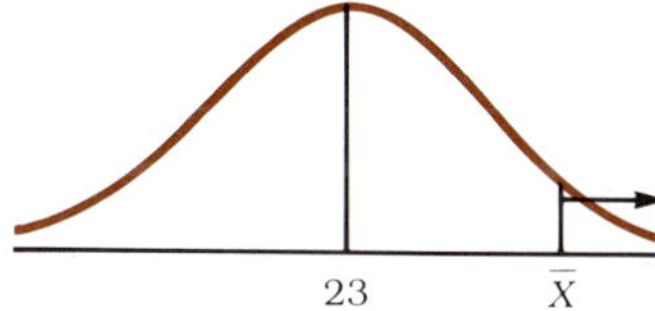

Figure 8.1

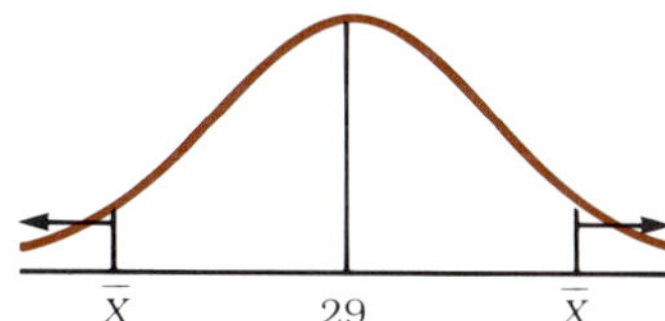

Figure 8.2

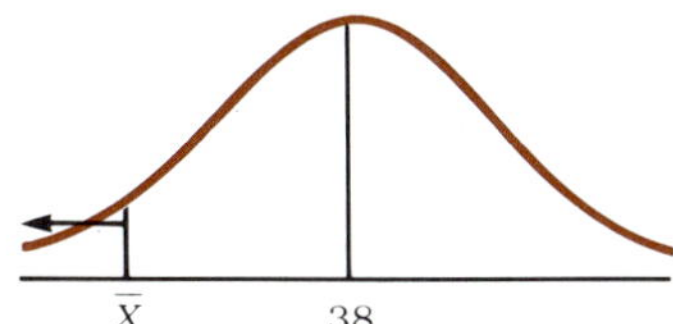

Figure 8.3

Next we need criteria for making a decision about whether or not we can reject H_0. We test μ in each case. To do that we will gather sample data, calculate $\bar{X}$, and compare it to the H_0 value for μ. If $\bar{X}$ is far enough from the claimed value, we will reject the claim. Since we are dealing with $\bar{X}$, we know from the central limit theorem that sample means are normally distributed. Fuel efficiency may not be normally distributed, but sample means are: $\mu_{\bar{X}}$ equals μ in the null hypothesis under the assumption that H_0 is true unless we can show otherwise.

Let us review each of the examples and determine where the sample result $\bar{X}$ should fall in order to justify our motivation. In Example 8.1 we believe that the company's mileage result is less than 38. This means that we would like for $\bar{X}$ to fall somewhere to the left of 38, as in Fig. 8.1. On the other hand, in Example 8.2, we believe that the mileage is greater than 23 and so would like the sample results to fall somewhere to the right of 23, as in Fig. 8.2. Finally, if we believe that μ is not 29 as in Example 8.3, we should be satisfied if $\bar{X}$ were to fall far from 29 on either side, as in Fig. 8.3.

We have divided the diagrams into two areas: We have said that if the sample results fall sufficiently far from μ, we will reject H_0; if not, we will not be able to reject H_0. Thus we have identified what is known as the **rejection region,** sometimes called the **test region,** or the **critical region.**

> ***rejection region (test region or critical region)*** ■ The portion of a distribution that provides values for the sample results causing the rejection of H_0.

Our work is summarized in Fig. 8.4. Cases (a) and (b), from Examples 8.1 and 8.2, are called **one-tail** or **directional** tests

Figure 8.4

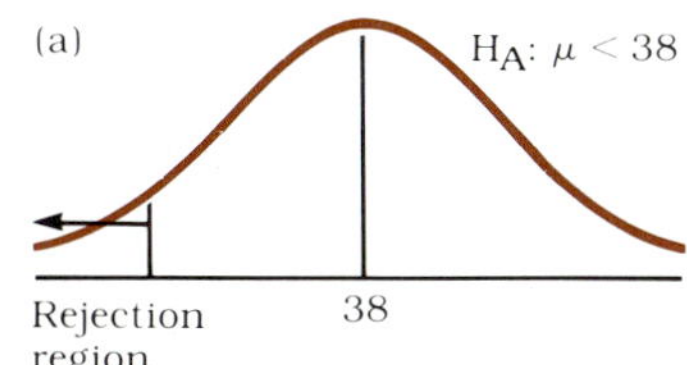

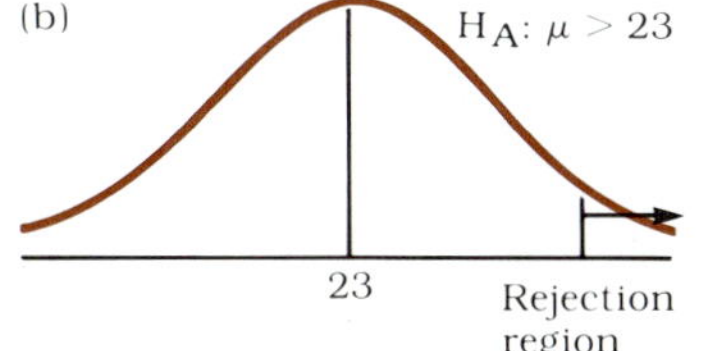

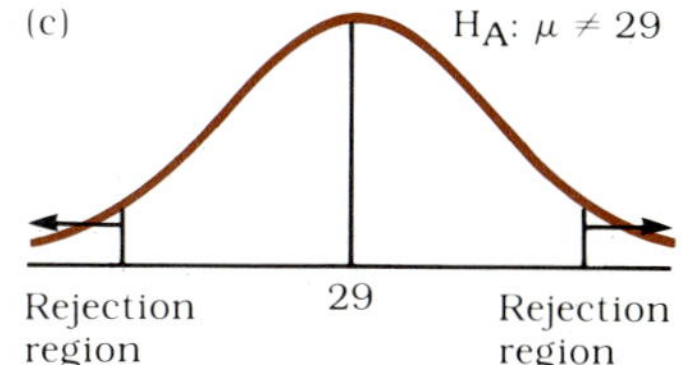

because their rejection regions are only in one tail of the normal distribution. Case (c), from Example 8.3, is a **two-tail** or **nondirectional** test since its rejection region is split between the two tails. Note that the symbol (< or >) in H_A points to the rejection region in the case of a directional test. The roles of the hypotheses should now be clear: *H_0 gives us a value to put in the center of the distribution, and H_A tells us where to look in order to be able to reject H_0.*

Our next goal is to locate the dividing point that tells us where the critical region begins. This is called the **critical value.**

> ***critical value(s)*** ▪ The value(s) that determines the rejection region.

When we find the critical value(s), we are almost done. (In a nondirectional test there are two critical values, equidistant from the mean.) We can compare our sample results to the critical value(s). If $\bar{X}$ falls in the rejection region, our conclusion is to reject H_0 and accept H_A. If $\bar{X}$ does not fall in the rejection region, we cannot reject H_0. We do not have to accept H_0, but the least we can say is that we do not reject it. This latter statement allows for the possibility of rejecting H_0 in the future, perhaps by gathering more data (taking a larger sample size if our original one was small) and trying again.

In order to find the critical value for Example 8.1, let us again recall the central limit theorem and sampling distributions. If H_0 is true and 38 really is the mean of this distribution, then Fig. 8.5 represents the distribution of all possible sample means that would spread out around 38. If we cut off part of this distribution in our rejection region, as in Fig. 8.6, we are actually cutting off legitimate $\bar{X}$'s, which are associated with 38 *if* H_0 *is true.*

38 $\bar{X}$

Figure 8.5

Thus if we obtain an $\bar{X}$ in the rejection region (causing us to reject H_0), we would be making an error *if* H_0 *is true.* What type of error is this? Rejecting a true null hypothesis is a Type I error. Associated with a Type I error is a probability α. Where do we place probabilities in diagrams of a normal distribution? Under the curve. So α represents the area under the curve above the rejection region. Alpha is usually called the **level of significance** of the test.

Figure 8.6

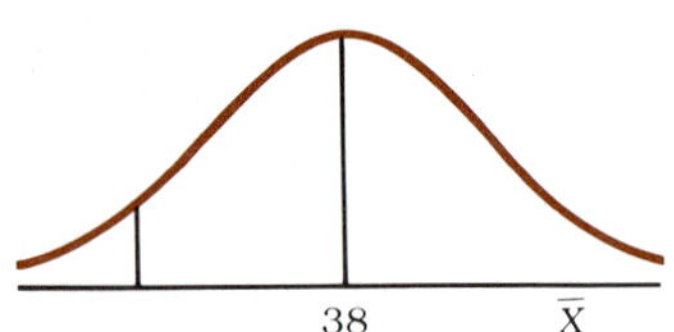

> ***level of significance*** ▪ The probability α of a Type I error.

When the results of a study are said to be statistically significant, this means that an α was chosen and a hypothesis test

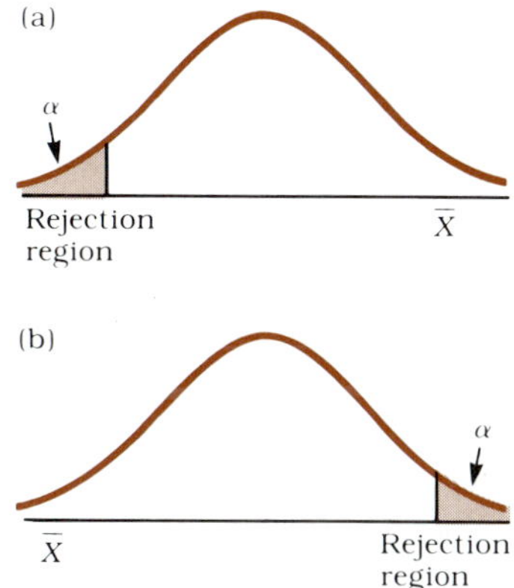

Figure 8.7

performed that led to the rejection of H_0. In News & Views 8.1, note the word *significantly*. Its use indicates that standard statistical procedures for hypothesis testing were used.

For a given α, we will always know the area under the curve above the rejection region. For a one-tail test, α may be located as shown in Fig. 8.7(a) *or* (b). If we have a two-tail test, we must be careful. If we put α in each tail, the total area in the rejection region is $\alpha + \alpha = 2\alpha$ (Fig. 8.8), and we have doubled the probability of a Type I error. Instead, we put only one-half of α, or $\alpha/2$, in each tail, and now the total area in the rejection region is $\alpha/2 + \alpha/2 = \alpha$ (Fig. 8.9). Knowing an area under the curve in either a normal distribution or a t distribution allows us to find the z or t score associated with the dividing point.

Figure 8.8

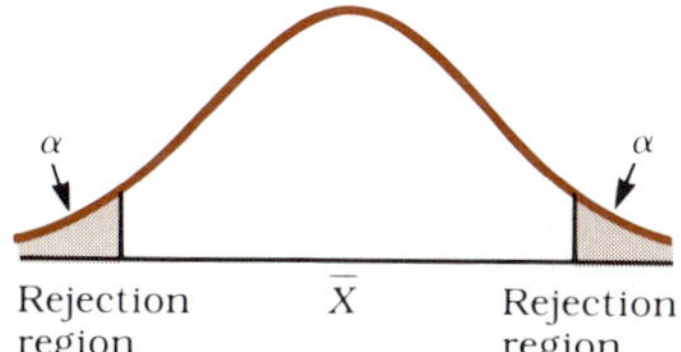

Figure 8.9

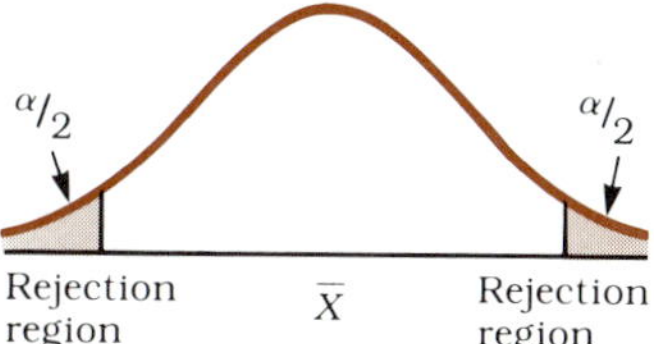

NEWS & VIEWS 8.1

Smoking may cancel helpful results of exercise

Cigarette smoking may cancel out a benefit of exercise, according to the preliminary findings of a study on cholesterol by the University of Louisville.

The research indicates that tobacco may inhibit the production of high-density cholesterol, a substance that helps remove other harmful kinds of cholesterol from the body, said Dr. Bryant Stamford, director of the university's exercise physiology laboratory.

Since last July, university researchers have tested about 500 volunteers to study the relationship between smoking and cholesterol. Those who had smoked more than a pack of cigarettes a day for at least five years had "significantly" less high-density cholesterol than the non-smokers, Dr. Stamford said in a recent interview.

Smokers had less high-density cholesterol, even if they exercised regularly. Exercise normally helps the body produce high-density cholesterol, Dr. Stamford said.

The study has not tested the effects of light or occasional smoking, he said.

Medical research indicates that a high intake of low-density cholesterol, part of animal fat, contributes to the risk of heart disease.

Source: *Baltimore Sun*, February 15, 1982.

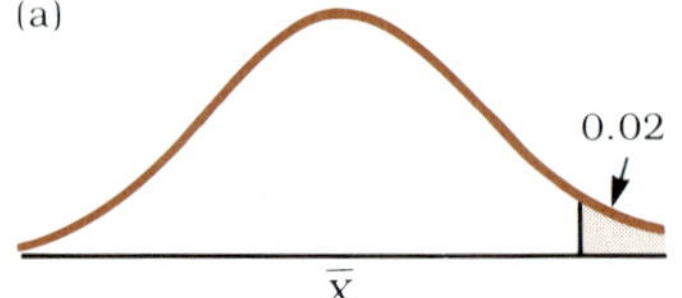

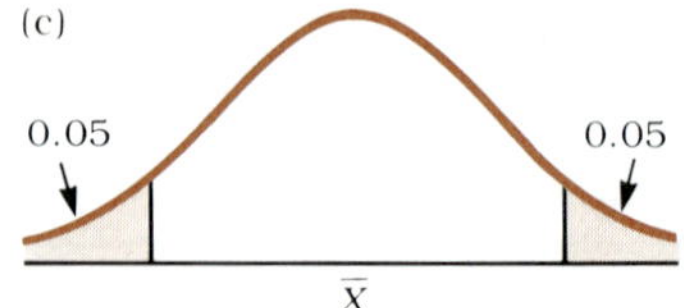

Figure 8.10

EXAMPLE 8.4 Find the z scores that are critical values for each diagram in Fig. 8.10.

Solution

a) From Chapter 6 recall that if 0.02 is the area in the tail, $0.5 - 0.02 = 0.48$ is the area bounded by the mean (Fig. 8.11). Looking for this area in the body of Table D.2, we find that $z = 2.05$.
b) Similarly if 0.10 is the area in the tail, then $0.5 - 0.10 = 0.40$ is the area bounded by the mean in Fig. 8.12. The z score found by looking up 0.40 in the body of the table is 1.28; the area is to the left of the mean, so we complete the problem by putting in a minus sign to get $z = -1.28$.
c) We can utilize symmetry here. In Fig. 8.13, $0.5 - 0.05 = 0.45$. The z score found by looking up 0.45 is 1.65. Symmetry tells us that the two z scores are $+1.65$ and -1.65. □

EXAMPLE 8.5 Find the critical values as t scores that go with the diagrams in Fig. 8.14.

Figure 8.11

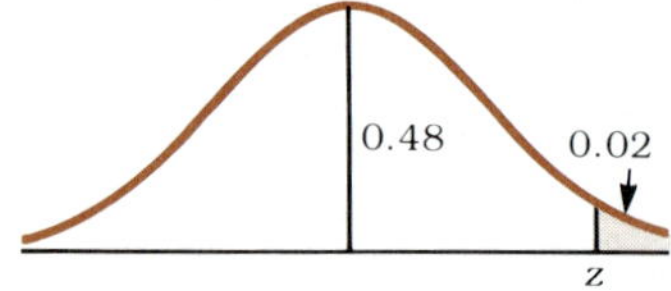

Figure 8.12

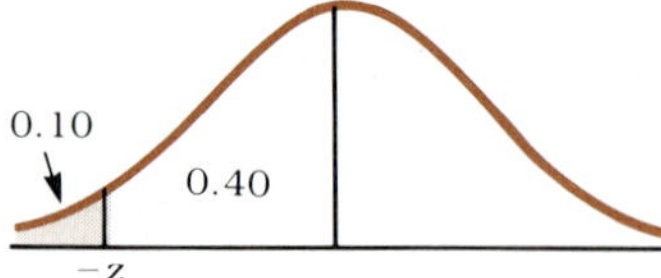

Figure 8.13

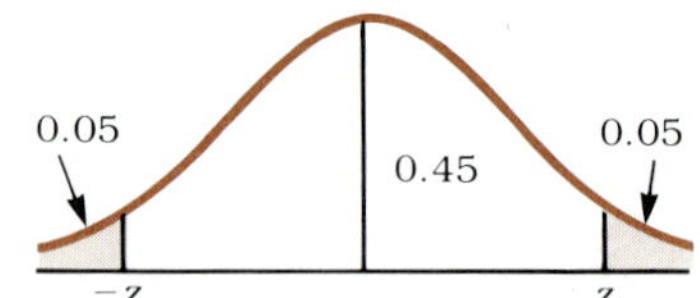

Figure 8.14

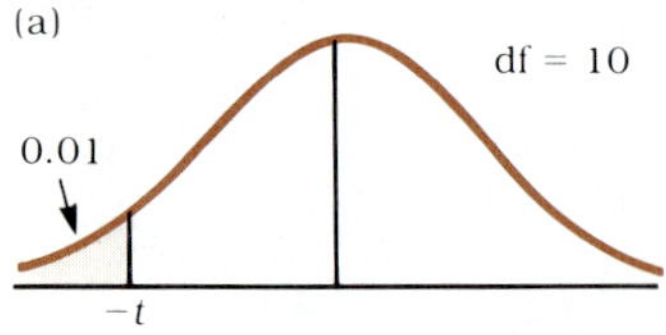

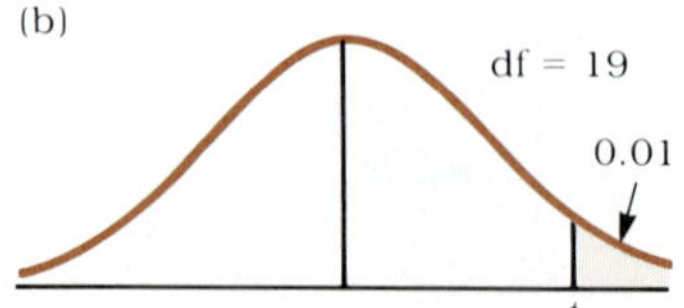

Solution

a) Looking in Table D.3, we find that for 10 degrees of freedom and 0.01 in the tail, the critical value is 2.76. Since the desired t is below the mean, we get $t = -2.76$.
b) Nineteen degrees of freedom and 0.01 for α give us $t = 2.54$.
c) Symmetry again aids us. With df = 15 and $\alpha = 0.025$, $t = 2.13$ and our critical values are +2.13 and −2.13. □

We are now able to find the decision points in terms of z scores or t scores. It becomes a simple matter to convert the sample result $\bar{X}$ to a z value, if we know σ, or to a t value, if we know only s. The z score or t score of the sample results is called the **test statistic.**

test statistic ■ A calculated value using sample results to be compared to a tabular value, such as a z score or t score of the sample results.

Now compare the test statistic to the test region and draw your conclusion.

REMARK

The null hypothesis is the one we are testing. It may be the status quo, something to be changed, past experience, or the path of least resistance. If you analyze problems from this point of view, you should make few errors in stating hypotheses.

This should aid in stating H_A. It is important that H_A be expressed correctly, since the critical region depends upon it. If H_A is one-tailed and backwards, for example, the rest of the test will be wrong.

EXERCISES/Section 8.3

1. Find the z scores that are critical values for each part of Fig. 8.15.
2. Find the critical values as t scores that go with each part of Fig. 8.16.
3. Find the appropriate z-score critical values using the following information.

a) H_A: $\mu \neq 5$m, $\alpha = 0.01$
b) H_A: $\mu > 10$ ft, $\alpha = 0.01$
c) H_A: $\mu < 103$ lb, $\alpha = 0.01$
d) H_A: $\mu < 0.03$ ml, $\alpha = 0.05$
e) H_A: $\mu \neq 5'8''$, $\alpha = 0.05$
f) H_A: $\mu > 10.2$ miles, $\alpha = 0.05$

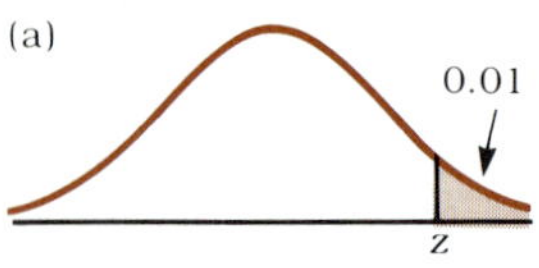

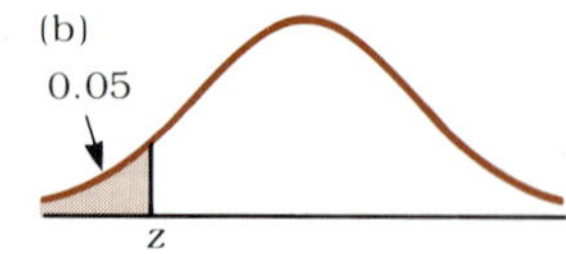

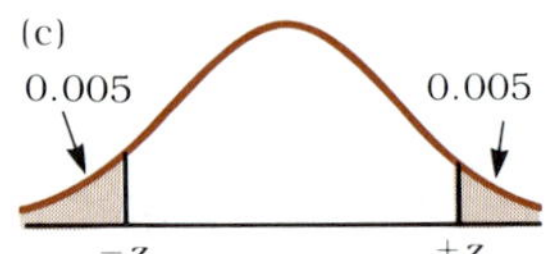

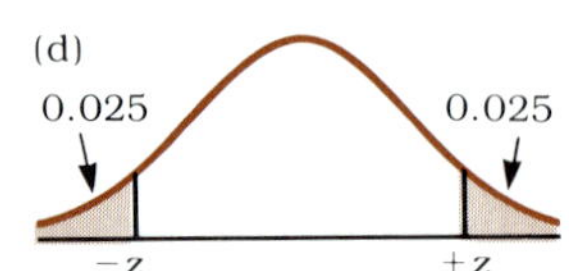

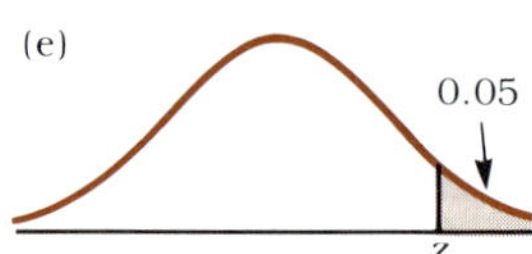

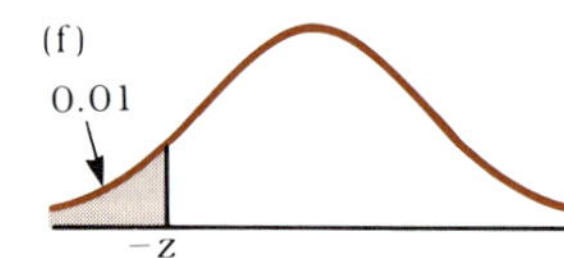

Figure 8.15

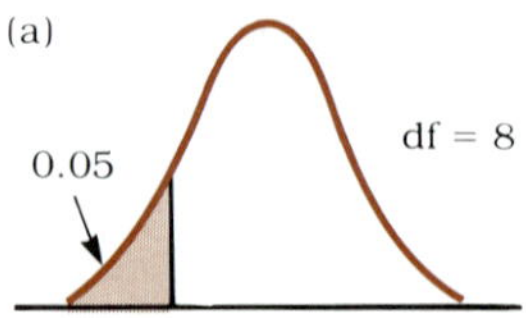

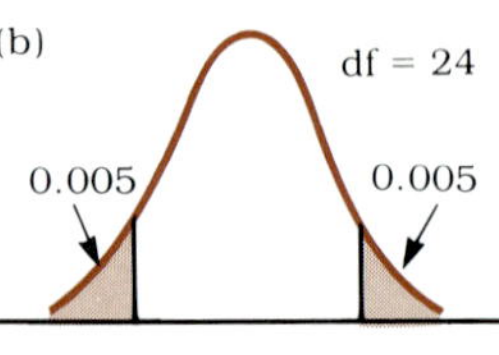

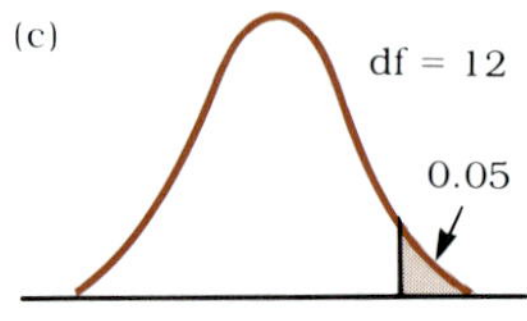

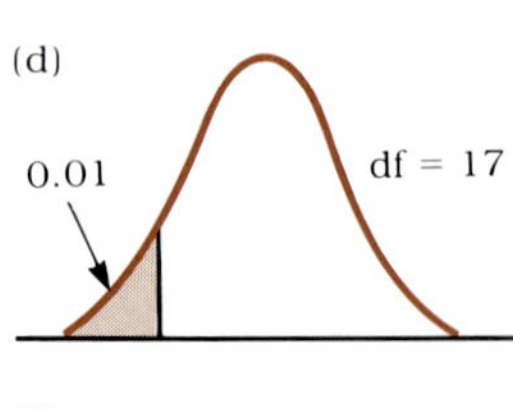

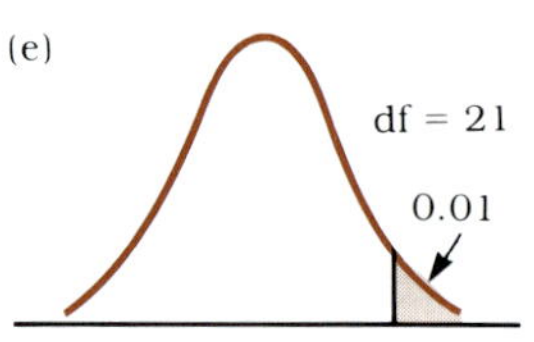

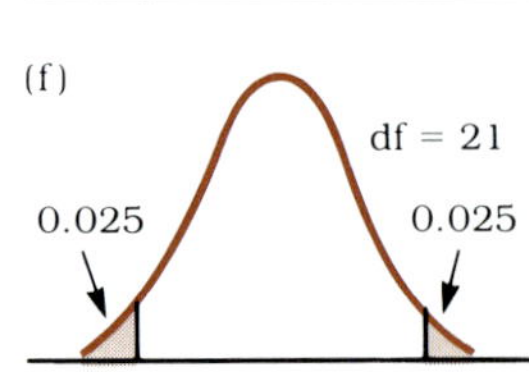

Figure 8.16

4. Find the appropriate *t*-score critical values using the following information.

a) H_A: $\mu < 6$, $\alpha = 0.05$, df = 10	**d)** H_A: $\mu \neq 44.2$, $\alpha = 0.01$, df = 29
b) H_A: $\mu > 21$, $\alpha = 0.01$, df = 15	**e)** H_A: $\mu \neq 0.06$, $\alpha = 0.05$, df = 24
c) H_A: $\mu > 91.7$, $\alpha = 0.05$, df = 17	**f)** H_A: $\mu < 1.578$, $\alpha = 0.01$, df = 7

5. State the null and alternative hypotheses you would use to test the following claims. (*Remember: Test the claim* makes the claim H_0.)
   a) The average weight of large dogs is 60 lb.
   b) The average commuting distance to work is at least 9 miles.
   c) The mean age at death among residents of City A is 55.3 yrs.
   d) The mean age for first pregnancies is no more than 20.
   e) On the average, bottles of beer contain 15.78 oz.
   f) Of those who walk, elementary-school children live within 0.8 miles of their school.
   g) There are no more than 2 min of commercials in every ½ hr of television.
   h) The average person eats more than 200 lb of meat per year.
   i) The mean length of time in a job is greater than 12 yrs.
   j) The average winter snowfall in Maryland is 10 in.

8.4 Further Examples

We are ready to work through some examples in detail. These examples include predetermined levels of significance α, but at the end of this section we introduce the *p*-value concept, which allows expression of an α that is achieved by the data.

EXAMPLE 8.6 In order to test the metropolitan traffic control office's claim that traffic lights are red, on the average, no more than 1 minute each cycle, an experimenter timed 36 traffic lights chosen throughout the city. She found that the mean time for red was 1.2 minutes. She tests the office's claim at the 0.05 level of significance, assuming that $\sigma = 0.06$ minutes.

Solution The null hypothesis is the claim we are to test.

$$H_0: \mu = 1.$$

The alternative hypothesis should contradict *no more than* in the claim.

$$H_A: \mu > 1.$$

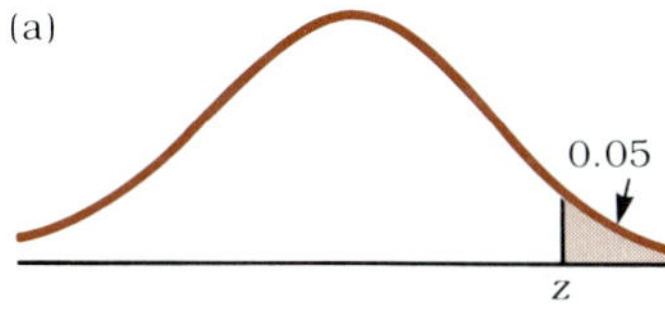

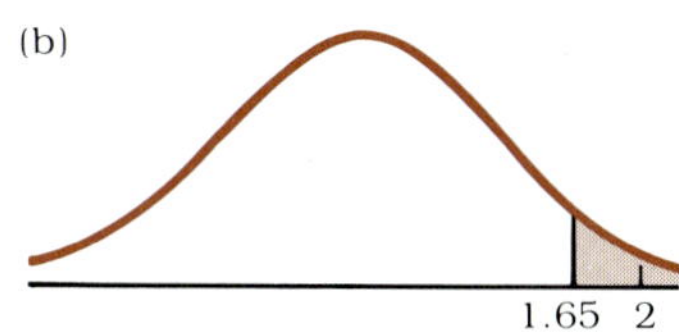

Figure 8.17

An α of 0.05 allows the experimenter to set up a one-tail test region (Fig. 8.17a). Since she is given the standard deviation of the population, σ, she uses z scores. The z score that goes with Fig. 8.17(a) is 1.65. She can now say that, if the z score of the sample result is greater than 1.65 and thus falls in the critical region, she will reject H_0. She calculates this z score as follows and shows the result in Fig. 8.17(b).

$$z^* = \frac{\bar{X} - \mu}{\sigma/\sqrt{n}} = \frac{1.2 - 1}{0.6/\sqrt{36}} = 2.$$

The z score of the sample result, the test statistic, is usually labeled with an asterisk (z^*) to indicate that it was calculated and is not a z score found in a table.

We will always need a value from the null hypothesis to help us calculate the test statistic; here the claimed value for μ was used. Since z^* falls in the critical region ($z > 1.65$), the experimenter rejects H_0 and accepts H_A: the traffic lights are red more than 1 minute per cycle.

Summary

H_0: $\mu = 1$.

H_A: $\mu > 1$.

Test region: We reject H_0 if $z^* > 1.65$.

Test statistic: $z^* = \dfrac{1.2 - 1}{0.6/\sqrt{36}} = 2.$

Conclusion: Reject H_0 in favor of H_A at the 5% level of significance and conclude that the average is greater than 1 minute. □

EXAMPLE 8.7 Cereal manufacturers claim that children watching Saturday morning cartoons are subjected, at most, to 120 seconds of commercials in each half hour of programing. Sarah Jones, a mother of three small children, cannot stand their whining for products in the grocery store and decides to monitor their TV viewing. She finds a sample mean of 140 seconds of cereal ads per half hour when she watches 25 half-hour cartoon programs. Do her data refute the cereal companies' claim for $\alpha = 0.01$? Assume that $\sigma = 55$ s.

Solution Sarah wants to try to reject the companies' claim. That tells her the value for H_0.

$$H_0: \mu = 120.$$

She wants to contradict the claim of *at most*.

$$H_A: \mu > 120.$$

She has σ and so will use z scores.

As in Example 8.6, the critical value is 1.65 (Fig. 8.18a), and Sarah will reject H_0 when $z^* > 1.65$. Her test statistic is

$$z^* = \frac{\bar{X} - \mu}{\sigma/\sqrt{n}} = \frac{140 - 120}{55/\sqrt{25}} = 1.82.$$

This is illustrated in Fig. 8.18(b), and thus Sarah rejects the cereal manufacturers' claim.

(a)

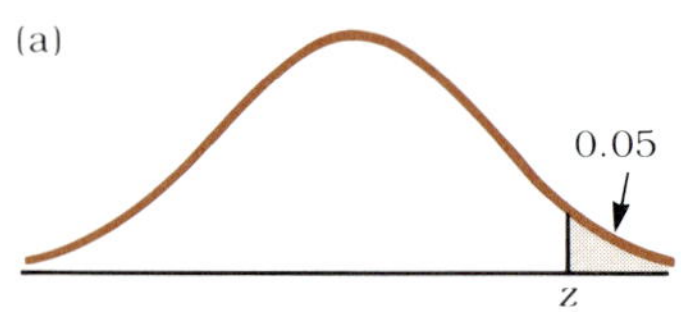

(b)

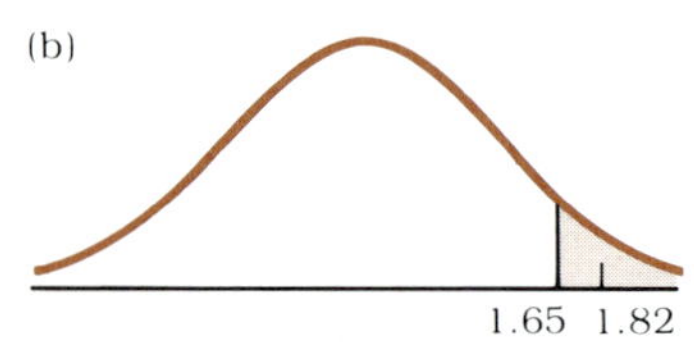

Figure 8.18

Summary

$H_0: \mu = 120.$

$H_A: \mu > 120.$

Test region: Reject H_0 if $z^* > 1.65$.

Test statistic: $z^* = \dfrac{140 - 120}{55/\sqrt{25}} = 1.82.$

Conclusion: Reject H_0 in favor of a mean greater than 140 seconds at the 0.05 level of significance. □

EXAMPLE 8.8 Customers complained to a grocery-store manager that the house-brand ketchup bottles were not adequately filled. The customers doubted the accuracy of the labels, which claimed a weight of 44 ounces. The manager decided to test the ketchup manufacturer's claim at the 0.01 level of significance by checking the accuracy of its bottling procedures. He randomly selected 48 bottles and weighed the contents. He obtained a sample mean weight of 42.6 ounces. What could he conclude? Assume that $\sigma = 4.75$ ounces.

Solution Here the customers are challenging the status quo as offered on the ketchup label. This becomes the null hypothesis.

$$H_0: \mu = 44.$$

The store manager can find out whether the customer complaints are justified by rejecting $\mu = 44$ in favor of inadequately filled jars, or

$$H_A: \mu < 44.$$

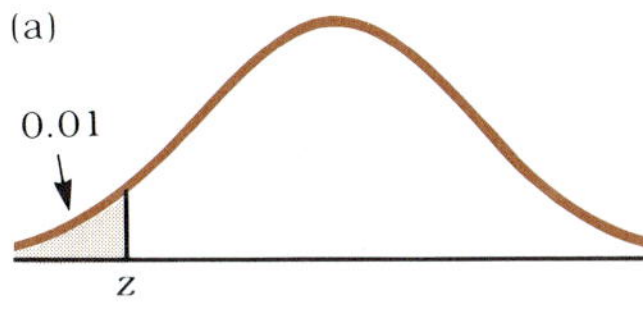

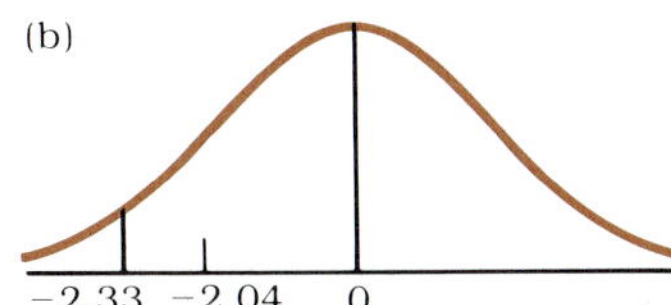

Figure 8.19

This H_A indicates a critical region that is directional to the left and contains all of $\alpha = 0.01$ in Fig. 8.19(a). Since the store manager knows the population standard deviation σ, he will use z scores; now the critical value is −2.33. (Do not forget the minus sign, which is needed because the critical region is below the mean.) The store manager will reject H_0, the manufacturer's claim, if $z^* < -2.33$.

$$z^* = \frac{\bar{X} - \mu}{\sigma/\sqrt{n}} = \frac{42.6 - 44}{4.75/\sqrt{48}} = -2.04.$$

This is shown in Fig. 8.19(b). Since z^* does not fall in the rejection region, the store manager cannot reject the manufacturer's posted weight in favor of the complaints.

Summary

$H_0: \mu = 44.$

$H_A: \mu < 44.$

Test region: Reject H_0 if $z^* < -2.33$.

Test statistic: $z^* = \dfrac{42.6 - 44}{4.75/\sqrt{48}} = -2.04.$

Conclusion: Do not reject H_0. □

EXAMPLE 8.9 A recent newspaper article claimed that the average fuel economy figure for autos traveling at 55 mph was 25.7 miles per gallon. Two neighbors, one with a small car and one with a larger model, debated this claim and decided to test it at the 0.05 level of significance. They enlisted the aid of 11 friends who did most of their driving on highways, averaging 55 mph. These 11 drivers averaged 23.6 miles per gallon with a standard deviation of 3.5 miles per gallon. Were the neighbors correct in challenging the newspaper's claim?

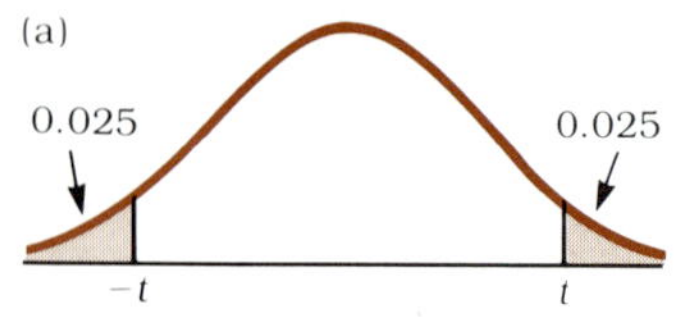

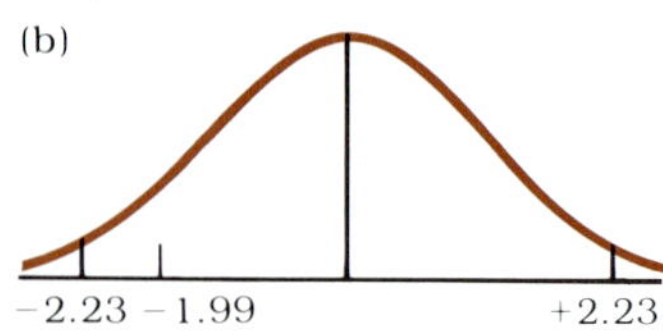

Figure 8.20

Solution The newspaper's claim to be tested provides the null hypothesis.

$$H_0: \mu = 25.7.$$

The neighbors are disputing the claim, but they haven't said whether they think the claim is too high or too low, so the alternative hypothesis is two-tailed.

$$H_A: \mu \neq 25.7.$$

This indicates a split critical region with one-half of α in each tail, as in Fig. 8.20(a).

They must now find two critical values. Looking for the standard deviation to tell them whether to use z or t scores, they find that the standard deviation occurs in the sentence giving sample results; thus they have s, not σ, and will use t scores. The degrees of freedom are $n - 1 = 11 - 1 = 10$, and the critical values from Table D.3 are $+2.23$ and -2.23. The neighbors will reject H_0 if $t^* > 2.23$ or $t^* < -2.23$.

$$t^* = \frac{\bar{X} - \mu}{s/\sqrt{n}} = \frac{23.6 - 25.7}{3.5/\sqrt{11}} = -1.99.$$

This is illustrated in Fig. 8.20(b). Since t^* does not fall in the test region, the neighbors cannot reject the newspaper's claim.

Summary

$H_0: \mu = 25.7.$

$H_A: \mu \neq 25.7.$

Test region: Reject H_0 when $t^* < -2.23$ or $t^* > 2.23$.

Test statistic: $t^* = \dfrac{23.6 - 25.7}{3.5/\sqrt{11}} = -1.99.$

Conclusion: Do not reject H_0. □

EXAMPLE 8.10 A national poll reported that average wages had reached a record high of \$21,300. In order to determine how the city compared to the national mean, the chamber of commerce gathered data from the 25 largest employers and obtained an average of \$22,750 with a standard deviation of \$2750. Can the chamber support the claim that the city is significantly different from the national average at the 0.01 level?

Solution The chamber of commerce is challenging the national average, so it provides the null hypothesis.

$$H_0: \mu = \$21{,}300.$$

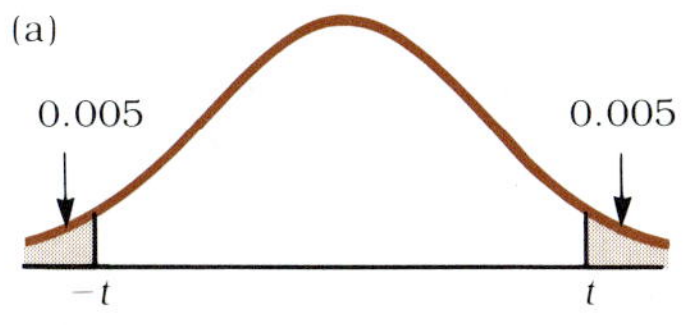

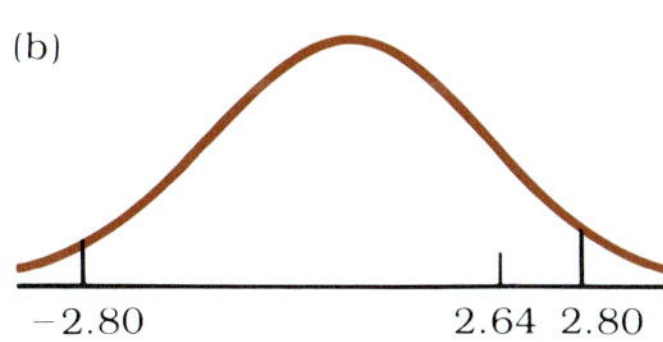

Figure 8.21

The claim that the city is different suggests an alternative hypothesis that is nondirected.

$$\mathrm{H_A}: \mu \neq \$21{,}300.$$

The critical region is then two-tailed with α split (Fig. 8.21a).

The standard deviation in the problem comes from sample results and so is s, not σ. That means that the chamber must use t scores with $n - 1 = 25 - 1 = 24$ df. The two critical values needed are thus $+2.80$ and -2.80. The chamber will reject the national average for the city if $t^* < -2.80$ or $t^* > 2.80$.

$$t^* = \frac{\bar{X} - \mu}{s/\sqrt{n}} = \frac{22{,}750 - 21{,}300}{2750/\sqrt{25}} = 2.64.$$

The test statistic is shown in Fig. 8.21(b). The calculated t^* is less than 2.80 and so does not fall in the rejection region. The chamber of commerce cannot reject the national poll results.

Summary

$\mathrm{H_0}: \mu = \$21{,}300.$

$\mathrm{H_A}: \mu \neq \$21{,}300.$

Test region: Reject $\mathrm{H_0}$ if $t^* < -2.80$ or $t^* > 2.80$.

Test statistic: $t^* = \dfrac{22{,}750 - 21{,}300}{2750/\sqrt{25}} = 2.64.$

Conclusion: Do not reject $\mathrm{H_0}$. □

NOTE

Use z scores when σ is known.

Use t scores when σ is unknown but s is given for an approximately normal population. If the sample size is large enough (greater than 30), we will drop to the bottom of the t table and use z scores.

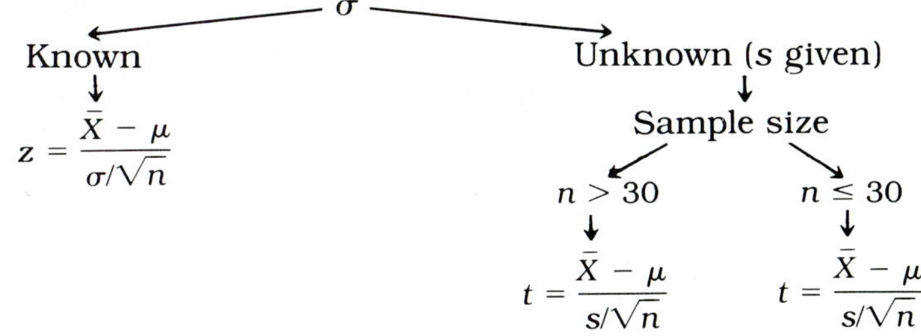

$$z = \frac{\bar{X} - \mu}{\sigma/\sqrt{n}} \qquad t = \frac{\bar{X} - \mu}{s/\sqrt{n}} \qquad t = \frac{\bar{X} - \mu}{s/\sqrt{n}}$$

In general:

$$\text{Test statistic} = \frac{\text{Sample result} - \text{Parameter in } \mathrm{H_0}}{\text{Appropriate standard deviation}}$$

Thus far we have chosen a level of significance before solving a problem. This method allows for a clearcut decision but only for that specific value of α. Many authors of journal articles and other professional writings use ***p* values** to express a level of significance achieved by the data. Rather than limiting conclusions with a preselected α, the author allows the reader to draw a personal conclusion based on a level of significance acceptable to the reader.

The p value is defined as the probability of obtaining a test statistic more extreme than z^* if H_0 is true; that is, p value $= P(z > |z^*|)$, where $|z|$ is the absolute value of z^*.

> ***p* value** ■ The level of significance achieved by the data; that is, the probability of obtaining a test statistic more extreme than z^* if H_0 is true.

Let us demonstrate this concept using the test statistics from Examples 8.6–8.8. In Example 8.6, the data gave us a test statistic $z^* = 2$. The p value associated with $z^* = 2$ is $P(z > 2) = 0.5 - 0.4772 = 0.0228$. The α chosen was 0.05. Thus for the selected α, p value $< \alpha$, and we rejected H_0. However, if we had chosen $\alpha = 0.01$, the p value would have been greater than α, and we could not have rejected H_0.

In Example 8.7 we got $z^* = 1.82$. Its associated p value is $P(z > 1.82) = 0.5 - 0.4656 = 0.0344$. Thus we can see that the result in Example 8.6 is more significant than the result in Example 8.7; that is, it has a smaller p value, or a smaller area in the tail to the right of z^*.

Example 8.8 with its $z^* = -2.04$ has a p value $= P(z < -2.04) = 0.5 - 0.4793 = 0.0207$. We can apply the p-value concept to either tail or to both tails in the case of a two-tail test; for the latter we double the probability determined by using z^* and apply z^* to both tails in order to obtain the p value.

After examining the p value, the reader can judge whether the calculated p value is acceptable for rejecting H_0 and can obtain a measure of the degree of significance of a specific set of data results. Thus p values are regularly published with experimental results.

EXERCISES/Section 8.4

1. With a mean snowfall of 24″ at 10 locations around the state by the end of February, this winter seems worse than those in recent years. Since observations can be exaggerated or misleading, John decided to test this hypothesis at the 0.05 level of significance by digging into National Weather Service records. For the previous 10 years, he found $\bar{X} =$

22.7″ and $s = 2.3″$. Can John conclude that recent winters had less snowfall than the current year?

2. Jane parks at the same meter every day when she goes to work. Lately she has been getting parking tickets because the time has expired on the meter. She has been putting in the same amount of money but feels that she has been getting less time on the meter. Before she complains to the city, she decides to collect some data to try to convince the city to fix the meter. One quarter is supposed to give her 2 hr, or 120 min. After timing the meter for 2 weeks (10 work days), she finds that the meter expires after a mean of 110 min with a standard deviation of 10.5 min. At the 0.01 level of significance, should Jane be able to convince the city that the meter is broken?

3. In an effort to make comparisons with last year's sales, a car manufacturer wishes to determine if there has been a change from last year's average sales of 2050 cars per day during July. Using the 31 days in July for this year, he obtains a mean for daily sales of 1965. At the 0.05 level of significance, has there been a change? Assume that $\sigma = 250$.

4. A new brand of fertilizer claims to produce larger tomatoes. An agricultural experiment station attempted to test this claim using a variety of tomatoes that is normally 8.6 cm in diameter with a standard deviation of 2.4 cm. The sample of 50 tomatoes raised using the new fertilizer had a sample mean of 9.2 cm. Can the fertilizer manufacturer's claim be supported at the 0.01 level of significance? What is the p value for this problem?

5. In an effort to justify additional budget requests to the city council, the chief of police gathered data to show that increased police presence on the streets deters crime. Last year an average of 6 assaults per day occurred in a 20-block area. This year, special extra funds were used to station additional police officers visibly in this area during July and August. Using this 2-month (62-day) period as his sample, the chief found an average assault rate of 5.1 per day with a standard deviation of 2.31. At the 0.05 level of significance, can the police chief make his case for a decrease in the assault rate?

6. A quality control supervisor is trying to determine whether the jars on his assembly line are being properly filled. They are supposed to contain 118 ml with $\sigma = 7.1$. If the jars are too full, the company is giving out more of its product than necessary and is losing money. If the jars are not full enough, the company will be subject to complaints and claims of false advertising. The supervisor pulls 50 jars off the line at different times of the day and on different days of the week. He finds that $\bar{X} = 120$. At the 0.05 level, is the difference significant; that is, should the company adjust its bottle filling machine? What is the p value for this problem?

7. There is discussion in the media of grade inflation; that is, average grades rising over time. To determine if grade averages are really higher now than in the past, Professor Smith gathered records from 90 students to compare to the mean of 2.54 that existed 10 years ago. The 90 students produced a grade average of 2.69 with a standard deviation of 0.96. At the 0.05 level of significance, are grade averages higher than in the past?

8. Have heights changed? Based on examples of clothing in museums, it is estimated that the average height of adult males 100 years ago was 5′7″ with $\sigma = 3.4″$. A random sample of 36 adult males produced a mean of 5′8½″. At the 0.01 level, is this difference significant? What is the p value?

9. The mean reaction time for a certain species of pigs when given a stimulus is 0.9 s. After 18 pigs were given 3 oz of alcohol, their reaction time slowed to 1.1 s with a standard deviation of 0.26 s. At the 0.05 level, did intake of alcohol significantly increase reaction times?

10. Tests that result in too high an average score are considered too easy. Similarly, test results with too low a mean score may be too difficult. On a 100-point test, a mean of 75 is reasonable. If Professor Hill's test for a class of 17 students produced a mean of 67 with a standard deviation of 15.6, should it be judged a reasonable test at the 0.05 level of significance?

8.5 Estimation

The second question that we set out to answer at the beginning of the chapter is: What is the value of a parameter? As before we will discuss the mean while formulating procedures. If I asked you to tell me the average income of electricians in your area, I am asking you for the value of the parameter μ, the mean income of the population of electricians in your area. What would you do to find this information? If you live in a large metropolitan region, you would be forced to take a sample and calculate $\bar{X}$. Suppose you found that $\bar{X} = \$24{,}200$. What would you tell me was the μ for all electricians? If your reply were $\mu = \$24{,}200$, you have used a **point estimate.** You directly transferred the value of the sample mean to the parameter.

> ***point estimate*** ■ The sample statistic given as the estimate of the corresponding parameter.

This is the simplest procedure. As a matter of fact, we used point estimates when we substituted s for the unknown σ in calculating t.

I hope that you would not be dogmatic and insist that μ was *exactly* \$24,200. You are more likely to say that μ is close to \$24,200 or about \$24,200 or in the neighborhood of \$24,200. When you do this, you are giving an **interval estimate.** You are hoping that μ will be in a group of numbers that cluster around $\bar{X}$.

> ***interval estimate*** ■ An interval of numbers surrounding the sample statistic that is used to estimate the corresponding parameter.

Figure 8.22

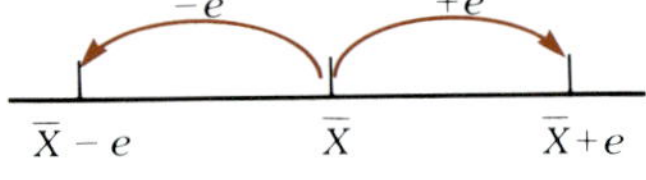

In order to find this interval we will begin by using $\bar{X}$ as the center of the interval and then both add something to it and subtract something from it, as shown in Fig. 8.22.

We now have an interval that, we hope, contains

$$\bar{X} - e < \mu < \bar{X} + e.$$

This says that μ lies between $\bar{X} - e$ and $\bar{X} + e$. It is greater than $\bar{X} - e$ and less than $\bar{X} + e$.

Figure 8.23

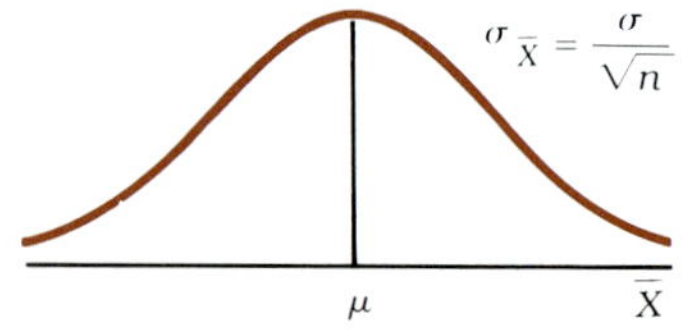

To discover what e is, let us return to the central limit theorem. It says that $\bar{X}$ is distributed approximately normally with $\mu_{\bar{X}}$ equal to μ. So $\bar{X}$ is distributed about μ, as shown in Fig. 8.23.

As with hypothesis testing, we can make some decisions about which part of the curve to use. If we cut off the tails of the

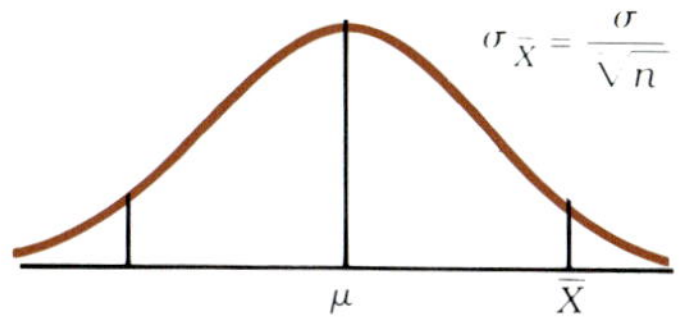

Figure 8.24

curve in Fig. 8.24, we can use z scores to find areas under the curve and determine the likelihood that the $\bar{X}$ from our particular sample is close to μ. The total area in the tails is always α. Now we want to concentrate on the area between the critical values, the area in the interval. This will tell us the probability that $\bar{X}$ is close to μ. This area is $1 - \alpha$, since α is the total area in the tails, and is called the **level of confidence.** It indicates how confident we are that our estimate is correct. If $\alpha = 0.05$, the level of confidence is 0.95 or 95%. Thus interval estimates are usually called **confidence intervals.**

> ***confidence interval*** ■ An interval estimate with which a level of confidence, $1 - \alpha$, is associated.

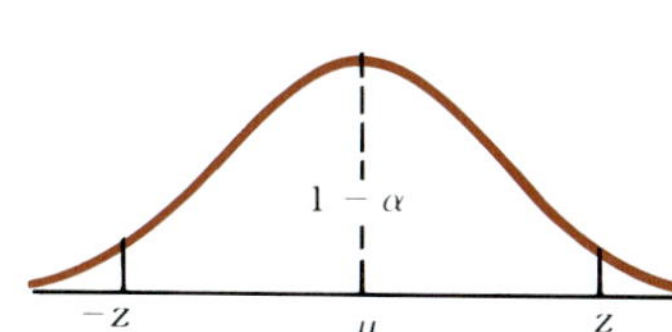

Figure 8.25

Now we can figure out what e has to be. The normality assumptions of the central limit theorem tell us that, if we have a confidence level, we can find a z score. We will consider z scores, but the discussion applies to t scores as well. Since z scores measure distance in standard deviations, we know from Fig. 8.25 that $\bar{X}$ is at most z standard deviations from μ with a $1 - \alpha$ level of confidence. If $\bar{X}$ is at most z standard deviations from μ, then μ is at most z standard deviations from $\bar{X}$. Think about that for a moment. It says the equivalent of: If you are three feet from a telephone pole, the telephone pole is three feet from you. Rather than putting μ in the middle between z scores, as in Fig. 8.25, we can put $\bar{X}$ in the middle, as in Fig. 8.22.

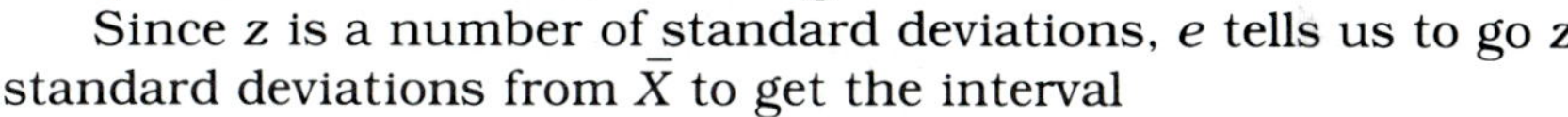

Since z is a number of standard deviations, e tells us to go z standard deviations from $\bar{X}$ to get the interval

$$e = z \cdot \text{Standard deviation},$$

which in this case is

$$e = z \cdot \frac{\sigma}{\sqrt{n}}.$$

Thus the confidence interval $\bar{X} - e < \mu < \bar{X} + e$ becomes

$$\boxed{\bar{X} - \frac{z\sigma}{\sqrt{n}} < \mu < \bar{X} + \frac{z\sigma}{\sqrt{n}}.}$$

If σ is unknown and $n \leq 30$, we need to use t instead of z, so

$$\bar{X} - t\frac{s}{\sqrt{n}} < \mu < \bar{X} + t\frac{s}{\sqrt{n}}.$$

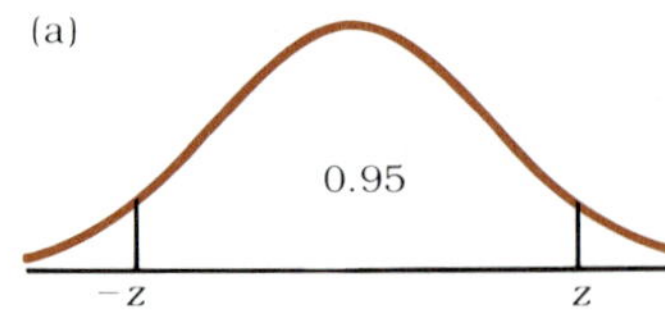

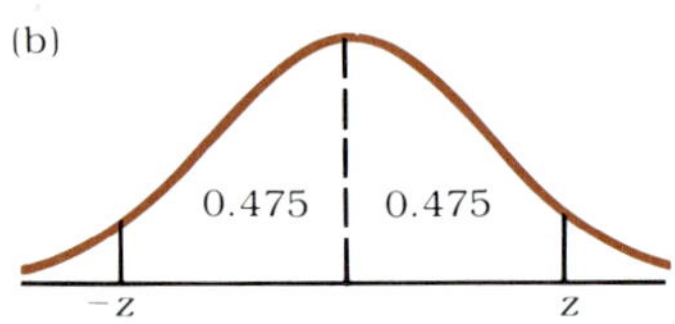

Figure 8.26

EXAMPLE 8.11 Find the 95% confidence interval estimate for the mean IQ score if $\bar{X} = 110$, as determined from a random sample of 50 people. Assume that $\sigma = 15$.

Solution The 95% confidence level tells us that 0.95 is the area in the central part of the normal distribution in Fig. 8.26(a). We want to use Table D.2 to find a z score. Remember that the values in this table represent the area under the curve from the mean to the z score. Therefore we divide $1 - \alpha$ by 2, as shown in Fig. 8.26(b), and use 0.475 as the value to look up in Table D.2. We get $z = 1.96$.

We go out 1.96 standard deviations in both directions from the sample mean of 110:

$$110 + 1.96 \cdot \frac{15}{\sqrt{50}} = 110 + 4.16 = 114.16.$$

$$110 - 1.96 \cdot \frac{15}{\sqrt{50}} = 110 - 4.16 = 105.84.$$

We are 95% confident that

$$105.84 < \mu < 114.16. \quad \square$$

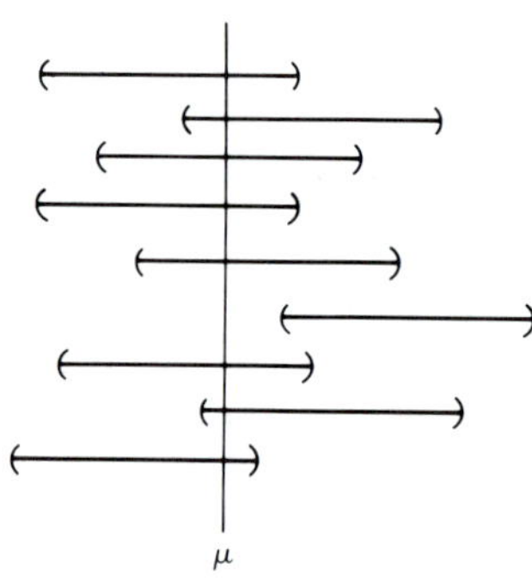

Figure 8.27

Before going on to further examples, let us look at what this procedure has given us. If we were to take 100 samples and construct 100 confidence interval estimates of the true mean IQ score, we would come up with 100 slightly different intervals, but approximately 95 of them should contain the true μ, as depicted in Fig. 8.27.

EXAMPLE 8.12 Find the 99% confidence interval estimate for the mean price of gasoline in the state if $\bar{X} = \$1.582$ from a sample of size 60. Assume that $\sigma = \$0.20$.

Figure 8.28

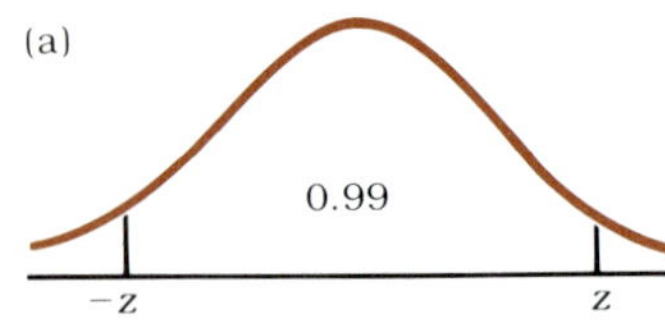

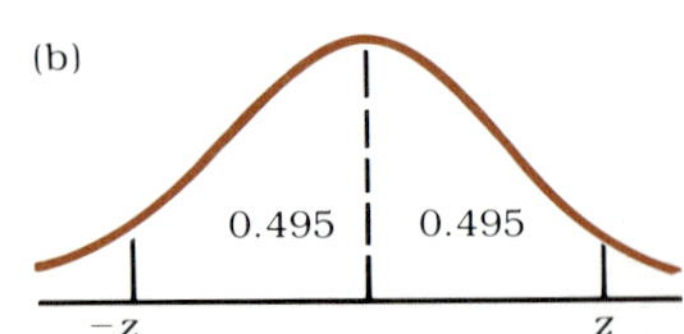

Solution A 99% confidence level tells us that the area under the curve between $-z$ and z is 0.99, as shown in Fig. 8.28(a). Again, we use $1 - \alpha/2$ (Fig. 8.28b) as the value to look up in Table D.2, which yields $z = 2.58$. We go out 2.58 standard deviations in both directions from the mean.

$$1.582 + 2.58 \cdot \frac{0.20}{\sqrt{60}} = 1.582 + 0.067 = 1.649.$$

$$1.582 - 2.58 \cdot \frac{0.20}{\sqrt{60}} = 1.582 - 0.067 = 1.515.$$

Thus $\$1.515 < \mu < \$1.649$ with 99% confidence. $\square$

Since confidence intervals are formed by both adding and subtracting from the sample mean, they are always two-tailed. If σ is unknown, the only procedural changes needed are to use t for small samples and s for σ.

EXAMPLE 8.13 Find the 95% confidence interval estimate for the mean hourly wage for part-time employees of ABC Corporation if $\bar{X}$ = \$4.21 and s = \$0.43 from a sample of 20 payroll records.

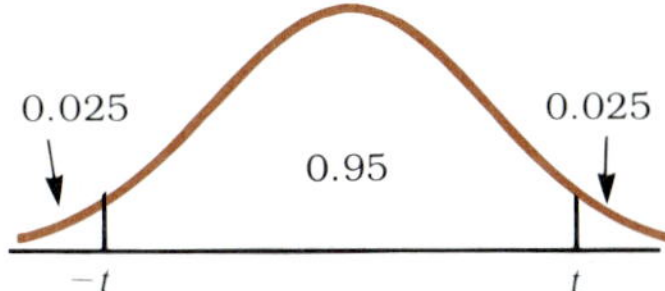

Figure 8.29

Solution The confidence level is the area under the curve in Fig. 8.29, as before. Since σ is not known and $n < 30$, we use Table D.3. Now $n = 20$ and df = 19; $t = 2.09$. Adding and subtracting 2.09 standard deviations gives

$$4.21 + 2.09 \cdot \frac{0.43}{\sqrt{20}} = 4.21 + 0.20 = 4.41;$$

$$4.21 - 2.09 \cdot \frac{0.43}{\sqrt{20}} = 4.21 - 0.20 = 4.01.$$

Thus the 95% confidence interval estimate is

$$4.01 < \mu < 4.41. \quad \square$$

EXAMPLE 8.14 Find with 99% confidence the mean miles per gallon for a pickup truck, if $\bar{X} = 16.7$, and $s = 2.5$ based on a test of 12 models.

Figure 8.30

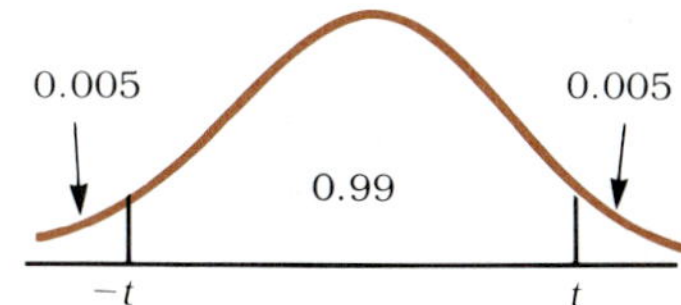

Solution We now use Fig. 8.30. Since σ is not known and $n <$ 30, we use Table D.3 again, but with df = 11. Since t is 3.11,

$$16.7 + 3.11 \cdot \frac{2.5}{\sqrt{12}} = 16.7 + 2.24 = 18.94;$$

$$16.7 - 3.11 \cdot \frac{2.5}{\sqrt{12}} = 16.7 - 2.24 = 14.46;$$

$14.46 < \mu < 18.94$ with 99% confidence. □

What is the effect of changing the confidence level? If we repeat Example 8.14 using a 95% confidence level, the area under the center of the curve (Fig. 8.31) will be smaller; $t = 2.20$, and

Figure 8.31

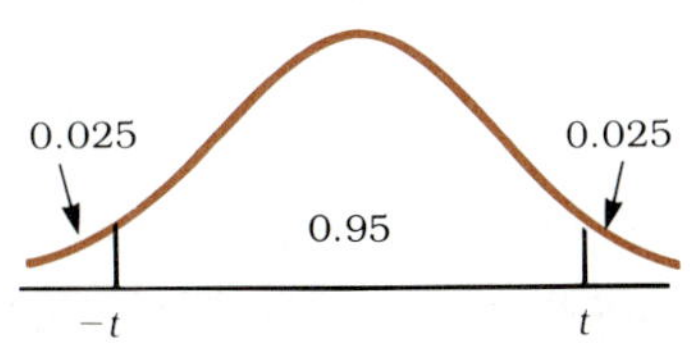

$$16.7 + 2.20 \cdot \frac{2.5}{\sqrt{20}} = 16.7 + 1.23 = 17.93;$$

$$16.7 - 2.20 \cdot \frac{2.5}{\sqrt{20}} = 16.7 - 1.23 = 15.47;$$

$15.47 < \mu < 17.93$ with 95% confidence.

NOTE

> If you want to be correct more often, you must select a wider interval. If you want the precision of a narrower interval, you must settle for a lower level of confidence.

However, there is another way to affect the size of the interval without changing the confidence level. We can control only two of the numbers in these problems; we pick α or $1 - \alpha$. Both $\bar{X}$ and σ are determined by the sample and the population. What is left? The sample size n, which we can usually control. In Chapter 7 some of the examples showed what happens when n is changed: In general, since we are dividing by $\sqrt{n}$, as n becomes larger the amount added or subtracted (previously called $e = z\sigma/\sqrt{n}$) becomes smaller. That observation provides us with a useful tool.

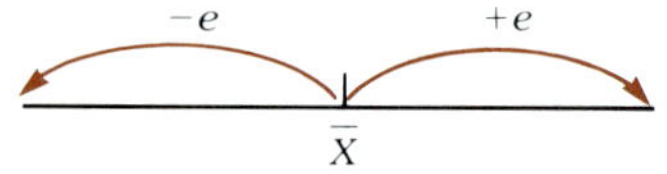

Figure 8.32

We began the discussion of confidence intervals by setting up an e (Fig. 8.32) and assuming a probability of, for example, 95%. We said that μ lies in the interval from $\bar{X} - e$ to $\bar{X} + e$ 95% of the time for such intervals. This also says that $\bar{X}$ lies at most a distance of e from μ. We call this e the **maximum error of the estimate.** Since we do not know μ exactly, there is some error in our estimate of it, but we can attach a probability and a maximum size to that error.

> ***maximum error of the estimate*** ■ One-half the length of the confidence interval. When estimating the mean,
>
> $$e = z\frac{\sigma}{\sqrt{n}}.$$

Remember the verbal definition. When we come to examples involving another parameter, the definition in words will be more meaningful than the equation; definitions as concepts are easier to generalize when we confront other applications.

We can decide in advance of an experiment how closely we wish to estimate μ, say to within 1 lb, 3 cm, or 5 ml, with a specified degree of confidence. We know e and can easily find z from Table D.2. If σ is also known from previous research or prior experience, then we have to find only the proper sample size n to satisfy our conditions for accuracy and degree of confidence. If σ is not known, we may be able to take a small preliminary sample to get an estimate for σ and then calculate the needed sample size before we begin the main study.

The equation $e = z\sigma/\sqrt{n}$ is not in the most convenient form for solving for n. Let us transform it algebraically. Multiplying both sides by $\sqrt{n}$,

$$\sqrt{n} \cdot e = z \frac{\sigma}{\cancel{\sqrt{n}}} \cancel{\sqrt{n}}.$$

Dividing by e,

$$\frac{\sqrt{n}\cancel{e}}{\cancel{e}} = \frac{z\sigma}{e} \quad \text{or} \quad \sqrt{n} = \frac{z\sigma}{e}.$$

Squaring both sides,

$$\boxed{n = \left(\frac{z\sigma}{e}\right)^2}$$

EXAMPLE 8.15 Determine the size of the sample needed to estimate the mean age at marriage for women if we want to be within ½ year of the true age with 95% confidence. Assume that $\sigma = 4$ years.

Solution

$$n = \left(\frac{z\sigma}{e}\right)^2.$$

With 95% confidence, $z = 1.96$. We want to be within ½ year, so the value of e is 0.5.

$$n = \left(\frac{1.96 \cdot 4}{0.5}\right)^2 = 245.86.$$

Since we are determining the number of people to be questioned, n must be a whole number. We also want to be sure that we have satisfied the conditions in the problem, so we always *round up*, or $n = 246$. □

EXAMPLE 8.16 What size sample is needed to estimate the average weight of newborn babies with 99% confidence if we want to be correct to within 6 ounces? $\sigma = 10$ ounces.

Solution

$$n = \left(\frac{z\sigma}{e}\right)^2.$$

With 99% confidence, $z = 2.58$. We want to be correct to within 6 ounces, so $e = 6$.

$$n = \left(\frac{2.58 \cdot 10}{6}\right)^2 = 18.49.$$

Rounding up (always), we get $n = 19$. □

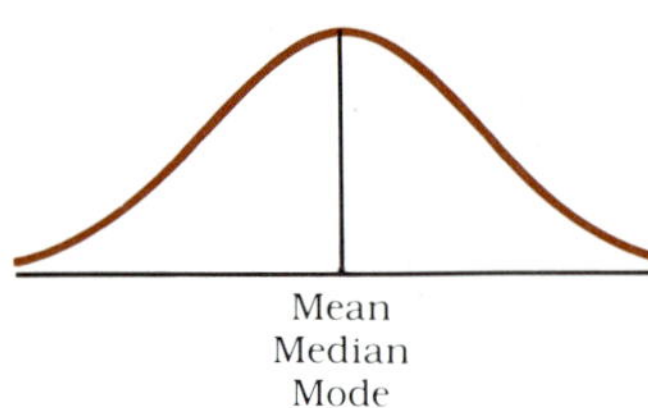

Figure 8.33

Table D.3 cannot be used to determine sample size for this type of problem. Why not? Think for a moment about what you need to know to use that table, and you can answer this question for yourself.

Before leaving this section, let us consider the normal distribution. From Chapter 4 we know that the mean, median, and mode all lie at the same point in the middle (Fig. 8.33). Why not use the sample median or the sample mode to estimate μ? In general for a normal distribution, $\bar{X}$ is the best estimate of μ. Theoretical analysis has determined that $\bar{X}$ is what is known as an *unbiased estimator.* The central limit theorem applied to the distribution of sample means tells us about the variance of $\bar{X}$; $\bar{X}$ has a smaller variance than other estimators of μ and thus is more reliable. Generally, we will use unbiased estimators when trying to determine values of parameters.

And finally, before we begin practicing with the exercises, note the general pattern for confidence intervals:

Sample results ± Critical value · Appropriate standard error.

If we remember the pattern as we encounter examples of confidence intervals in the coming chapters, we will not be confused by the different equations presented. This single pattern will help us get through a variety of situations.

EXERCISES/Section 8.5

1. Find the 95% confidence interval estimate for μ if $\bar{X} = 45.2$, $\sigma = 3$, and $n = 50$.
2. Find the 99% confidence interval estimate for μ if $\bar{X} = 106.7$, $\sigma = 15$, and $n = 45$.
3. Find the 90% confidence interval estimate for μ if $\bar{X} = 61.4$, $\sigma = 12$, and $n = 120$.
4. Find the 95% confidence interval estimate for μ if $\bar{X} = 231$, $\sigma = 25$, and $n = 300$.
5. Find the 95% confidence interval estimate for μ if $\bar{X} = 14.2$, $s = 0.36$, and $n = 25$.
6. Find the 99% confidence interval estimate for μ if $\bar{X} = 64.5$, $s = 21.3$, and $n = 14$.
7. Find the 90% confidence interval estimate for μ if $\bar{X} = 73.5$, $s = 13.2$, and $n = 20$.
8. Find the 95% confidence interval estimate for μ if $\bar{X} = 205$, $s = 19.9$, and $n = 75$.
9. A manufacturing company wishes to estimate the average life μ of its television sets in order to revise its warranty period. After questioning 50 purchasers of a specific model, the company determined that the sample mean $\bar{X}$ was 5.2 yrs.
   a) What is the point estimate for μ?
   b) Find the 95% confidence interval estimate for the mean life of that model; $\sigma = 1.6$ yrs.
   c) What is the 99% confidence interval estimate for μ?
10. A stereo cartridge manufacturer wants to estimate the average lifetime of the cartridge needles it manufactures. Testing a sample of 60, the company finds an $\bar{X} = 932$ hr.
    a) What is the point estimate for μ?
    b) Find the 95% confidence interval for μ if $\sigma = 200$ hr.

c) What is the 99% confidence interval estimate for μ?

11. An ecology class of 25 students wished to determine the average distance that a specific model car would travel on 1 gal of gas under various road conditions. They each drove the car, and the class efforts resulted in $\bar{X} = 30.2$ miles and $s = 2.4$ miles.
   a) What is the point estimate for μ?
   b) Find the 95% confidence interval estimate for μ?
   c) What is the 99% confidence interval estimate for μ?

12. A group of 100 victims of heart attacks had a mean diastolic blood pressure of 97 with a standard deviation of 10.
   a) What is the point estimate for μ, the mean DPB for all heart-attack victims?
   b) Find the 95% confidence interval estimate for μ?
   c) What is the 99% confidence interval estimate for μ?

13. Based on a sample of 15 games, the mean number of points scored by Player A on the Central High School basketball team was 10 with $s = 4$. Find the 95% confidence interval estimate for his final season average.

14. A fire chief wants to know how quickly his men respond to fires. During the month of July they responded to 24 fires, getting out of the firehouse in trucks in an average $\bar{X}$ of 4.2 min and $s = 2.3$ min. Find the 99% confidence interval estimate of mean reaction time.

15. Achievement tests given to 90 second graders produced a mean of 77 with a standard deviation of 11.2. Find the 95% confidence interval estimate for the mean of all second graders.

16. A group of concerned parents wants to estimate the number of hours of television watched by children each week. Surveying 120 children, they determined that $\bar{X} = 9$ hr and $s = 4.1$ hr. Find the 90% confidence interval estimate of mean viewing time.

17. In an attempt to estimate state climatic conditions, a weather observer obtained the following mean readings in different parts of the state during June: 73°, 78°, 82°, 75°, 80°, 70°, 71°, 69°, 75°, and 81°. Find the 95% confidence interval estimate of the mean June temperature for the state.

18. Determine the sample size necessary to estimate the true mean weight of adult females to within 5 lb with 95% confidence. Assume that $\sigma = 25$ lb.

19. What size sample is needed to estimate the true average miles per gallon for subcompacts driven in city traffic if you want to be within 2.1 mpg with 99% confidence? Assume that $\sigma = 5.1$.

20. Find the sample size necessary to estimate the mean wearing life of Brand A tires if you want the estimate to be within 500 miles of the true mean with 95% confidence. Use $\sigma = 1200$ miles.

21. What sample size is necessary to estimate the mean driving time to work in the morning of commuters in a large southern city if you want to be within 10 min of the true μ with 95% confidence? Use $\sigma = 33$ min.

22. Determine the sample size needed to estimate the mean size of mortgages if you want to be accurate to within $2000 with 99% confidence. Use $\sigma = \$22{,}000$.

Study Notes

KEY TERMS

alternative hypothesis ■ A statement that is to be accepted if the null hypothesis is rejected. It utilizes the same parameter as the null hypothesis but gives motivation for the rest of the test procedure. This is the statement we want to accept.

confidence interval ■ An interval estimate with which a level of confidence, $1 - \alpha$, is associated.

critical value(s) ■ The value(s) that determines the rejection region.

interval estimate ■ An interval of numbers surrounding the sample statistic that is used to estimate the corresponding parameter.

level of confidence ■ $1 - \alpha$.

level of significance ■ The probability α of a Type I error.

maximum error of the estimate ■ One-half the length of the confidence interval. When estimating the mean,

$$e = z\frac{\sigma}{\sqrt{n}}.$$

null hypothesis ■ A statement about a population parameter that is being tested by the use of sample results and a decision-making process.

p value ■ The level of significance achieved by the data; that is, the probability of obtaining a test statistic more extreme than z^* if H_0 is true.

point estimate ■ The sample statistic given as the estimate of the corresponding parameter.

rejection region (test region or critical region) ■ The portion of a distribution that provides values for the sample results causing the rejection of H_0.

test statistic ■ A calculated value using sample results to be compared to a tabular value, such as a z score or t score of the sample results.

Type I error ■ Rejecting a true null hypothesis.

Type II error ■ Failing to reject a false null hypothesis.

OUTLINE

I. Hypothesis testing
  A. Set up the null hypothesis.

$$H_0: \mu = \underline{\qquad}.$$

  B. Set up the alternative hypothesis.

$$H_A: \mu > \underline{\qquad}.$$
$$\mu < \underline{\qquad}.$$
$$\mu \neq \underline{\qquad}.$$

  C. Test region:
    1. Based on α.
    2. Based on H_A.

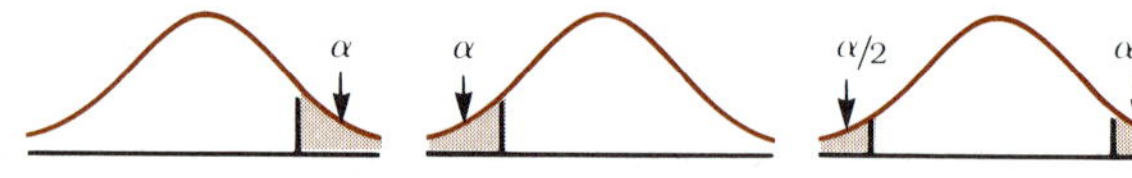

    3. Use a standard normal (z) distribution if σ is known or if $n > 30$.
    4. Use a t distribution if σ is unknown (s is known) and $n \leq 30$.
  D. Test statistic:
    1. Calculate the z or t value of the sample results.
    2. $z^* = \dfrac{\bar{X} - \mu}{\sigma/\sqrt{n}}$. $\quad t^* = \dfrac{\bar{X} - \mu}{s/\sqrt{n}}$.

    In general,

    Test statistic

$$= \frac{\text{Sample results} - \text{Parameter in } H_0}{\text{Appropriate standard error}}.$$

  E. Conclusion—drawn by comparing test statistic to critical region.
  F. Errors:
    1. Type I—rejecting a true hypothesis.
    2. Type II—accepting a false hypothesis.

II. Estimation
  A. Point estimate.
  B. Interval estimate:
    1. Associated with a level of confidence.
    2. Puts the sample result in the center of a group of numbers.
    3. $\bar{X} - \dfrac{z\sigma}{\sqrt{n}} < \mu < \bar{X} + \dfrac{z\sigma}{\sqrt{n}}$.

    or

$$\bar{X} - \frac{ts}{\sqrt{n}} < \mu < \bar{X} + \frac{ts}{\sqrt{n}}.$$

    Sample result ± Critical value × appropriate standard error.

  C. Sample size calculations involving z:
    1. $e = \dfrac{z\sigma}{\sqrt{n}}$ and $n = \left(\dfrac{z\sigma}{e}\right)^2$.
    2. e = maximum error of the estimate.
    3. Always round up for n.

REVIEW PROBLEMS

1. A company gives a manual-dexterity test to all new employees to measure how well they can assemble small parts. If the mean score in previous years was 72 with a standard deviation of 7.5, is a new group of 40 employees superior with $\bar{X} = 74.5$ at the 0.05 level of significance?
2. A local brewery has been spending different amounts each month on television advertising. In order to plan a budget for the year ahead, the president of the company wants an estimate of the average expense per month for this form of advertising. If the company spent an average $\bar{X}$ of $12,250 per month over the past 12 months with $\sigma = \$2000$, find the 99% confidence interval estimate for its average ad costs.
3. A seedsman claims that a certain variety of tomato plant will bear fruit no more than 68 days after seedlings are planted. A local gardener wonders about this because the 20 seedlings of this variety that she planted required, on the average, 75 days to mature with $s = 12$ days. Does she have sufficient evidence to reject the seedsman's claim at the 5% level of significance?
4. To soften the blow of "sticker shock," a car dealer claims that the mean price per model of new cars sold by him is no more than $8000. A skeptical consumer activist group dispatches some of its members to the dealer's showroom. They return with the following model prices: $9500; $6300; $10,800; $7800; $11,900; $8500; $7100; $9900. Is this sufficient evidence to reject the dealer's claim? Use $\alpha = 1\%$.
5. A certain professor is fond of claiming that students in her class spend no more than the length of a class period doing homework that she assigns. The classes she teaches last 75 min each. On her annual teacher evaluation form, she asked 50 students how much time each spent doing homework each night. She learned from this sample that $\bar{X} = 80$ min. Assuming that $\sigma = 15$ min, is this result significant at the 5% level?
6. The manager of a video game arcade claims that the average length of time spent by adolescents at the machines in his establishment is no more than 30 min. A survey by a subcommittee of the local PTA disclosed that a random sample of 12 such customers had a mean time of 40 min and $s = 13$ min. At the 1% level, is this sufficient evidence to reject the manager's claim?
7. The state inspector who verifies the accuracy of scales used in the local butcher shop weighed a standard 1 lb weight 10 times on the butcher's scale and came up with the following weights: 15.9 oz, 16 oz, 15.8 oz, 16.1 oz, 16.1 oz, 16 oz, 16.2 oz, 16.1 oz, 15.9 oz, and 16.3 oz. At the 0.01 level of significance, can the inspector say that the scale does not weigh accurately enough?
8. An automobile manufacturer claims that one of his models averages at least 450 miles on a tank of gas with a standard deviation of 25 miles. An investigative reporter asked ten owners of that particular car to keep careful records of mileage traveled for each of the next four tanksful. The mean number of miles traveled for these 40 tanksful was found to be 435. Does this supply sufficient evidence to reject the manufacturer's claim at the 5% level of significance?
9. Assuming that $\sigma = \$20$, determine the 90% confidence interval estimate for the mean monthly utility bill if a sample of 100 customers revealed an average monthly bill of $98.50.
10. A time-and-motion study at a certain office disclosed that a sample of 50 employees spent an average of 12.25 min on break. Assuming that $\sigma = 1.5$ min, determine the 95% confidence interval estimate for the mean time spent on break by all the employees.
11. A sample of ten apple trees of the same variety reveals that the trees yield a mean of 5.6 bu with $s = 0.75$ bu. Use this information to determine the 99% confidence interval estimate for μ.

12. A sample of ten buyers of new houses in a certain town disclosed a mean price of \$98,000. If that same sample also had s = \$4500, determine the 95% confidence interval estimate of the mean new house price for all the houses in that town.

13. A randomly chosen sample of eight students in a statistics course had a mean grade on a recent quiz of 71 with $s = 9.5$. Determine the 90% confidence interval estimate for the value of μ for all the students who took the test.

14. Find the size of the sample necessary to estimate the mean life of light bulbs manufactured by the Edison Bulb Company if the company wants to be within 50 hr of the true mean with 95% confidence. Past experience has shown that $\sigma = 160$ hr.

15. Determine the sample size necessary to estimate the mean height of two-week-old seedlings for a new variety of tomato if the estimate is to be within one-half inch of the true height with 99% confidence. This variety of tomato in the past has had $\sigma = 1.2''$.

16. What size sample is needed to estimate the average viewing time, in hours, of TV by preschoolers if we want the estimate to be within 15 min (0.25 hr) of the true mean with 95% confidence? Assume that $\sigma = 1.3$ hr.

17. How many companies should we question in order to determine the mean dollar value per company of exports of photographic equipment if we want the estimate to be within \$10,000 with 95% confidence? Assume that σ = \$50,000.

18. A newspaper article claimed that the *average American* spends 25½ hr per week watching TV. A sociology professor thinks that this figure is too high. A survey of 100 students discloses that the average is 21 hr per week. Assuming that $\sigma = 2$ hr, is this sufficient evidence to reject the paper's claim? Use $\alpha = 0.05$.

19. A sales manager, new in town, is curious about the cost of dinner at restaurants that serve fine continental cuisine. He asks 30 acquaintances and colleagues how much they pay for such a meal. He finds the average cost to be \$27.75 (with drinks). Assuming that σ = \$3, determine the 99% confidence interval estimate of the average cost of a continental dinner.

20. The dean of students at a local community college claims that the number of absences averages 3.2 days per student per semester. A certain professor does not believe the claim and selects a random sample of ten names from his class list. This sample had 28 absences with $s = 0.6$ day. Is this sufficient evidence to reject the dean's claim at the 1% level?

21. A random sample of eight buyers of new cars disclosed an average purchase price of \$9750 with s = \$1200. Use this information to construct the 95% confidence interval estimate for the purchase price of a new car.

Statistical inference concerning the means of two populations

9

9.1 Differences between Means—Large Samples

At this point we need to extend our work to consider hypothesis testing and confidence intervals for comparing the means of two populations. We will turn our attention first to samples that are independent. **Independent samples** can be described simply as samples that do not contain the same members.

> ***independent samples*** ▪ Samples that do not contain the same members and produce unrelated (unpaired) data.

The data they produce cannot be paired and can be compared only by analyzing each sample separately to produce means, say $\bar{X}_1$ and $\bar{X}_2$, and standard deviations, say s_1 and s_2, which are then compared.

In this section we will examine how to treat the results from large samples; Section 9.2 presents small-sample analysis; and, finally, Section 9.3 shows us how to deal with **dependent samples.**

> ***dependent samples*** ▪ Samples that contain the same members and/or produce matched pairs of data.

Dependent samples produce data that can be paired and thus have some convenient properties for sample-design problems and statistical analysis.

The basic techniques of Chapter 8 apply here as well. We will extend to two samples what we learned about the central limit theorem and the t distribution.

What do we need to know? We will be testing or estimating the difference between two population means, so we will be subtracting $\mu_1 - \mu_2$ and $\bar{X}_1 - \bar{X}_2$. If independent samples of size n_1 and n_2 are drawn from populations with means μ_1 and μ_2, standard deviations σ_1 and σ_2, and n_1 and n_2 both greater than 30, the sampling distribution $\bar{X}_1 - \bar{X}_2$ (instead of $\bar{X}$, as before) has the following properties:

1. It is approximately normally distributed.
2. Its mean is $\mu_1 - \mu_2$.
3. Its standard deviation is

$$\sqrt{\frac{\sigma_1^2}{n_1} + \frac{\sigma_2^2}{n_2}}.$$

If σ_1 and σ_2 are unknown, then its standard error can be approximated by

$$\sqrt{\frac{s_1^2}{n_1} + \frac{s_2^2}{n_2}},$$

using sample variances.

We can use this information to modify the z score.

$$z = \frac{\text{Sample results} - \text{Parameter value}}{\text{Appropriate standard error}}$$

$$= \frac{(\bar{X}_1 - \bar{X}_2) - (\mu_1 - \mu_2)}{\sqrt{\frac{s_1^2}{n_1} + \frac{s_2^2}{n_2}}}.$$

If we know σ_1 and σ_2, we should use them. This is frequently not the case, so we will continue to use s_1 and s_2.

Hypothesis Tests

We need to consider how to state the null hypothesis H_O. It will usually state that there is no difference between the two population means.

$$H_0\colon \mu_1 - \mu_2 = 0.$$

This is algebraically equivalent to $\mu_1 = \mu_2$, but remember that we take the number from H_0 to substitute into the z calculation; thus it is obvious that utilizing a 0 is desirable. The equation then becomes

$$z = \frac{\bar{X}_1 - \bar{X}_2}{\sqrt{\frac{s_1^2}{n_1} + \frac{s_2^2}{n_2}}}.$$

The alternative hypothesis H_A also requires some consideration. We must think about what $\mu_1 - \mu_2$ means. If we think that μ_1 is larger than μ_2, then $\mu_1 - \mu_2$ is positive and $H_A\colon \mu_1 - \mu_2 > 0$. If we think that μ_1 is smaller than μ_2, then $\mu_1 - \mu_2$ is negative and $H_A\colon \mu_1 - \mu_2 < 0$. It no longer suffices to read the problem and automatically translate words into $>$ or $<$. Now let us consider an example.

EXAMPLE 9.1 Last year a sample of 100 new cars averaged 33.8 miles per gallon of gasoline. This year a sample of 100 new cars averaged 35.1 mpg. If each sample had a standard deviation of 10 mpg, can the auto manufacturers say that they have significantly

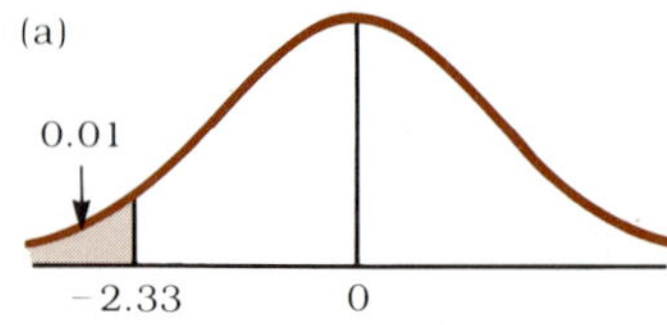

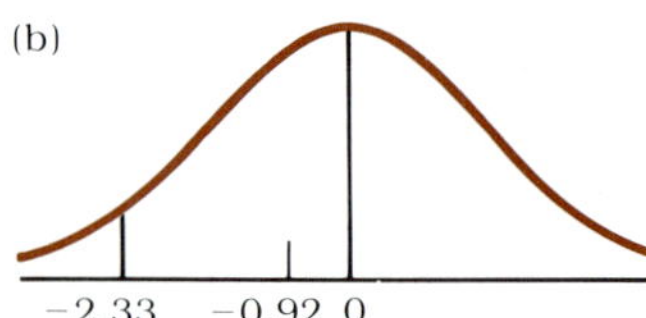

Figure 9.1

increased the mileage for their new fleet at the 0.01 confidence level?

Solution We assume that there is no difference and require the manufacturers to prove their counterclaim. Let μ_1 be the true mean for last year's models and μ_2 the true mileage for this year's. Then H_0: $\mu_1 - \mu_2 = 0$. If the manufacturers believe they have increased the mileage, they think that μ_2 is larger than μ_1 and thus H_A: $\mu_1 - \mu_2 < 0$.

The order of subtraction must be consistent because H_A contributes to the critical region and H_0 determines the order for the z critical value. Since $\alpha = 0.01$ and H_A has a < symbol, we get the z critical value as shown in Fig. 9.1(a).

We will reject H_0 if $z^* < -2.33$. The test statistic is

$$z^* = \frac{\bar{X}_1 - \bar{X}_2}{\sqrt{\dfrac{s_1^2}{n_1} + \dfrac{s_2^2}{n_2}}} = \frac{33.8 - 35.1}{\sqrt{\dfrac{10^2}{100} + \dfrac{10^2}{100}}} = -0.92.$$

Since −0.92 lies in the acceptance region (Fig. 9.1b), we cannot reject H_0.

Summary

H_0: $\mu_1 - \mu_2 = 0$.

H_A: $\mu_1 - \mu_2 < 0$.

Test region: Reject H_0 if $z^* < -2.33$.

Test statistic:

$$z^* = \frac{33.8 - 35.1}{\sqrt{\dfrac{10^2}{100} + \dfrac{10^2}{100}}} = -0.92.$$

Conclusion: The manufacturers did not substantiate their claim. □

EXAMPLE 9.2 A sampling of 50 women and 40 men produced a mean age at death for women of 75.5 years and for men of 67.9 years with standard deviations of 16.2 years for women and 18.3 years for men. At the 0.05 level of significance, do women live longer than men?

Solution We assume no difference; H_0: $\mu_w - \mu_m = 0$. The question as asked indicates a predisposition for μ_w to be larger than

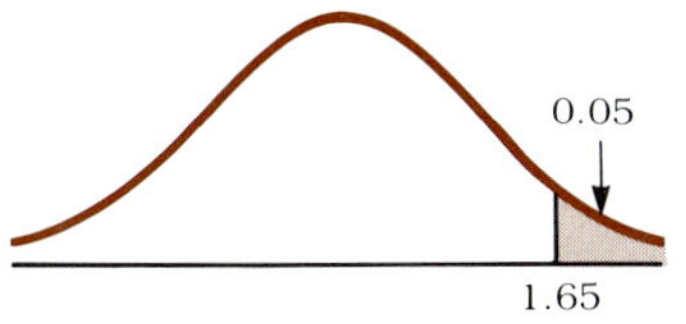

Figure 9.2

μ_m, or H_A: $\mu_w - \mu_m > 0$. Since $\alpha = 0.05$ we get a critical value of 1.65 (Fig. 9.2).

$$z^* = \frac{\bar{X}_w - \bar{X}_m}{\sqrt{\frac{s_w^2}{n_w} + \frac{s_m^2}{n_m}}} = \frac{75.5 - 67.9}{\sqrt{\frac{16.2^2}{50} + \frac{18.3^2}{40}}} = 2.06.$$

Since $2.06 > 1.65$, we can reject H_0.

Summary

H_0: $\mu_w - \mu_m = 0$.

H_A: $\mu_w - \mu_m > 0$.

Test region: Reject H_0 if $z^* > 1.65$.

Test statistic:

$$z^* = \frac{75.5 - 67.9}{\sqrt{\frac{16.2^2}{50} + \frac{18.3^2}{40}}} = 2.06.$$

Conclusion: Reject H_0. Women outlive men. □

Confidence Intervals

We can now use confidence intervals to estimate the difference between the means of two populations by revising the pattern identified in Chapter 8.

Sample results ± Critical value · Appropriate standard error

becomes

$$\bar{X}_1 - \bar{X}_2 \pm \text{Critical value} \cdot \sqrt{\frac{s_1^2}{n_1} + \frac{s_2^2}{n_2}}.$$

Again, if you know σ_1^2 and σ_2^2, it is preferable to use them instead of s_1^2 and s_2^2, but usually you will not know them.

An additional suggestion is helpful here. Since we have no null hypothesis in a confidence-interval problem to determine the order of subtracting the sample means, it is generally better to subtract the smaller $\bar{X}$ from the larger $\bar{X}$. This produces a positive result and avoids the possible mistake of losing minus signs. If we were to subtract the larger from the smaller, we would obtain the same numerical answer but with the signs reversed.

EXAMPLE 9.3 The same standardized test is given to students selected at random from two schools. The following table summarizes the results.

	n	X	s
School A	50	68.7	6.0
School B	40	65.1	6.4

Construct a 95% confidence interval estimate for the difference in the mean level of performance for the two schools.

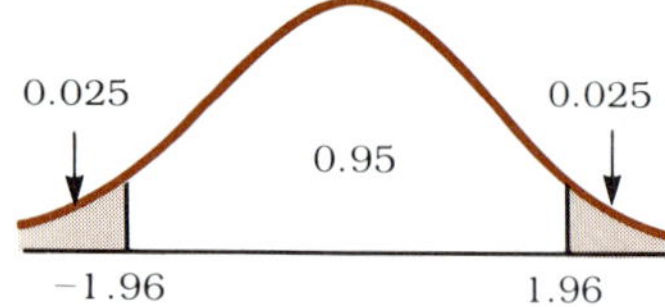

Figure 9.3

Solution We begin by finding the point estimate of the difference: $\bar{X}_A - \bar{X}_B = 68.7 - 65.1$. (We use $\bar{X}_A$ minus $\bar{X}_B$ because $\bar{X}_A$ is larger.) Then we expand this into an interval estimate by adding and subtracting the number of standard deviations determined by the critical value (Fig. 9.3).

$$68.7 - 65.1 \pm 1.96\sqrt{\frac{6.0^2}{50} + \frac{6.4^2}{40}} = 3.6 \pm 2.59.$$

$$1.01 < \mu_A - \mu_B < 6.19.$$

Thus the 95% confidence interval estimate for the difference in mean performance level goes from 1.01 to 6.19.

We know that we are estimating $\mu_A - \mu_B$ because we subtracted the sample means in that order. If we had used $\bar{X}_B - \bar{X}_A$, we would have obtained $-6.19 < \mu_B - \mu_A < -1.01$, a negative result caused by subtracting the larger from the smaller. □

Besides obtaining two positive or two negative endpoints when determining the confidence interval estimate of a difference, it is possible to obtain one positive and one negative endpoint. This is illustrated in the next example.

EXAMPLE 9.4 Find the 99% confidence interval estimate of the difference between IQ scores for students in schools in two different ethnic neighborhoods. The results of the samples are:

Figure 9.4

	n	$\bar{X}$	s
School 1	75	105	10.2
School 2	60	108	12.4

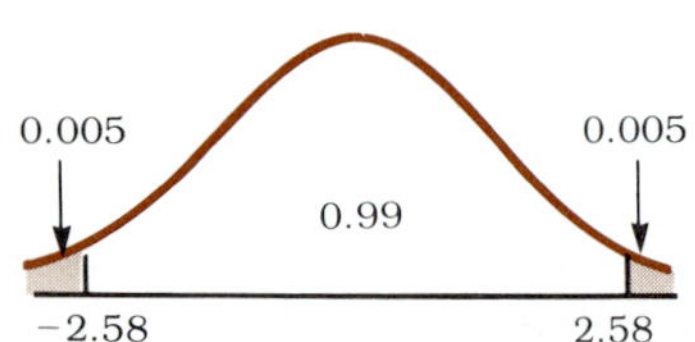

Solution We will use $\bar{X}_2 - \bar{X}_1$, since School 2 produced the larger mean. A 99% confidence level requires that we use 2.58 as our z

score (Fig. 9.4). So we calculate

$$108 - 105 \pm 2.58\sqrt{\frac{10.2^2}{75} + \frac{12.4^2}{60}} = 3 \pm 5.13.$$

$$-2.13 < \mu_2 - \mu_1 < 8.13. \quad \square$$

Figure 9.5

An interval that stretches from negative to positive numbers contains 0, as shown in Fig. 9.5. Therefore such an interval indicates that there is probably no significant difference between the means being estimated, just as a test statistic in an acceptance region indicates that there is no significant difference between the means when the null hypothesis contains 0.

EXERCISES/Section 9.1

1. A survey of a broadcasting year (52 weeks) indicated that in Great Britain an average of 71.1 hr per week is spent on educational programming. The same 52-week period in the United States produced a mean of 54.6 hr per week. Assume a standard deviation of 29.6 hr for each country. At the 0.05 level of significance, is there any difference in the amount of educational programming?
2. The Lifelong Lightbulb Company manufactures a bulb that, they claim, will burn longer than other brands. A sample of 60 Lifelong bulbs produced a mean of 1500 hr with a standard deviation of 275 hr. A sample of 60 bulbs from Lifelong's main competitor produced a mean of 1450 hr with $s = 179$ hr. Is Lifelong's claim justified at the 0.05 level of significance?
3. Samples of men and women from 50 hospitals produced the following results with regard to cancer death rate per 100,000 patients who were treated for any reason. Do men have a higher cancer death rate than women at the 0.01 level of significance?

	n	$\bar{X}$	s
Men	50	13.8	4.9
Women	50	11.3	3.8

4. In an age where regional differences are minimized by communication and transportation networks, is it more expensive to live on one of the coasts than in the midsection of the country? The budgets of 60 families of four were surveyed in each of two cities with the following results. Let $\alpha = 0.01$.

	$\bar{X}$	s
East Coast city	$19,200	$2451
Midwest city	$18,800	$1632

5. Daily sales records in two suburban branches of a major department store produced the following results, in thousands of dollars, averaged over a 45-day period. Is there a difference in sales performance at the two stores at the 0.05 level of significance?

	$\bar{X}$	s
West store	45	9.75
East store	52	17.52

6. Find the 99% confidence interval estimate of the difference between the mean price, in dollars, of Brandname ketchup and Storebrand

ketchup. Data were gathered on each once a week over a 52-week period.

	$\bar{X}$	s
Brandname	1.166	0.436
Storebrand	0.897	0.392

7. Estimate the difference in effectiveness in preventing tooth decay for two brands of toothpaste with 95% confidence. Adults were tested for a period of 2 yr, which produced the following results. Let $\bar{X}$ = mean number of cavities.

	n	$\bar{X}$	s
Brand A	100	0.92	0.94
Brand B	175	1.23	0.86

8. A large university decided to estimate the difference in students' ability to learn calculus by using two different teaching methods. One group of 125 used self-paced materials and studied on their own. Another group of 300 attended a lecture three times per week with one problem-solving session each week available to the individual student. Both groups were given the same final exam. Using the results in the following table, find the 95% confidence interval estimate of the difference in accomplishment.

	n	$\bar{X}$	s
Self-paced	125	76.3	12.6
Lecture	300	74.2	9.2

9. To estimate the effect of peer support, a researcher examined weight losses by two groups of people. The first group reported only to their doctors for supervision, whereas the second also met weekly as a group to compare notes and discuss problems. Find the 95% confidence interval estimate of the difference in average weight loss, in pounds, over a six-month period for these two groups.

	n	$\bar{X}$	s
Peer support group	40	24.7	2.6
No peer support group	42	21.6	4.7

10. Find the 99% confidence interval estimate of the difference in mean programming time for educational viewing between the United States and Great Britain using the information given in Exercise 1.

9.2 Differences between Means—Small Samples

When we utilized inferential procedures to make judgments about small-sample results for single samples, we assumed that the sample was taken from a normal population and then we used a t statistic to help us draw conclusions or make estimates. We will extend that discussion to work with the difference between the means of two populations from which two small samples ($n \leq 30$) are taken. We will assume that

1. two independent random samples are taken;
2. the original populations yielding these samples are each approximately normal; and
3. the two population variances or standard deviations are equal. (They will probably also be unknown.)

This last assumption may seem too restrictive, but in many instances populations have the same general spread but different centers (means). If the population variances are unequal,* the only difference in the analysis is in determining which degrees of freedom to use when reading the t table. Unequal population variances that occur when we test or estimate the difference between means using small samples require use of the Fisher-Behrens formula:

$$\text{df} = \frac{\left(\dfrac{s_1^2}{n_1} + \dfrac{s_2^2}{n_2}\right)^2}{\left(\dfrac{s_1^2}{n_1}\right)^2\left(\dfrac{1}{n_1 - 1}\right) + \left(\dfrac{s_2^2}{n_2}\right)^2 \dfrac{1}{n_2 - 1}}.$$

The answer is always rounded to a whole number. This formula is presented here only to complete the subject. We will proceed, using our assumption of equal variances.

We need to make a comment about assumption 2. If the original populations from which the samples are drawn are not normal, or their normality is unknown, large samples should be used in order to take advantage of the central limit theorem, as applied to two populations.

Hypothesis Tests

Performing a hypothesis test to compare results from two small samples requires us to use the t distribution again. The only new piece of information we need is the test statistic, which we calculate. The t test statistic,

$$t = \frac{\text{Sample results} - \text{Parameter claim}}{\text{Appropriate standard error}},$$

in this case becomes

$$t = \frac{(\bar{X}_1 - \bar{X}_2) - (\mu_1 - \mu_2)}{s_p\sqrt{\dfrac{1}{n_1} + \dfrac{1}{n_2}}}.$$

* We will be able to determine this in Chapter 10.

If the null hypothesis is $\mu_1 - \mu_2 = 0$, as it usually will be, we can shorten the preceding equation to

$$t = \frac{\bar{X}_1 - \bar{X}_2}{s_p\sqrt{\frac{1}{n_1} + \frac{1}{n_2}}}.$$

The s_p in the denominator results from our assumption that the population variances are equal. Ideally, the appropriate standard deviation for comparing two population means is

$$\sqrt{\frac{\sigma_1^2}{n_1} + \frac{\sigma_2^2}{n_2}}.$$

If σ_1^2 and σ_2^2 are unknown (the usual case), they can be estimated by s_1^2 and s_2^2. When samples are large, we substitute these values as we did in Section 9.1. When samples are small, the relationship between s_1^2 and s_2^2 becomes more important because we are trying to make judgments with less information. We are assuming that s_1^2 and s_2^2 are just different estimates of that common value for σ^2, assuming that $\sigma_1^2 = \sigma_2^2 = \sigma^2$. Thus we "pool" these two sample variances to estimate better the common σ^2 with a pooled variance s_p^2:

$$s_p^2 = \frac{(n_1 - 1)s_1^2 + (n_2 - 1)s_2^2}{n_1 + n_2 - 2}.$$

This equation gives a value between s_1^2 and s_2^2, weighted in the direction of the standard deviation derived from the larger sample. The larger sample (more information) contributes more to the calculation because of multiplication by its greater degrees of freedom in the numerator. Note that the denominator is the sum of the individual degrees of freedom,

$$\text{df} = (n_1 - 1) + (n_2 - 1) = n_1 + n_2 - 2.$$

and provides df when using the t table.

If we now substitute the pooled estimate of the common variance into

$$\sqrt{\frac{\sigma_1^2}{n_1} + \frac{\sigma_2^2}{n_2}}.$$

we get

$$\sqrt{\frac{s_p^2}{n_1} + \frac{s_p^2}{n_2}} = \sqrt{s_p^2\left(\frac{1}{n_1} + \frac{1}{n_2}\right)} = s_p\sqrt{\frac{1}{n_1} + \frac{1}{n_2}}$$

as the appropriate standard error for this type of problem.

EXAMPLE 9.5 A local coach has a theory, based on his experience, that participants in sports which stress individual achievement (such as track or swimming) do better academically than those who participate in team sports. He believes that the skills required for individual success, such as motivation and self-discipline, carry over to academic areas. To test his theory, he collected the following data on cumulative grade-point averages of female athletes.

	n	$\bar{X}$	s
Team participants (T)	7	2.56	0.871
Individual achievers (I)	8	3.04	0.621

Is his assumption justified at the 0.05 level of significance?

Solution The null hypothesis states that there is no difference, or H_0: $\mu_T - \mu_I = 0$. He wants to reject this in order to show that the theory is justified. Since the coach's theory is that individual achievers do better, μ_I should be larger than μ_T and thus H_A: $\mu_T - \mu_I < 0$. Since both samples are small, we have a t test with $n_T + n_I - 2 = 7 + 8 - 2 = 13$ degrees of freedom, and we use Fig. 9.6.

Figure 9.6

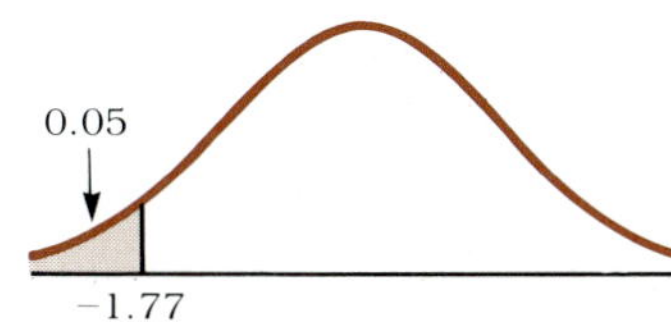

We need to calculate s_p before we can find t^*.

$$s_p = \sqrt{\frac{(7-1)(0.871)^2 + (8-1)(0.621)^2}{13}} = 0.747.$$

Now

$$t^* = \frac{\bar{X}_T - \bar{X}_I}{s_p\sqrt{\frac{1}{n_T} + \frac{1}{n_I}}} = \frac{2.56 - 3.04}{0.747\sqrt{\frac{1}{7} + \frac{1}{8}}} = -1.24.$$

Since -1.24 does not lie in the rejection region, these data do not support the coach's theory.

Summary

H_0: $\mu_T - \mu_I = 0$.

H_A: $\mu_T - \mu_I < 0$.

Test region: Reject H_0 when $t^* < -1.77$.

Test statistic:

$$t^* = \frac{2.56 - 3.04}{0.747\sqrt{\frac{1}{7} + \frac{1}{8}}} = -1.24.$$

Conclusion: We cannot reject H_0. The coach's theory is unproven. □

Confidence Intervals

Confidence intervals are created as before:

Sample results ± Critical value · Appropriate standard error.

Now we transform this into

$$\bar{X}_1 - \bar{X}_2 \pm \text{Critical value} \cdot s_p \sqrt{\frac{1}{n_1} + \frac{1}{n_2}},$$

remembering to subtract the smaller from the larger for $\bar{X}_1 - \bar{X}_2$.

EXAMPLE 9.6 As an incentive to pay cash and thus save processing charges for credit card usage, a major oil company ran an experiment in some of its stations. Ten stations sold gasoline at a cash-only pump with a lower price per gallon, while 25 stations had a uniform price regardless of method of payment. The results are as follows for a six-week period. Let $\bar{X}$ = mean price per gallon in dollars.

	n	$\bar{X}$	s^2
A: Discount for cash stations	10	1.182	0.162
B: Uniform-price stations	25	1.264	0.189

Figure 9.7

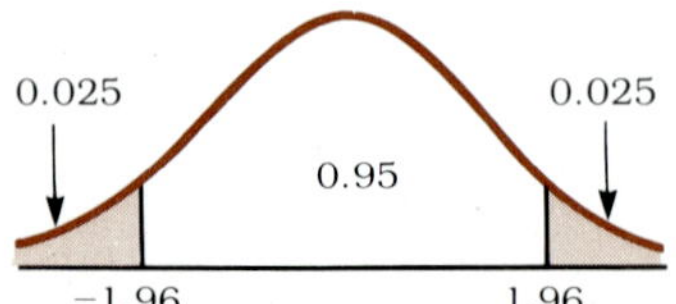

Find the 95% confidence interval estimate for the difference in the mean prices of gasoline.

Solution The small sample sizes tell us to use Table D.3 (the t table). For Fig. 9.7, df = $n_A + n_B - 2 = 10 + 25 - 2 = 33$. Remember that such a large number of degrees of freedom will take us to the z scores at the bottom of the t table. We will also need s_p.

$$s_p = \sqrt{\frac{(n_A - 1)s_A^2 + (n_B - 1)s_B^2}{n_A + n_B - 2}}$$

$$= \sqrt{\frac{9 \cdot 0.162 + 24 \cdot 0.189}{33}} = 0.0426.$$

Note the following:

1. The table of data gave us s_A^2 and s_B^2, so we did not need to square the terms in our calculations. Be careful to read the form in which data are given.
2. If we calculate the standard deviations,

$$s_A = 0.0402 \qquad s_B = 0.0435,$$

we can see that $s_p = 0.0426$ lies between them and closer to the standard deviation for the larger sample. The B sample had more influence on the value of s_p because of its larger size.

To continue, our confidence interval

$$\bar{X}_B - \bar{X}_A \pm \text{Critical value} \cdot s_p \sqrt{\frac{1}{n_A} + \frac{1}{n_B}} =$$

$$(1.264 - 1.182) \pm 1.96 \cdot 0.0426 \cdot \sqrt{\frac{1}{10} + \frac{1}{25}} =$$

$$0.082 \pm 0.0312.$$

$$\$0.0508 < \mu_B - \mu_A < \$0.1132.$$

The 95% confidence interval estimate for the difference in the mean prices of gasoline is from $0.0508 to $0.1132. □

EXERCISES/Section 9.2

1. One method of ascertaining the size of a wildlife population is the capture-recapture method that involves tagging. Animals are caught, tagged, and then later trapped again; careful records are kept of previously tagged and untagged animals. Over time this method can provide birth and death rates. At 15 different locations on two days, six months apart, the following data were collected on leopards. Was there a decline in population size during this six-month period at the 0.05 level of significance? Let $\bar{X}$ = the mean number of leopards caught.

	n	$\bar{X}$	s
Day 1	15	12.6	3.1
Day 2	15	10.2	3.6

2. The amount of time required for a reaction to a drug indicates information about its effectiveness. A group of 20 men and a group of 25 women were each given the same dosage of an experimental drug, with the following results. Do men react significantly faster than women at the 0.05 level of significance? Let $\bar{X}$ = the mean reaction times in minutes.

	n	$\bar{X}$	s^2
Men	20	21.2	26.9
Women	25	24.9	27.2

3. Complaints are being received that the mean amount of liquid in bottles from bottling line A is greater than that for bottling line B. A sampling of bottles from each produced the following results. Do the machines on line A fill bottles with significantly more liquid than those on line B for $\alpha = 0.01$? Let $\bar{X}$ = the mean amount of liquid per bottle (in ounces).

	n	$\bar{X}$	s
A	10	33.5	5.2
B	15	31.9	5.8

4. A study was conducted to determine whether tires wear faster on large cars than on small cars. Identical tires were installed on all cars and driven until at least one tire on a car was deemed unsafe by an independent inspector. The following data below gives the mean life of a tire in thousands of miles. Do these data indicate a difference in wearability based on car size at the 0.05 level of significance?

	n	$\bar{X}$	s
Large cars	25	11.6	2.8
Small cars	25	11.1	2.2

5. To determine the effect of price on merchandising, a store owner decided to feature his store-brand paper towels during a monthly sale. He wanted to see if price were a sufficient factor to draw customers away from a popular national brand on which he did not have much markup. For six consecutive monthly sales, he featured the store brand at a reduced price and obtained the following data. Do these data indicate that sales of the store-brand towels significantly exceeded those of the national brand at the 0.01 level of significance? Let $\bar{X}$ = mean monthly sales (in cartons).

	n	$\bar{X}$	s^2
Store-brand paper towels	6	104	16.7
National-brand paper towels	6	96	20.9

6. Using the data given in Example 9.5, estimate with 99% confidence the difference in mean grade-point averages between team participants and individual sport achievers.

7. Two popular computer programming manuals reported the following sales figures for a seven-year period. With 95% confidence, estimate the difference in mean sales for these two manuals. Let $\bar{X}$ = the mean number of manuals, in hundreds, sold per year.

	$\bar{X}$	s
Manual A	42.5	9.18
Manual B	19.4	6.54

8. In surveying, a residual tells how far from the true value a specific measurement is. As a means of rating his surveying students, a civil engineering professor kept track of their residuals for ten test measurements of a predetermined length. Estimate the difference in the means of these two students with 95% confidence. Let $\bar{X}$ = the mean residual in feet.

	n	$\bar{X}$	s
Student A	10	0.026	0.008
Student B	10	0.012	0.004

9. For US air carriers, the National Safety Council reports civil aviation deaths resulting from accidents. Six randomly selected years provide the following data. Determine the 95% confidence interval estimate for the difference in mean number of deaths for domestic flights versus those for international flights. Let $\bar{X}$ = the mean number of deaths per year.

	$\bar{X}$	s
Domestic flights	129.5	74.3
International flights	55.0	65.5

10. To estimate the difference in yield provided by two of its best-selling fertilizers, a large fertilizer company planted 20 plots of the same size in different locations with the same variety of vegetable. It treated one-half of every plot with one fertilizer and one-half of every plot with the other, with a barrier between the halves to prevent runoff from one to the other. The results are given in the following table. Find the 95% confidence interval estimate for the difference in mean yield. Let $\bar{X}$ = mean yield in bushels.

	n	$\bar{X}$	s^2
Fertilizer 1	20	12.3	13.9
Fertilizer 2	20	14.6	10.4

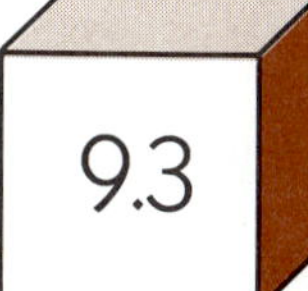

9.3 Differences between Means—Paired Samples

Now it is time to turn to data produced by dependent samples. As you will recall, *dependent* samples contain the same members and thus produce data that can be paired. For example, we can gather blood pressure information from a group of 50 people, then administer a medicine designed to reduce blood pressure, and again take blood pressures after a suitable time has passed. We would have two sets of data—*before* and *after*, as shown in Table 9.1—but both sets would have come from the same sample members and could be paired. If we treat the two sets of data independently, that is, calculate $\bar{X}_b$, s_b, $\bar{X}_a$, and s_a, we lose the extra information that goes with pairing. For example, if we separate the 80 from the 84 in the second row, we lose the fact that both numbers are for the same person (Person 2).

Since we ultimately want to find the difference between the two sets of data, rather than finding two sample means and then subtracting as we did before, we can now subtract within pairs and preserve that pairing. This produces a condensed set of data that is usually labeled difference, or d, as shown in the right-hand column of Table 9.1. We must continue to be careful and consistent when subtracting and recording the sign of the result. Here we subtracted the number in the *after* column from the number in the *before* column for each pair.

TABLE 9.1

Diastolic Blood			
Person 1	110	106	+4
Person 2	80	84	−4
Person 3	88	76	+12
Person 4	75	68	+7
.	.	.	.
.	.	.	.
.	.	.	.
Person 50	90	82	+8

Now we have one set of data, the d column, representing one set of people. What started as two sets can be treated as one, and we are back to the type of problem we solved in Chapter 8. We find

$$\overline{X}_d = \frac{\Sigma d}{n}$$

and

$$s_d = \sqrt{\frac{n\Sigma d^2 - (\Sigma d)^2}{n(n - 1)}}.$$

which are the equations we used in Chapter 4, except that d is now used instead of X.

Hypothesis Tests

We can perform hypothesis tests with H_O: $\mu_d = 0$ (no difference) and estimate μ_d with a confidence interval. If samples are large, z scores are used; if samples are small, t scores are used, just as before. Frequently these hypothesis tests are called *matched t tests.*

Why are dependent samples so useful and how do we obtain them? *Before* and *after* experiments are classic cases involving dependent samples. For the one just presented, many other factors influence blood pressure: weight, age, heredity, diet, occupation, smoking habits, and so on. These are confounding variables, which we discussed in Chapter 2. If we had used two totally different groups of people, taken blood pressures in one without giving the medication, and taken blood pressures in the other after giving the medication, we would have no way to determine whether these other factors influenced the results. Comparison of the groups would not tell us what we wanted to know about the medication. However, by using the same people to obtain both sets of data, these other variables were controlled; they have the same effect in both sets of data. Thus dependent samples are quite helpful in solving design problems such as those involving confounding variables.

In order to obtain dependent, matched samples, we must control for confounding variables. We can do this by using the same sample members and having them participate in such a way that aspects of an experiment likely to produce confounding variables are treated equally. How should two brands of shoe soles be tested for wearability? If 100 people participate, should we give 50 of them shoes with brand A soles and the other 50 shoes with brand B? If we do, we will end up with independent samples and have to analyze the data as we did in Section 9.1. Is it satisfactory to give everyone two pairs of shoes, one with sole A and one with sole B, and tell them to wear the A shoes for six weeks and then to switch to the B shoes for 6 weeks? This is better than the first suggestion but is still not ideal. What if it rains during much of the first six weeks and not during the second? Perhaps the best idea is to give everyone one pair of shoes, one shoe with an A sole and one shoe with a B sole. That way both soles would be subjected to the same external conditions that affect wear. The shoes should consist of 50 pairs with sole A on the right shoe and 50 pairs with sole A on the left shoe, distributed randomly to the 100 participants. Why do we do this? Because wearing occurs unevenly between right and left feet; we are right- or left-footed as surely as we are right- or left-handed. Anyone in doubt of this should watch toddlers climbing stairs.

EXAMPLE 9.7 To determine the value of a remedial math program, eight students were chosen at random and given a standardized math aptitude test. After completion of the program the same students were given an equivalent test. The grades of the students on both tests were:

Student	1	2	3	4	5	6	7	8
Pretest score	60	65	50	58	56	62	52	55
Posttest score	75	88	65	73	56	59	69	70

Have the eight students benefited from the program; that is, have their test scores increased significantly? Use $\alpha = 0.01$.

Solution We begin by deciding how to generate d. A quick glance at the data tells us that we will have more positive answers (and thus fewer minus signs to keep track of) if we subtract the pretest from the posttest results. Performing the subtractions, we get the following d values.

$$d = \text{posttest results} - \text{pretest results}$$
$$= 15 \quad 23 \quad 15 \quad 15 \quad 0 \quad -3 \quad 17 \quad 15$$

Now we can ignore the original data and concentrate on the d values. Hence

$$\bar{X}_d = \frac{\Sigma d}{n} = \frac{97}{8} = 12.125,$$

and

$$s_d = \sqrt{\frac{n\Sigma d^2 - (\Sigma d)^2}{n(n-1)}} = \sqrt{\frac{8 \cdot 1727 - 97^2}{8 \cdot 7}} = 8.87.$$

The null hypothesis always says that the status quo is preserved, or H_0: $\mu_d = 0$. We are looking for improvement from the pretest score to the posttest score, so H_A: $\mu_d > 0$. There are only eight members of the sample and we have only the calculated standard deviation s_d, so this is a t test. We use Fig. 9.8. Thus

Figure 9.8

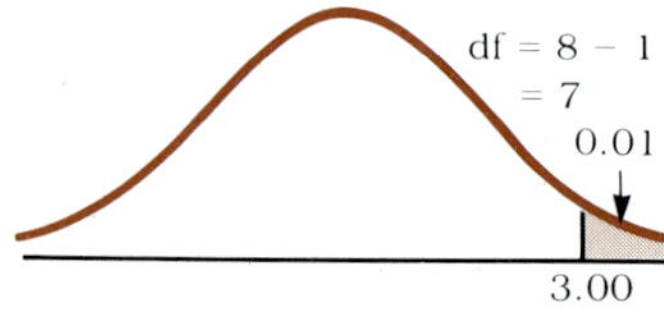

$$t^* = \frac{\bar{X} - \mu}{s/\sqrt{n}},$$

which becomes, in the case of matched pairs,

$$t^* = \frac{\bar{X}_d - \mu_d}{s_d/\sqrt{n}},$$

and since $\mu_d = 0$ by H_0,

$$t^* = \frac{\bar{X}_d}{s_d/\sqrt{n}}.$$

Substituting, we obtain

$$t^* = \frac{12.125}{8.87/\sqrt{8}} = 3.87.$$

Since 3.87 does lie in the rejection region, we can reject H_0.

Summary

H_0: $\mu_d = 0$.

H_A: $\mu_d > 0$.

Test region: Reject H_0 when $t^* > 3.00$.

Test statistic:

$$t^* = \frac{12.125}{8.87/\sqrt{8}} = 3.87.$$

Conclusion: We can reject H_0 and thus say that the program has made a significant difference with these eight students. □

Confidence Intervals

EXAMPLE 9.8 The following table gives fuel economy differences, in miles per gallon, based on two driving speeds for ten automobiles randomly selected from each model.

	55 mph	60 mph	d
Chevrolet Chevette	22.5	21.7	0.8
Honda Civic	32.5	28.7	3.8
Ford Pinto	25.3	23.7	1.6
Plymouth Horizon	30.6	27.9	2.7
Ford Fairmont	22.5	21.1	1.4
Olds Cutlass	24.3	22.6	1.7
Chevrolet Caprice	17.8	16.2	1.6
Olds Omega	28.0	26.4	1.6
Toyota Corolla	27.4	24.1	3.3
Olds Cutlass (diesel)	27.0	24.6	2.4

Find the 95% confidence interval estimate of the mean difference in fuel economy from 55 mph to 60 mph.

Solution We need to find the sum of the differences by adding the numbers in the d column: $\Sigma d = 20.9$.

$$\bar{X}_d = \frac{\Sigma d}{n} = \frac{20.9}{10} = 2.09.$$

$$s_d = \sqrt{\frac{n\Sigma d^2 - (\Sigma d)^2}{n(n-1)}} = \sqrt{\frac{10 \cdot 51.55 - (20.9)^2}{10 \cdot 9}} = 0.94.$$

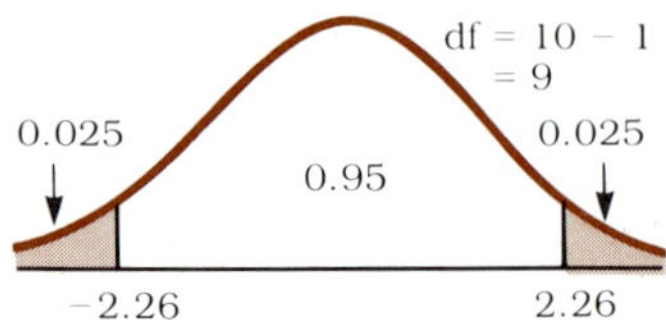

Figure 9.9

Since we have a small sample and only a sample standard deviation, we use Table D.3 and Fig. 9.9. The confidence interval pattern is

$$\text{Sample results} \pm t \cdot \text{Appropriate standard error},$$

which becomes

$$\bar{X}_d \pm t \cdot \frac{s_d}{\sqrt{n}}.$$

Substituting, we get

$$2.09 \pm 2.26 \cdot \frac{0.94}{\sqrt{10}} = 2.09 \pm 0.67.$$

$$1.42 \text{ mpg} < \mu_d < 2.76 \text{ mpg}.$$

The 95% confidence interval estimate for the mean difference in fuel economy is from 1.42 mpg to 2.76 mpg. □

EXERCISES/Section 9.3

1. In order to test whether a hypertension drug reduces performance anxiety (stage fright), 18 professional and student musicians were asked to give two performances: one after taking the drug and one after taking a placebo. The mean difference $\bar{X}_d$ in heart rate was 19 beats per minute with $s_d = 20.4$ beats per minute. Did the drug produce a significant decrease at the 0.05 level of significance?

2. In an attempt to judge extrasensory perception ten subjects were put through a series of tests while awake and then again while hypnotized. The following scores were obtained.

Subject	Waking Score	Hypnotized Score
1	23	23
2	12	23
3	21	21
4	15	24
5	19	21
6	15	27
7	17	22
8	21	19
9	18	19
10	23	23

At the 0.05 level of significance, did hypnotism improve the ESP of these subjects?

3. The following are election returns for 12 randomly selected precincts by percent of votes cast for the Democratic candidate during two consecutive presidential elections. Did the vote substantially decrease at the 0.01 level of significance?

Precinct	1	2	3	4	5	6
19X4	64	72	38	29	51	32
19X8	62	69	39	28	55	28
Precinct	7	8	9	10	11	12
19X4	97	63	13	44	91	83
19X8	75	57	22	43	86	79

4. The following data come from a study that followed 16 veterans over a ten-year period. It gives their diastolic blood pressures at the beginning and again at the end of the test period. Does blood pressure change significantly with age, at the 0.01 level of significance?

Patient	1	2	3	4	5	6	7	8	9	10	11	12	13	14	15	16
Initial DBP	80	110	88	94	74	80	80	80	90	75	80	70	90	80	82	110
End DBP	84	58	76	88	90	68	58	76	96	68	66	82	78	78	80	84

5. A study to determine the effect of nursing intervention on behavior modification by patients produced the following test results. Has the nursing intervention produced a significant change in behavior at the 0.05 level of significance?

	Social Maturity	
Subject	Initial	Terminal (2 Months Later)
1	31	32
2	40	39
3	41	41
4	118	134
5	57	73

6. Using the data given in Example 9.7, find the 95% confidence interval estimate of the change in mathematics skills evidenced during the remedial math program.

7. With 95% confidence, estimate the mean difference in wearability between Brand A and Brand B tires. The data given below were produced by using ten different car models with each car having 2 tires of A and 2 tires of B in diagonally opposite positions.

A ╳ B, B ╳ A or B ╳ A, A ╳ B

	Life of Tire (thousands of miles)	
Model	Brand A	Brand B
1	10.0	11.0
2	12.0	12.1
3	13.0	12.2
4	9.7	9.7
5	10.9	11.1
6	14.0	15.1
7	11.1	12.4
8	10.1	12.3
9	9.7	11.2
10	14.3	15.3

8. Use the data given in Exercise 1 to estimate with 99% confidence the mean decrease in heart rate.

9. A quality-control inspector is determining whether lotion bottles are properly filled. He weighs 12 randomly selected bottles full (gross weight) and then empty (tare weight). Determine the 95% confidence interval estimate for the mean weight, in grams, of the contents.

Gross Weight	Tare Weight
416.12	106.98
414.10	104.38
415.32	106.24
414.54	105.78
413.88	106.14
415.28	104.04
416.66	104.98
414.90	104.66
416.30	104.80
416.04	105.64
416.66	104.92
415.74	105.08

10. With 95% confidence, estimate the mean difference in votes cast for the Democrat in two consecutive presidential elections using the data in Exercise 3.

Study Notes

KEY TERMS

dependent samples ■ Samples that contain the same members and/or produce matched pairs of data.

independent samples ■ Samples that do not contain the same members and produce unrelated (unpaired) data.

OUTLINE

I. Differences between means—large samples
   A. Independent samples:
      1. Samples that do not contain the same members.
      2. Analyze them separately to produce $\bar{X}_1$, $\bar{X}_2$, s_1, s_2.
   B. If independent samples of size n_1 and n_2 are drawn from populations with means μ_1 and μ_2 and standard deviations σ_1 and σ_2 and $n_1 > 30$ and $n_2 > 30$, then the sampling distribution of $\bar{X}_1 - \bar{X}_2$ has the following properties.
      1. It is approximately normal.
      2. Its mean is $\mu_1 - \mu_2$.
      3. Its standard deviation is

$$\sqrt{\frac{\sigma_1^2}{n_1} + \frac{\sigma_2^2}{n_2}},$$

which can be approximated by

$$\sqrt{\frac{s_1^2}{n_1} + \frac{s_2^2}{n_2}},$$

if σ_1 and σ_2 are unknown.

   C.
$$z = \frac{(\bar{X}_1 - \bar{X}_2) - (\mu_1 - \mu_2)}{\sqrt{\frac{s_1^2}{n_1} + \frac{s_2^2}{n_2}}}.$$

   D. $H_0: \mu_1 - \mu_2 = 0$.
   E. H_A:
      1. $\mu_1 - \mu_2 < 0$ means that μ_1 is smaller than μ_2.
      2. $\mu_1 - \mu_2 > 0$ means that μ_1 is larger than μ_2.
   F. Confidence intervals:

$$\bar{X}_1 - \bar{X}_2 \pm z\sqrt{\frac{s_1^2}{n_1} + \frac{s_2^2}{n_2}}.$$

II. Differences between means—small samples
   A. Assumptions:
      1. Two independent random samples are taken.
      2. The original population yielding these samples are each approximately normal.
      3. The two population variances or standard deviations are equal and probably unknown.
   B. $$t = \frac{(\bar{X}_1 - \bar{X}_2) - (\mu_1 - \mu_2)}{s_p\sqrt{\frac{1}{n_1} + \frac{1}{n_2}}}.$$
   C. $$s_p = \sqrt{\frac{(n_1 - 1)s_1^2 + (n_2 - 1)s_2^2}{n_1 + n_2 - 2}}.$$
   D. $df = n_1 + n_2 - 2$.
   E. Confidence interval:

$$\bar{X}_1 - \bar{X}_2 \pm ts_p\sqrt{\frac{1}{n_1} + \frac{1}{n_2}}$$

III. Differences between means—paired samples
   A. Dependent samples contain the same members and produce matched pairs of data.
   B. $$\bar{X}_d = \frac{\Sigma d}{n}.$$
   C. $$s_d = \sqrt{\frac{n\Sigma d^2 - (\Sigma d)^2}{n(n - 1)}}.$$

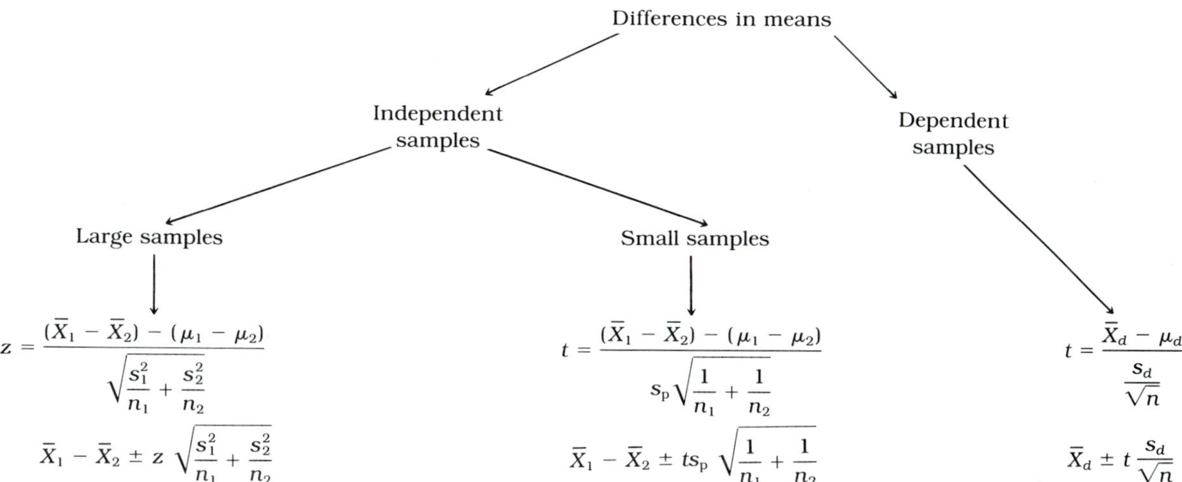

Figure 9.10
How to decide which formulas to use

D. $H_0: \mu_d = 0$.

E. Dependent samples control for confounding variables.

F. $t^* = \dfrac{\bar{X}_d}{s_d/\sqrt{n}}$.

G. Confidence interval:

$$\bar{X}_d \pm t\frac{s_d}{\sqrt{n}}.$$

IV. How to decide which formulas to use (see Fig. 9.10).

REVIEW PROBLEMS

1. Over a six-month period a major manufacturing concern kept records of absenteeism for men and women. Estimate the difference in absenteeism between men and women with 95% confidence. Let $\bar{X}$ = mean number of days absent.

	n	$\bar{X}$	s
Men	200	7.42	5.75
Women	75	5.21	2.36

2. Ten married employees comparing notes on the number of years of college attended. At the 0.05 level of significance, is there any difference in educational level between spouses?

Husband	Wife
4	4
1	0
7	2
6	4
5	4
4	0
7	6
2	5
2	4
4	6

3. Two record companies, A and B, are competitors for the same market. A random sample of 50 records produced by A yielded a mean playing time (for both sides) of 51.5 min with $s = 5$ min. A random sample of 75 records produced

by B yielded a mean playing time of 48.25 min with $s = 3.5$ min. Is this sufficient evidence to reject the claim that there is no difference between the mean playing times of the two companies' records at the 0.05 level of significance?

4. The parents of a child about to graduate from elementary school must decide between two local high schools, HS_1, a public school and HS_2, a private school. They decide to send their child to the school that has the better record of senior performance on SAT exams. A random sample from HS_1 indicates that 100 seniors had a mean combined SAT score of 1050 with $s = 50$, whereas a random sample of 50 seniors from HS_2 indicates a mean combined test score of 1100 with $s = 100$. At the 0.01 level of significance, is this sufficient evidence to conclude that HS_2 is preferable by the standards the parents had established?

5. Two accounting teachers had a disagreement about which of them was the better teacher. They agreed to settle the argument statistically. Each selected 100 names at random from the other's lifetime class list. M's sample yielded an average course grade of 73.2 with $s = 9.8$, while N's sample yielded an average course grade of 76.4 with $s = 13.5$. At the 0.10 level of significance, is this sufficient evidence to conclude that there is a difference between their teaching abilities?

6. Determine the 95% confidence interval estimate of the difference between the mean price, in cents per pound, of generic macaroni and the house brand. Data was gathered for each once a week over a 52-week period as follows:

	$\bar{X}$	s
Generic	30.25	10
House brand	34.12	15

7. Determine the 99% confidence interval estimate for the difference between the mean life span, in years, of two different video cassette recorders.

	n	$\bar{X}$	s
Brand B	45	8.75	1.50
Brand V	60	9.50	1.25

8. Determine the 90% confidence interval estimate for the difference between the mean price, in dollars, of 30 math and 60 English textbooks.

	n	$\bar{X}$	s
Math	35	19.50	5.00
English	60	17.25	6.25

9. A guidance counselor at a local high school often gets requests to recommend a speed-reading course. Two such courses are available in the area. To determine the more effective program, she asked ten graduates of each course to read the same short story. The results of this study are summarized in the accompanying table. Can the counselor conclude that there is no difference between the two courses at the 5% level of significance? Let $\bar{X}$ = average time, in minutes, needed to read the assigned short story.

	n	$\bar{X}$	s
Course A	10	2.25	0.8
Course B	10	2.50	1.0

10. Paul Jameson decides he must lose some weight quickly. He consults two friends, one a doctor and the other a physical education instructor. Both recommend plans that include diet and exercise, though in different proportions. To judge effectiveness, he asked each to provide data on mean weight loss in one month's time. The data are summarized in the following table. Can Paul conclude that there is no difference at the 10% level of significance? Let $\bar{X}$ = the mean weight loss, in pounds, in one month.

TABLE 9.2

Employee	1	2	3	4	5	6	7	8	9	10
X with music	28	32	30	29	27	29	31	33	35	30
X without music	25	31	30	27	26	31	25	30	34	25

	n	$\bar{X}$	s
Plan D	15	12.5	3.0
Plan I	10	14.0	4.5

11. Two regional airlines claim to be the most reliable ones serving their region. They base their claims on mean delays in departure time. A consumer activist organization collects the following data. Can the consumer group conclude that Y is the more reliable airline at the 1% level of significance? Let $\bar{X}$ = mean delay, in minutes, in departure times.

	n	$\bar{X}$	s
Airline X	20	30	5.0
Airline Y	15	28	7.5

12. Two remedial math programs reported post-test scores on standardized achievement tests, as indicated in the following table. With 95% confidence, estimate the difference in mean scores for these two programs.

	n	$\bar{X}$	s
Program A	18	76.5	6.8
Program B	26	79.2	9.1

13. The frequency of repairs for two samples (one foreign, one domestic) of automobiles is shown in the accompanying table. With 99% confidence, estimate the difference in mean number of repairs over a five-year period. Let $\bar{X}$ = the average number of repairs over a five-year period.

	n	$\bar{X}$	s
Domestic	20	7.2	1.5
Foreign	15	5.9	1.2

14. Two brands of light bulb were tested, with the following results regarding average life span in hours of use. With 90% confidence, estimate the difference in mean life span for these two brands.

	n	$\bar{X}$	s
Brand M	10	495	20
Brand N	20	480	25

15. "Things Go Better with Music" is the advertising slogan of a local easy-listening-music radio station. It justifies this claim on the basis of the figures in Table 9.2. Do these figures bear out the slogan's claim at the 5% level of significance? Let X represent the number of units produced per workhour in a certain factory.

16. Fluoride is supposed to reduce the incidence of caries (tooth decay). To test this hypothesis, a producer of toothpaste chose 12 willing participants at random and had each of them use a fluoridated tooth paste for a year and then use a nonfluoridated toothpaste for a year. Table 9.3 records the number of cavities per patient per year. Do these figures support the hypothesis at the 10% level of significance?

Table 9.3

Nonfluoridated	4	6	2	3	0	4	2	3	4	1	2	5
Fluoridated	3	2	1	2	1	3	0	1	1	1	2	3

17. Find the 95% confidence interval estimate of the difference in mean production using the data given in Problem 15.

18. With 99% confidence, estimate the mean difference between incidence of caries using the data in Problem 16.

Statistical inference concerning proportions, standard deviations, and variances

10

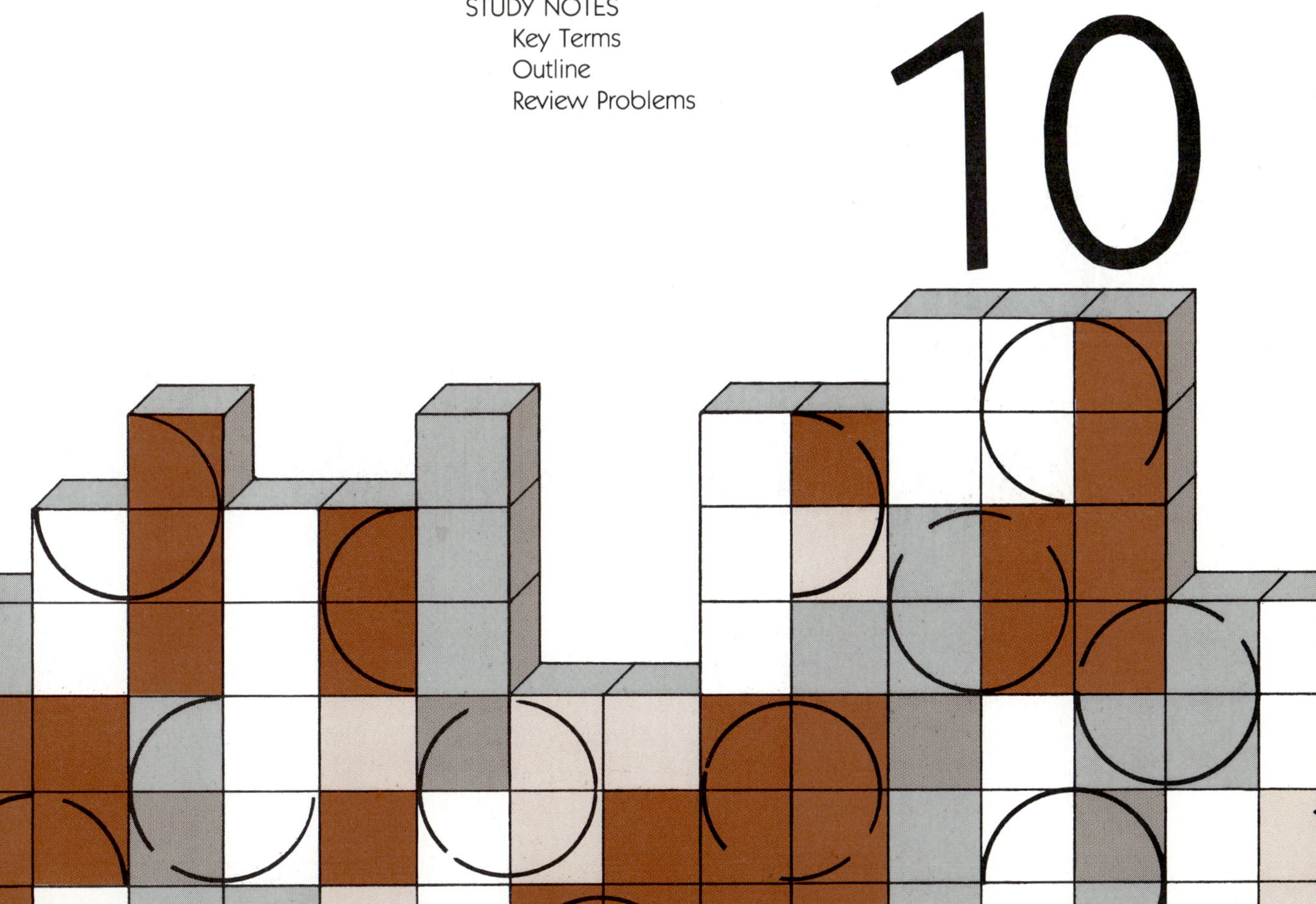

10.1 Proportions—One Population

Population proportions, or percents, give information based on counting: the number of people or objects with a certain characteristic divided by the total number considered, producing a decimal or a percent. Thus we are back to discrete data and relative frequencies. In this chapter we concentrate on binomial probabilities, assessing situations where the population can be divided into two categories; later, in Chapter 13, we will examine cases where more than two categories are required, such as for hair color or eye color. Recall that, with the binomial distribution, p is the parameter giving the probability for an individual success or the proportion of the population having the desired characteristic. Similarly, $q = 1 - p$ is the probability of failure or the proportion of the population that does not have the desired characteristic. Finally, n is the number of independent trials conducted; however, we will now use n for sample size.

Note that we are digressing from the usual practice of using Greek letters for parameters. The Greek letter for p is π, which has a specific meaning in mathematics; we will not use it here in order to avoid confusion. We will use p for the population parameter and use $\hat{p}$ (read "p hat") for a sample result. This should not be too confusing because we have added extra marks to letters before to indicate calculated sample results (such as $\bar{X}$) and will do so again. The law of large numbers tells us that $\hat{p}$, a relative frequency from a sample, provides us with a good estimate of p, the population proportion in which we are interested.

We also need to recall that for a binomial distribution

$$\mu = np \quad \text{and} \quad \sigma = \sqrt{npq},$$

which are in units of X and X = the number of successes. We can convert these equations into forms that allow us to work directly with $\hat{p}$ in units of proportion, or percent, by dividing both by n. Thus $\mu_{\hat{p}} = p$ as the law of large numbers states; the theoretical, expected number is the true proportion, and

$$\sigma_{\hat{p}} = \frac{\sqrt{npq}}{n} = \sqrt{\frac{npq}{n^2}} = \sqrt{\frac{pq}{n}}.$$

NOTE

The sampling distribution of $\hat{p}$ has the following useful characteristics:

1. It is approximately normal, provided that $np > 5$ and $nq > 5$.
2. Its mean (center) is p; $\mu_{\hat{p}} = p$.
3. Its standard deviation or standard error is $\sigma_{\hat{p}} = \sqrt{pq/n}$.

Hypothesis Tests

We are ready to perform hypothesis tests or find confidence intervals; we know that we use the z table if np and nq are both greater than 5. We can easily convert

$$z = \frac{\text{Sample results} - \text{Parameter value}}{\text{Appropriate standard error}}$$

to the proper z score:

$$\boxed{z = \frac{\hat{p} - p}{\sqrt{\dfrac{pq}{n}}}.}$$

EXAMPLE 10.1 The state of New Jersey has a high cancer death rate. A 20-year study found a death rate for white males of 205 per 100,000. The national rate is 174 per 100,000. Does New Jersey have a significantly high death rate for cancer, at the 0.01 level of significance?

Solution We are told the national rate:

$$p = \frac{174}{100{,}000}.$$

Thus

$$H_0: p = \frac{174}{100{,}000} = 0.00174.$$

The problem asks whether New Jersey's rate is higher, so we have a one-tail test.

$$H_A: p > \frac{174}{100{,}000} = 0.00174.$$

Figure 10.1

(a)

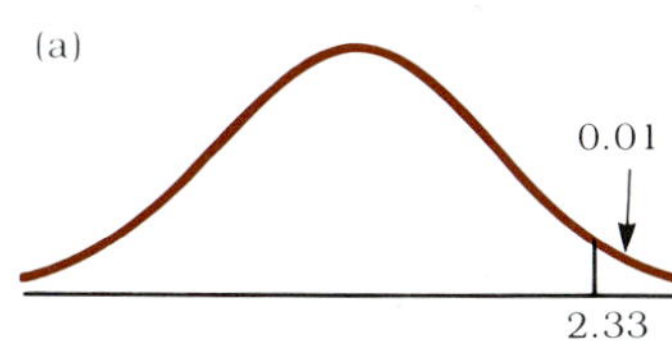

(b)

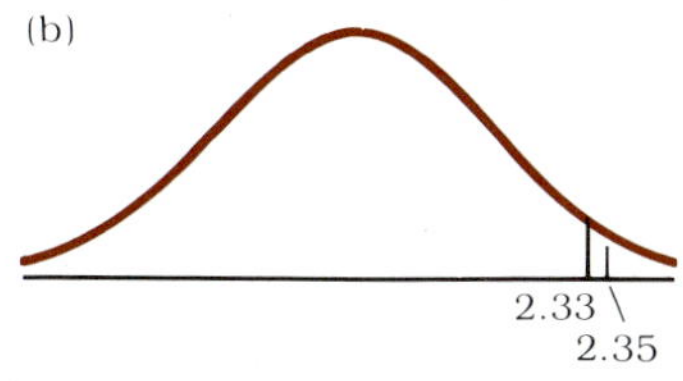

We use z scores in Fig. 10.1(a).

$$z^* = \frac{\hat{p} - p}{\sqrt{\dfrac{pq}{n}}} = \frac{0.00205 - 0.00174}{\sqrt{\dfrac{0.00174 \cdot 0.99826}{100{,}000}}} = 2.35.$$

Since z^* falls in the rejection region in Fig. 10.1(b), we can reject H_0.

Summary

$H_0: p = 0.00174.$

$H_A: p > 0.00174.$

Test region: Reject H_0 if $z^* > 2.33$.

Test statistic:

$$z^* = \frac{0.00205 - 0.00174}{\sqrt{\dfrac{0.00174 \cdot 0.99826}{100{,}000}}} = 2.35.$$

Conclusion: Reject H_0. New Jersey has a cancer death rate significantly higher than the nation as a whole. □

EXAMPLE 10.2 The same research found a cancer death rate for women to be 148 per 100,000 as opposed to a national rate of 130 per 100,000. Are news reports and editorials that vilify New Jersey's environment as a link to cancer justified in the case of women at the 0.01 level of significance?

Solution Again H_0 uses the national rate,

$$H_0: p = \frac{130}{100{,}000},$$

and we have a one-tail alternative,

$$H_A: p > 0.00130.$$

The rejection region is the same as in Example 10.1.

$$z^* = \frac{\hat{p} - p}{\sqrt{\dfrac{pq}{n}}} = \frac{0.00148 - 0.00130}{\sqrt{\dfrac{0.00130 \cdot 0.99870}{100{,}000}}} = 1.58.$$

Figure 10.2

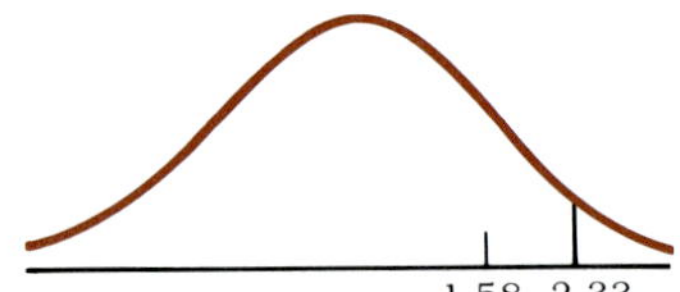

Since z^* does not fall in the rejection region in Fig. 10.2 we cannot reject H_0.

Summary

$H_0: p = 0.00130$,

$H_A: p > 0.00130$.

Test region: Reject H_0 if $z^* > 2.33$.

Test statistic:

$$z^* = \frac{0.00148 - 0.00130}{\sqrt{\dfrac{0.00130 \cdot 0.99870}{100{,}000}}} = 1.58.$$

Conclusion: We cannot reject H_0. For women, New Jersey does not have a significantly higher overall cancer death rate than the nation as a whole. □

Note the following points:

1. You may have difficulty verifying these calculations on a calculator unless it has an exponential key, but the problem is sufficiently thought-provoking to warrant examination even if you can't do all the numerical calculations.
2. The value for p from H_0 is used whenever possible in z; it shows up in the right-hand side of the numerator and is used for p and $q = 1 - p$ in the denominator. Be careful that you only use the sample result once, in the numerator. We assume validity for H_0 until it is rejected, so we must use its information.
3. Whites are singled out from other racial groups. Why? Men are recorded separately from women. Why?
4. Anytime a characteristic is rated—putting some values of the characteristic high and others low on a list—impressions are created. It is important to look at the size of the differences. Is the worst rating significantly different from the middle or the best rating? If Professor Jones makes $20,000 per year and each of her students earns $19,999, she ranks first, but obviously there is no significant difference in their salaries. In the case of New Jersey's health problems, further analysis revealed that the perception of a high cancer death rate for white men was justified; but that for women the perception is not justified.

Confidence Intervals

For confidence interval problems we must convert our pattern,

Sample results ± Critical value · Appropriate standard error,

using $\hat{p}$ and $\hat{q} = 1 - \hat{p}$, because we will have no value for p from H_0. When we substitute, we obtain

$$\hat{p} \pm z\sqrt{\frac{\hat{p}\hat{q}}{n}}.$$

Example 10.3 The Maryland State Board of Dental Examiners handles complaints against dentists. In an attempt to evaluate its effectiveness, it randomly selected 52 people from its files who had expressed dissatisfaction with dental care. Forty-six were not satisfied with the board's decision on their complaint. Find the 95%

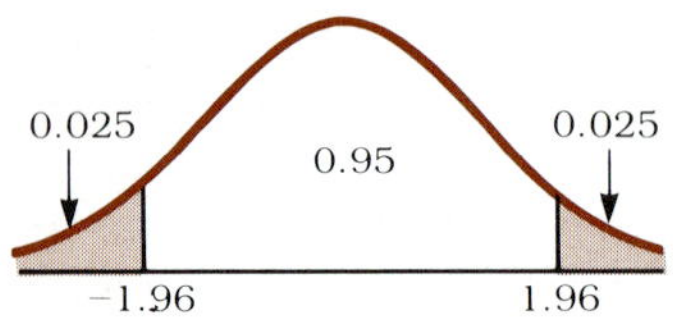

Figure 10.3

confidence interval estimate for the proportion of applicants who were displeased with the board's work.

Solution The sample proportions are: $\hat{p} = 46/52$ and $\hat{q} = 1 - \hat{p} = 6/52$. The z score for 95% confidence is 1.96 in Fig. 10.3.

$$\frac{46}{52} \pm 1.96 \sqrt{\frac{46/52 \cdot 6/52}{52}} = 0.885 \pm 0.087.$$

$$0.798 < p < 0.972.$$

We estimate the proportion of displeased petitioners to be between 79.8% and 97.2%, with 95% confidence. □

News & Views 10.1 contains a report of findings expressed as intervals. Usually newspaper articles attempt to simplify by using one number, but in this case the intervals produced by statistical procedures were actually reported.

Sample Size

We can use the concept of maximum error of the estimate from Chapter 8 to pose questions about the determination of sample size. Recall that the maximum error of the estimate e is one-half the length of the confidence interval. In this case it is

$$e = z \sqrt{\frac{pq}{n}}.$$

When asked to determine the sample size, we cannot use $\hat{p}$ or $\hat{q}$ since they are sample results—and we will not have sample results before taking the sample. A little algebra puts this equation into a more useful form for finding n.

$$\text{Square both sides:} \quad e^2 = z^2 \frac{pq}{n}.$$

$$\text{Multiply both sides by } n\text{:} \quad ne^2 = z^2 \frac{pq}{\cancel{n}} \cancel{n}.$$

$$\text{Divide both sides by } e\text{:} \quad \frac{n\cancel{e^2}}{\cancel{e^2}} = \frac{zpq}{e^2}.$$

$$n = \frac{z^2pq}{e^2}$$

NEWS & VIEWS 10.1

Ovary removal found to speed mineral loss

Women whose ovaries have been removed have an unexpectedly high rate of bone mineral loss, but estrogen therapy can prevent and even reverse the problem, researchers have reported.

The finding by University of California researchers may be significant for all women, since all go through natural or surgical menopause and have some degree of bone mineral loss.

The condition can cause curvature of the spine and leave bones susceptible to fracture. Millions of people in the United States, particularly post-menopausal women, are affected by it, and as many as 20,000 people die each year as a result of bone mineral loss, usually from complications following fractures of the hip.

Doctors who made the two-year study at the university's San Francisco campus expected a small percentage of bone mineral loss among the 37 premenopausal women whose ovaries were removed. Instead, they found an annual average loss of 7 percent to 9 percent, and a loss of as much as 15 percent to 20 percent in some women, according to Dr. Harry K. Genant, professor of radiology, medicine and orthopedics.

The women were treated with either a placebo, low-dose estrogen therapy or 0.6 milligrams of conjugated estrogen.

Of the six women in the last group, five showed no loss of bone mineral. The mean loss was less than 0.5 percent for that group, compared to 7 percent to 9 percent in the other two groups.

After two years, some of the women who had received placebos were treated with a 0.6-milligram dose of conjugated estrogen and regained bone mineral, Dr. Genant said.

The findings were reported recently in *Annals of Internal Medicine.*

The study, the first done on spinal mineral content after removal of ovaries, revealed far greater bone mineral loss than was found in previous research on the peripheral skeleton. The average bone mineral loss in those studies was only 1 percent to 2 percent a year.

Dr. Genant said the San Francisco researchers found a more severe problem because "the spine is a more sensitive measurement of mineral loss."

Source: *Baltimore Sun,* December 26, 1982.

As before, we determine z from the confidence level and e from the desired accuracy. However, we must also have something to substitute for p and q. The following two examples help us to solve this problem.

EXAMPLE 10.4 The state of Michigan is the nation's leading Christmas tree producer. Scotch pines seem to be the most popular tree. How many randomly selected trees need to be counted to determine the true proportion of Scotch pines sold, if we want to be accurate to within 4% with 95% confidence?

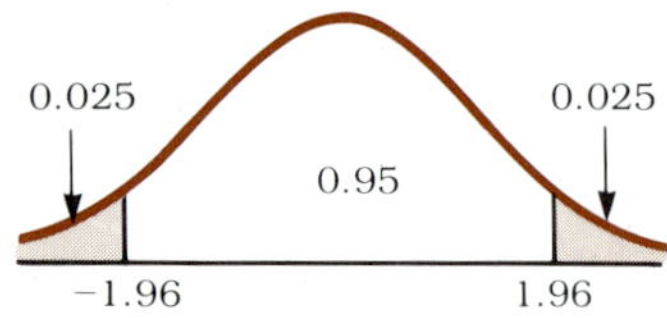

Figure 10.4

Solution The confidence level tells us that $z = 1.96$ in Fig. 10.4, and we know that $e = 4\% = 0.04$. In order to establish p and q, we consider the following: Is it better from a statistical point of view to have too large a sample or too small a sample? Large samples are always preferable, so let us examine pq and try to maximize it. A quick rundown of the following chart should convince you that pq reaches a maximum when $p = q = 0.5$.

p	q	pq
0.1	0.9	0.09
0.2	0.8	0.16
0.3	0.7	0.21
0.4	0.6	0.24
0.5	0.5	0.25

This is a manifestation of what the economist John Maynard Keynes called the Principle of Indifference: If you have no good reasons for supposing something to be true (a success) or false (a failure), then assign equal probabilities to each possibility.

Now

$$n = \frac{z^2pq}{e^2} = \frac{1.96^2 \cdot 0.5 \cdot 0.5}{(0.04)^2} = 600.25,$$

which rounds up to 601. □

EXAMPLE 10.5 A forestry professor at Michigan State University says that Scotch pines account for more than 80% of the trees sold. How many randomly selected trees must be counted to determine the true proportion of Scotch pines sold, if we want to be accurate to within 4% with 95% confidence?

Solution This problem is identical to the previous one, except that now we have an estimate for p (80%) based on previous research or experience.

$$n = \frac{1.96^2 \cdot 0.80 \cdot 0.20}{(0.04)^2} = 384.16,$$

which rounds to 385. □

Thus, for a desired degree of accuracy, the necessary sample size can be reduced by some preliminary research, or even a small test sample to provide an estimate for p.

EXERCISES/Section 10.1

1. A random sample of 500 voters were asked whether they approve of the president's economic policies and 219 said they do. Is this sufficient evidence to reject the claim of the president's party that at least one-half the voters approve of the president's policies, at the 1% level of significance?
2. A tourist in the British Isles had been warned to take rain gear because she could expect rain during at least one-half of her vacation. She experienced 19 rainy days during a 45-day trip. Is this sufficient evidence to counter the warning she was given, at the 0.05 level of significance?
3. The XYZ Industry, Inc., surveyed a random sample of 750 of its employees and found that 173 were white-collar workers. Can the company claim that 25% of its workers are white-collar, at the 0.01 level of significance?
4. The state police claim that extra manpower, and thus surveillance, lowered the highway death toll. Last year 8% of highway accidents resulted in at least one death. During the first four months of this year, 220 accidents occurred, of which 14 involved deaths. Should the legislature continue to fund the extra officers because of a significantly lower death rate on the state's highways? Let $\alpha = 0.01$.
5. A marketing analyst claimed that at least 45% of the population tint their hair. An independent researcher challenged this claim with results which indicated that, of 650 surveyed, 265 used a hair-color preparation. Was this challenge justified at the 0.05 level of significance?
6. A random sample of 100 students revealed that 63 own cars that are at least five-years old. Use this information to determine the 95% confidence interval estimate for p, the true proportion of owners of cars at least five years old.
7. A random sample of 1500 people produced 1140 who live in metropolitan areas. Determine the 95% confidence interval estimate for the percent of the population living in urban areas.
8. A survey of 900 people who are currently employed found 52 in the fields of agriculture, forestry, and fishing. Estimate with 98% confidence the true proportion of the population engaged in agriculture, forestry, and fishing.
9. A random sample of 1600 taken from voter registrations (in order to obtain a list of adults) produced 448 people who lived in apartments. With 99% confidence, estimate the true proportion of apartment dwellers in that jurisdiction.
10. Estimate with 95% confidence the percent of the population who suffer from severe headaches (severe enough to consult a doctor), if a sample of 250 people produced 35 such headache sufferers.
11. a) The traveler in Exercise 2 would like to know how many days she should spend in a country in order to estimate the percent of rainy days to within 10%, with 95% confidence. Find n.
    b) If her first week included 2 rainy days out of the 7, find n using the same level of confidence and accuracy requirements as in part (a).
12. a) What size sample is necessary to estimate to within 3% the current proportion of the work force employed in service professions, if you want a 95% level of confidence?
    b) If the Bureau of Labor Statistics tells you that last year 70% of the population was so employed, find n using the same requirements as in part (a).
13. a) How large a sample should be taken to determine the unemployment rates to within 0.5%, with 99% confidence?
    b) If last month's rate was 8.2%, find n using the requirements from part (a).
14. Determine the size of the sample necessary to estimate the proportion of TV tubes that do

not pass inspection, if you want to be within 1% with 99% confidence.

15. How many death certificates need to be examined for cause of death in order to estimate with 99% confidence the true percent of the U.S. population that dies of cancer, to within 1%, if the World Health Organization reports that 20% of deaths in industrialized countries are caused by cancer?

10.2 Comparison of Proportions—Two Populations

We will temporarily (until Chapter 13) complete our work with proportions and counted sample results by examining methods of dealing with two sets of sample data from two populations. Remember that we are considering situations in which characteristics can be divided into two categories (success, failure) and that sample sizes must be large. It will also help if we remember that the random variable X is the number of successes.

Hypothesis Tests

For a hypothesis test, the null hypothesis usually appears as it did for the difference of two means. We say that the proportions or percents in question for the two populations are the same and express this as no difference.

$$H_0: p_1 - p_2 = 0.$$

We approximate this population difference with the difference of sample results, $\hat{p}_1 - \hat{p}_2$. As before, we need to know the characteristics of the sampling distribution of $\hat{p}_1 - \hat{p}_2$.

The sampling distribution of $\hat{p}_1 - \hat{p}_2$ has the following characteristics.

1. It is approximately normal.
2. Its mean is $\mu_{\hat{p}_1 - \hat{p}_2} = p_1 - p_2$.
3. Its standard deviation is $\sigma_{\hat{p}_1 - \hat{p}_2} = \sqrt{(p_1 q_1/n_1) + (p_2 q_2/n_2)}$.

This tells us to use z scores to find the test region and helps us to convert a general z score,

$$z = \frac{\text{Sample results} - \text{Population parameter}}{\text{Appropriate standard error}},$$

to a specific z score,

$$z = \frac{(\hat{p}_1 - \hat{p}_2) - (p_1 - p_2)}{\sqrt{\dfrac{p_1 q_1}{n_1} + \dfrac{p_2 q_2}{n_2}}}.$$

The fact that H_0 claims $p_1 - p_2 = 0$ helps us simplify this to

$$z = \frac{\hat{p}_1 - \hat{p}_2}{\sqrt{\dfrac{p_1 q_1}{n_1} + \dfrac{p_2 q_2}{n_2}}}.$$

The null hypothesis also helps us to simplify the denominator. It says that $p_1 - p_2 = 0$, or $p_1 = p_2$, which is a common-value p (common to both populations); similarly, $q_1 = q_2 = q$. Thus

$$\sqrt{\frac{p_1 q_1}{n_1} + \frac{p_2 q_2}{n_2}} = \sqrt{\frac{pq}{n_1} + \frac{pq}{n_2}} = \sqrt{pq\left(\frac{1}{n_1} + \frac{1}{n_2}\right)}.$$

We do not know this common value p, but we can estimate it by pooling the sample results $\hat{p}_1$ and $\hat{p}_2$, in much the same way that we estimated a standard deviation common to two populations in Chapter 9. The equation we use for p depends on the way the data are presented.

Where X_1 and X_2 are actual, counted sample results, we estimate p by adding data from the two samples and call this estimate p':

$$p' = \frac{x_1 + x_2}{n_1 + n_2}.$$

However, if sample results are given only as sample percents or proportions (relative frequencies), we convert $\hat{p}_1 = (X_1/n_1)$ to $X_1 = n_1\hat{p}_1$ and calculate

$$p' = \frac{n_1\hat{p}_1 + n_2\hat{p}_2}{n_1 + n_2}.$$

Let us get back to z. We found p' as an estimate for an unknown p so

$$z = \frac{\hat{p}_1 - \hat{p}_2}{\sqrt{p'q'\left(\dfrac{1}{n_1} + \dfrac{1}{n_2}\right)}},$$

where p' is a pooled proportion and $q' = 1 - p'$.

EXAMPLE 10.6 Perceptions obviously affect opinions, particularly if information is one-sided as in this example. Professor French hears many excuses of illness from students just before a scheduled class exam. Thinking that he detects some insincerity, he decides to conduct a study. He finds that of a group of 150 students, 92 reported onset of the flu in the morning (before class); whereas of 200 of Professor German's students, 110 reported onset of the flu in the evening (after class). At the 0.05 level of significance, is there any difference between these groups as to when they become sick?

Solution The null hypothesis states that there is no difference.

$$H_0: p_F - p_G = 0.$$

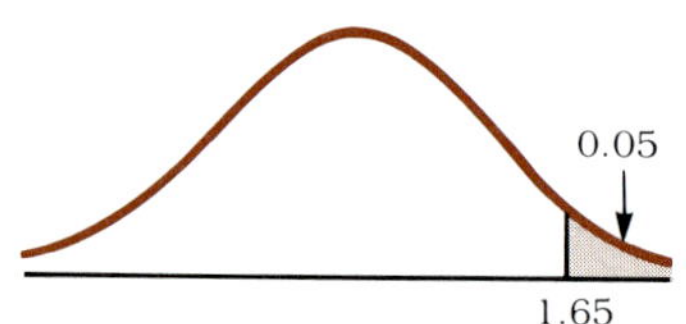

Figure 10.5

Professor French thinks his examinations motivate fake illness, so

$$H_A: p_F - p_G > 0.$$

We use z scores in Fig. 10.5.

$$z^* = \frac{\hat{p}_F - \hat{p}_G}{\sqrt{p'q'\left(\dfrac{1}{n_F} + \dfrac{1}{n_G}\right)}}.$$

$$p' = \frac{92 + 110}{150 + 200} = \frac{202}{350} \quad \text{and} \quad q' = 1 - \frac{202}{350} = \frac{148}{350}.$$

Substituting for p' and q', we find that

$$z^* = \frac{92/150 - 110/200}{\sqrt{\dfrac{202}{350} \cdot \dfrac{148}{350}\left(\dfrac{1}{150} + \dfrac{1}{200}\right)}} = 1.19.$$

Since 1.19 does not lie in the rejection region, the data do not support Professor French's perception.

Summary

$H_0: p_F - p_G = 0.$

$H_A: p_F - p_G > 0.$

Test region: Reject H_0 when $z^* > 1.65$.

Test statistic:

$$z^* = \frac{92/150 - 110/200}{\sqrt{\dfrac{202}{350} \cdot \dfrac{148}{350}\left(\dfrac{1}{150} + \dfrac{1}{200}\right)}} = 1.19.$$

Conclusion: We cannot reject H_0. Professor French heard only about the onset of illness in the morning and thus thought that it predominated. □

Confidence Intervals

In order to find confidence intervals, we need to go back to examine the standard deviation

$$\sqrt{\frac{p_1 q_1}{n_1} + \frac{p_2 q_2}{n_2}}.$$

We utilized the null hypothesis to simplify this expression, but we cannot do that now: there are no hypotheses in estimating. We have only sample results and so must modify the standard deviation in order to use them:

$$\sqrt{\frac{\hat{p}_1 \hat{q}_1}{n_1} + \frac{\hat{p}_2 \hat{q}_2}{n_2}}.$$

The pattern for confidence intervals,

Sample results + Critical value · Appropriate standard deviation,

becomes

$$\boxed{\hat{p}_1 - \hat{p}_2 \pm z\sqrt{\frac{\hat{p}_1 \hat{q}_1}{n_1} + \frac{\hat{p}_2 \hat{q}_2}{n_2}}.}$$

Remember to subtract the smaller from the larger for $\hat{p}_1 - \hat{p}_2$.

EXAMPLE 10.7 With 95% confidence, estimate the difference in the percent of the population below the poverty line between a rural county and a large city.

	n	$\hat{p}$
Rural county	400	16%
City	1500	21%

Solution The 95% confidence level gives us the critical values in Fig. 10.6.

Figure 10.6

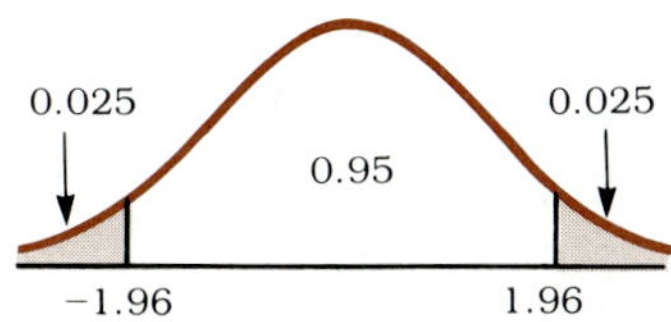

$$\begin{aligned}\hat{p}_1 - \hat{p}_2 &\pm z\sqrt{\frac{\hat{p}_1 \hat{q}_1}{n_1} + \frac{\hat{p}_2 \hat{q}_2}{n_2}} \\ &= 0.21 - 0.16 \pm 1.96\sqrt{\frac{0.21 \cdot 0.79}{1500} + \frac{0.16 \cdot 0.84}{400}} \\ &= 0.05 \pm 0.041.\end{aligned}$$

$$0.009 < p_C - p_R < 0.091$$

With 95% confidence, we estimate the difference in percent of the population between city and rural areas to be between 0.9% and 9.1%. □

EXERCISES/Section 10.2

1. A candidate for statewide political office wants to compare his appeal to urban voters with his appeal to suburban voters. A public opinion survey disclosed that, of 150 urban voters, 80 approved of his policies; of 100 suburban voters, 48 approved. Test $H_0: p_U - p_S = 0$ at $\alpha = 0.01$.

2. In order to compare the productivity of two plants that manufacture the same product, the parent corporation collected the following data for a one-week period. At the 0.05 level of significance, is there any difference between these two plants in the proportion of defective gadgets they produce? Let n = the number of gadgets made and X = the number of defectives produced.

	n	X
Plant A	200	32
Plant B	500	104

3. Almost 20% of the population dies from cancer. At the 0.01 level of significance, is there any difference in cancer death rates between men and women?

	n	Percent of Total Deaths
Men	1200	18.6
Women	1000	19.3

4. Should deodorant manufacturers aim more of their advertising at men or at women? Use $\alpha = 0.05$.

	n	Percent Using Deodorant Regularly
Men	200	82
Women	200	89

5. A group of 600 couples who are parents were asked to respond to the statement: Parents who can afford to pay for only one of their children to study should decide in favor of the girl, if she does better at school than the boy. Of the men, 348 agreed; of the women, 420 agreed. At the 0.01 level of significance, do more women agree with this statement than men?

6. Using the information in Exercise 5, find the 95% confidence interval estimate for the difference between the percent of men and women who agreed with the statement.

7. Find the 95% confidence interval estimate for the difference in deaths from heart disease and related diseases between men and women.

	n	Percent of Deaths from Heart Disease
Men	1500	51
Women	1500	56

8. Estimate, with 95% confidence, the difference between urban and rural voters in voter approval of a senatorial candidate's position on the economy.

	n	Percent Approval
Urban	550	62
Rural	300	39

9. Estimate the difference in percent of unemployment between white and black teenage youths, with 95% confidence.

	n	Percent Unemployed
White youth	600	18
Black youth	600	42

10. Find the 99% confidence interval estimate of the difference between the percent of germination of two varieties of bean seeds.

	n	Percent Germinating
Yellow Giant	100	92
Wonder Wax	110	73

10.3 Standard Deviation and Variance—One Population

Now is a good time to remind ourselves that:

1. Standard deviation $= \sqrt{\text{Variance}}$, or (Standard deviation)^2 = Variance.
2. Both the standard deviation and the variance measure the spread of the values of a random variable.

Standard deviation and variance are the last parameters we need to discuss. They are important because many people think only in generalities (averages) or portions of the whole population (percents). But it is also necessary many times for us to consider the spread of data about the mean. In manufacturing, for example, it is not sufficient to know that, on the average, everything turns out correctly. No mechanical process does everything exactly the same way every time. Each item produced varies a little from every other one; if these variations are minimal, no problems result. The amount of variation from the mean that can be tolerated is called, appropriately enough, the tolerance; if the tolerance is not met, parts will not fit together properly. A Japanese automaker advertised by showing a marble rolling down the gap where the hood and fender meet on one of its cars. The marble never waivers, demonstrating small tolerances and good quality control. It is quite easy to imagine that, if much variation were permitted along an assembly line, doors might not shut or windows might not close completely upon final assembly of a car. The average dimensions of a series of doors might indicate perfect fit, but the amount of individual variations is a measure of quality control.

This aspect of quality control can also be demonstrated in a grocery store. Choose several products that are contained in clear glass or plastic bottles. Do the contents of all the bottles come up to the same or approximately the same height? If not, be sure to buy one of the fullest bottles; the manufacturer needs to adjust the bottling equipment for consistency.

Standard deviation is important in the field of testing. For tests that require a high passing score to indicate competence, such as a nurses' licensing exam, a small variance is desirable. For broad, general tests given to measure knowledge or general aptitude, test designers do not want a small standard deviation. This would mean that scores were clustered, and it would be difficult to distinguish the truly knowledgeable or highly able from the average. Nor is a large variance desirable, because it would indicate a great deal of spread: the A-or-F, all-or-nothing syndrome. The CEEB (College Entrance Examination Board), which administers the SAT exam, aims for a standard deviation of 100 points on a test for which scores can range from 200 to 800. We will examine an example that involves a test, but first we need to learn the basics of hypothesis testing.

Hypothesis Tests

The null hypothesis may be expressed in terms of either σ or σ^2.

$$H_0\colon \sigma^2 = \underline{\qquad\qquad} \quad \text{or} \quad H_0\colon \sigma = \underline{\qquad\qquad}.$$

The alternative hypothesis matches H_0 except that a different symbol replaces the equals sign. Basically, we want to use the sample result s^2 and compare it with the σ^2 in H_0. This is done with a new test statistic, **chi square.**

$$\chi^2 = \frac{(n - 1)s^2}{\sigma^2}.$$

The χ is the Greek letter chi (pronounced kī). This statistic is a random variable; its value depends on the calculation of s^2 from a random sample of size n. As we did for $\bar{X}$, if we take many random samples, all of size n, from a normally distributed population with variance σ^2, we can generate a sampling distribution to give us probabilities for χ^2 (just as the normal distribution gave us probabilities for z). We get a family of curves, one curve for each n, just as we got a family of curves for the t distribution. We also use degrees of freedom when dealing with χ^2. If we examine the equation for χ^2, we can see some of the properties of these curves. Since n = sample size, $n - 1$ is always positive; s^2 and σ^2 are always positive because they are squares. Theoretically, it is possible, but not likely, for $s = 0$. Thus the χ^2 curve will always be

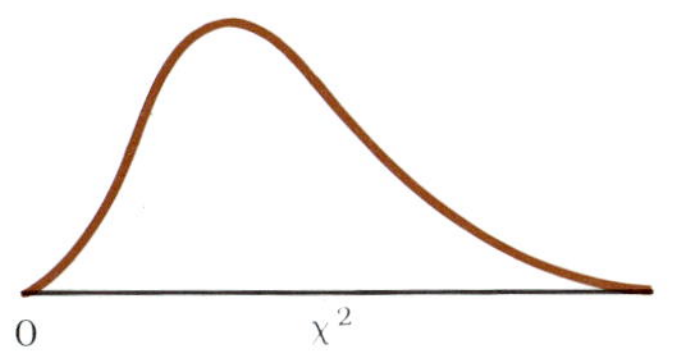

Figure 10.7

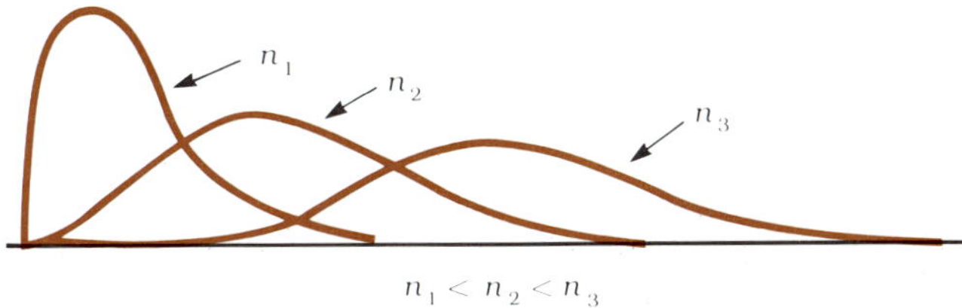

Figure 10.8

greater than or equal to zero; that is, it begins at 0. This is very different from both the standard normal distribution and Student's t distribution, both of which are centered about 0. In fact, the χ^2 distribution is not even symmetric, as shown in Fig. 10.7.

Again df = $n - 1$, and as n gets larger, the peak of the curve moves farther to the right of 0, as in Fig. 10.8. This distribution is said to be *skewed to the right.* Now let us summarize.

χ^2 distribution ■ If random samples of size n are taken from a normally distributed population with variance σ^2, then the random variable $\chi^2 = (n - 1)s^2/\sigma^2$ is used to study σ^2 or σ with probabilities from a χ^2 curve.

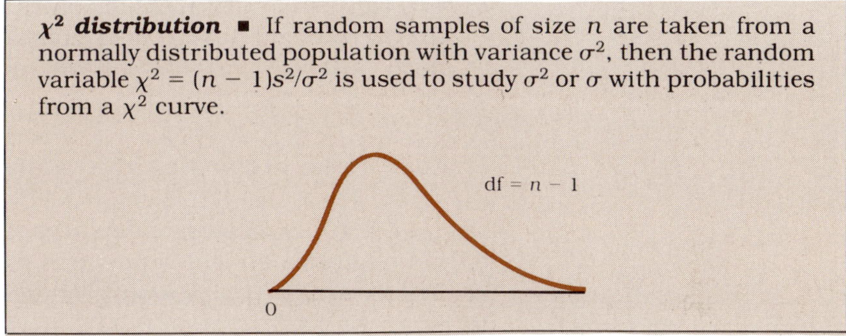

Note that, because this curve starts at 0, we can no longer just insert minus signs to obtain left-tail critical values. To obtain probabilities and critical values, we will use a new table, Table D.4 in Appendix D. As with the t table, df is listed down the side. The diagrams at the top show how the α row is utilized. They tell us that α is the area under the curve to the right of the χ^2 value; the body of the table gives us these values. We use the columns labeled 0.10, 0.05, 0.025, 0.01, and 0.005 as we did in Table D.3: Choose the proper column and the proper df row; read down and across to obtain a critical value for right tails. We use the first five columns in Table D.4 to obtain left-tail critical values after subtracting, $1 - \alpha$.

EXAMPLE 10.8 Find the χ^2 critical values that go with Fig. 10.9(a)–(c).

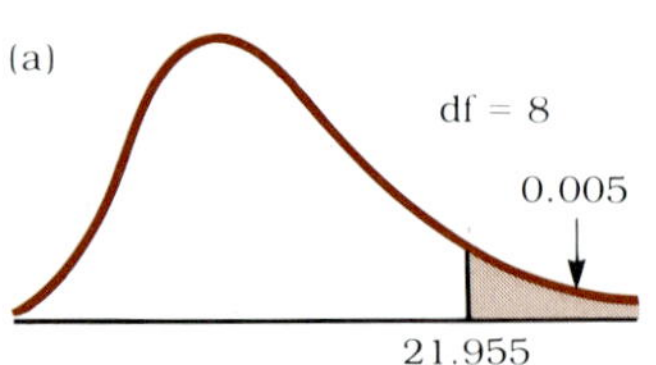

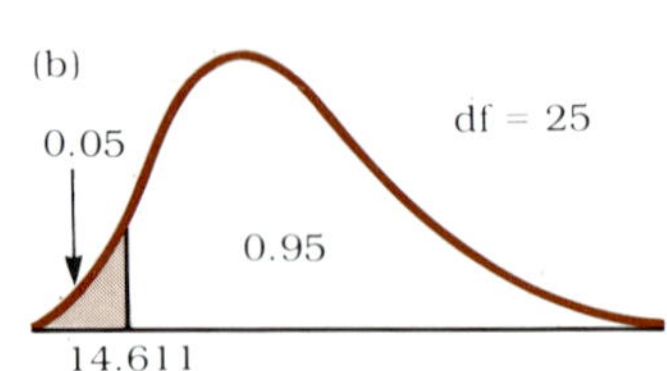

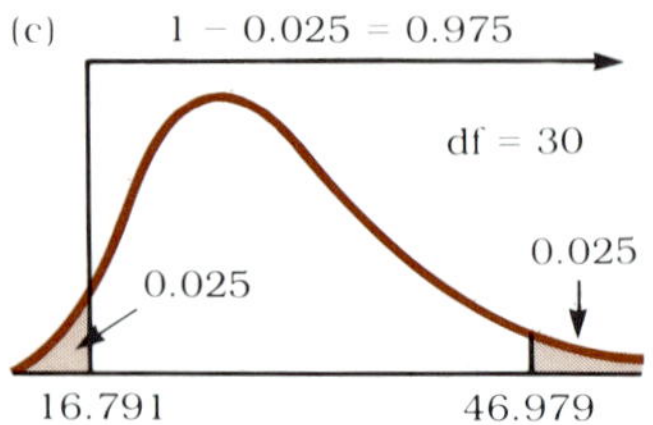

Figure 10.9

Solution

a) Since this diagram shows only a right-tail area, we go to the 0.005 column and the 8 row of Table D.4 to obtain 21.955.

b) We need to remind ourselves that in order to use the table, we must have an area to the *right* of the critical value. As with all probability distributions, the total area under the curve = 1, so we get 1 − 0.05 = 0.95 for the area to use in the table. We go to the 0.95 column and the 25 row to obtain 14.611.

c) The right-tail critical value comes from the 0.025 column and the 30 row and is 46.979. To obtain the left-tail value, we find the 1 − 0.025 = 0.975 column and row 30 value, or 16.791.

□

Note: There are no negative values.

EXAMPLE 10.9 Professor Spencer thinks that the standard deviation of grades on last semester's math final was no more than 12. The grades of 11 randomly chosen students had a standard deviation of 13.5. Use $\alpha = 0.01$ to determine whether the students' grades counter his perception.

Solution Professor Spencer is looking for a standard deviation of 12.

$$H_0\colon \sigma = 12.$$

We are asked to *counter his perception* of *no more than*, so

$$H_A\colon \sigma > 12.$$

Figure 10.10

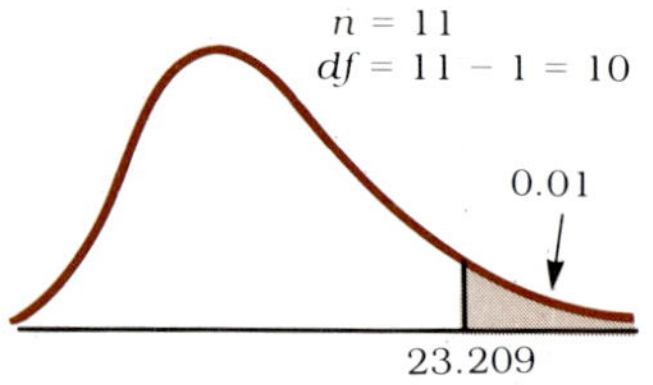

We find the test region using the 0.01 column and the 10 row for Fig. 10.10. We will reject H_0 if $\chi^{2*} > 23.209$.

$$\chi^{2*} = \frac{(n-1)s^2}{\sigma^2} = \frac{10 \cdot 13.5^2}{12^2} = 12.656.$$

The test statistic does not lie in the rejection region and thus the sample data support Professor Spencer's perception.

Note: As in Chapter 9, we must read carefully to determine whether the information given in the problem is squared (variance) or not (standard deviation) in order to know whether to square when calculating χ^{2*}.

Summary

H_0: $\sigma = 12$.

H_A: $\sigma > 12$.

Test region: Rejection H_0 when $\chi^{2*} > 23.209$.

Test statistic:

$$\chi^{2*} = \frac{10 \cdot 13.5^2}{12^2} = 12.656.$$

Conclusion: We cannot reject H_0. Professor Spencer's perception appears to be correct. □

Confidence Intervals

Confidence interval problems must be handled differently now. We can no longer form an interval by putting sample results in the middle and then going an equal distance in both directions from this value. Before, we had a symmetric distribution with which to work, but now we do not. Let us examine the χ^2 equation and see if we can transform it algebraically to help us.

$$\chi^2 = \frac{(n - 1)s^2}{\sigma^2}.$$

We are trying to estimate the parameter σ^2 (and later σ), so we get it out of the denominator by multiplying both sides by σ^2 and canceling:

$$\sigma^2\chi^2 = \frac{(n - 1)s^2\sigma^2}{\cancel{\sigma^2}}.$$

Dividing by χ^2 and canceling,

$$\frac{\sigma^2\cancel{\chi^2}}{\cancel{\chi^2}} = \frac{(n - 1)s^2}{\chi^2},$$

we get σ^2 to stand alone, or

$$\sigma^2 = \frac{(n - 1)s^2}{\chi^2}.$$

The fraction on the right-hand side of the equation will provide us with two values—an upper bound and a lower bound on

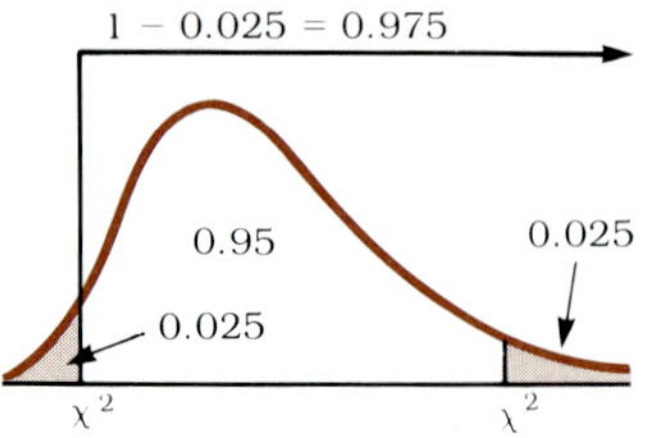

Figure 10.11

the confidence interval—when we find two values to substitute for χ^2. These two values are the critical values for the two tails. For a 95% confidence interval, we find in Fig. 10.11 the two critical values we need. Let us now look at an example.

EXAMPLE 10.10 A sample of 20 randomly chosen 42-ounce cans of tomato juice has a standard deviation of 1.2 ounces. Determine the 99% confidence interval estimate for the standard deviation of the weight of such cans of tomato juice.

Solution We first find the two χ^2 values we need for Fig. 10.12. Using the 19 row and the 0.995 column, we obtain χ^2 values of 6.844 and 38.582. We perform two calculations using $\sigma^2 = (n - 1)s^2/\chi^2$:

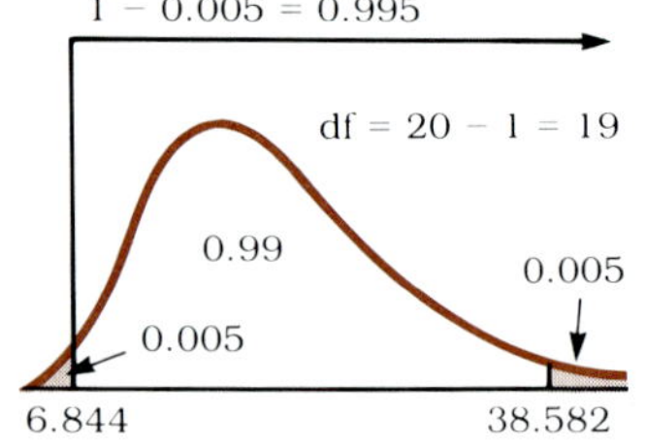

Figure 10.12

$$\frac{(20-1)\cdot 1.2^2}{6.844} = 3.998 \quad \text{and} \quad \frac{(20-1)\cdot 1.2^2}{38.582} = 0.709.$$

These are the upper and lower bounds of the interval:

$$0.709 < \sigma^2 < 3.998.$$

We were asked to estimate the *standard deviation*, so we finish the problem by taking the square root of the interval for σ^2 to obtain

$$0.842 < \sigma < 1.999.$$

The manufacturer can now examine this interval, compare it to acceptable tolerances, and decide whether adjustments need to be made. □

Let us summarize. Confidence intervals are found by using the confidence level as the area in the center of the distribution and cutting off two tails, each of area $\alpha/2$. Two χ^2 values are found in the χ^2 table and used in the denominator of $(n - 1)s/\chi^2$. Remember not to square the χ^2 values when substituting. The larger χ^2 value in the denominator produces the lower bound of the interval, and the smaller χ^2 value produces the upper bound.

To estimate σ^2:

$$\frac{(n-1)s^2}{\chi^2_{\alpha/2}} < \sigma^2 < \frac{(n-1)s^2}{\chi^2_{1-\alpha/2}}.$$

To estimate σ:

$$\sqrt{\frac{(n-1)s^2}{\chi^2_{\alpha/2}}} < \sigma < \sqrt{\frac{(n-1)s^2}{\chi^2_{1-\alpha/2}}}$$

EXERCISES/Section 10.3

1. Find the critical values for χ^2 using the diagrams in Fig. 10.13(a)–(e).
2. Find χ^2 values from Table D.4 using the diagrams in Fig. 10.14(a)–(e).
3. Based on past experience, a statistics teacher expects the average grade on a 125-point final exam to be 95 with a standard deviation of at least 12. This semester she discovered that, in one of her classes, $\bar{X} = 100$ and $s = 10$. Perform a hypothesis test at the 5% significance level to determine whether this class is more homogeneous (that is, has a significantly lower standard deviation) than all her other classes. Assume that $n = 31$.
4. A lumberman is willing to continue to use a certain saw so long as the boards cut by it are true to measure with a standard deviation of no more than 0.01″. A random sample of 100 boards disclosed $s = 0.015″$. Test the hypothesis that the standard deviation is no more than 0.01″, using $\alpha = 1\%$.
5. A battery manufacturer claims that the standard deviation of the life span of his batteries is no more than 2 hr. A random sample of 50 batteries taken from a recent production run yielded $s = 2.5$ hr. Test the manufacturer's hypothesis at the 0.5% level of significance.

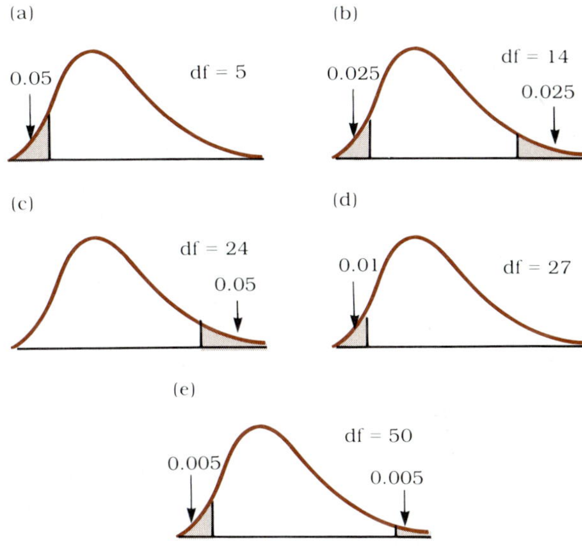

Figure 10.14

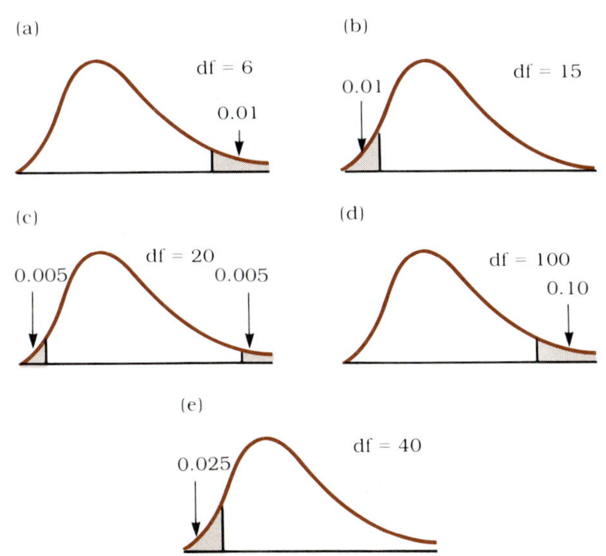

Figure 10.13

6. A fast-food restaurant chain uses a machine to form its hamburger patties. It is calibrated to produce quarter-pound patties with a standard deviation of no more than 0.1 oz. A sample of 25 patties recently formed by this machine yielded $s = 0.2$ oz. Is this evidence sufficient at the 1% level of significance to reject the claim that the standard deviation for this machine is no more than 0.1 oz.? Does the machine need adjustment?
7. The management at a certain airport claims that its work crews are so efficient that the standard deviation of the amount of time needed to refuel aircraft landing there is no more than 200 s. One of the airlines noted that for ten flights on a recent day, $s = 210$ s. Is this evidence sufficient to reject the airport management's claim? Let $\alpha = 0.05$.
8. A home gardener kept a diary to record the exact time required for the seeds of a certain variety of tomato to germinate. The standard

deviation for a packet of 25 seeds turned out to be 36 hr. Use this information to determine the 95% confidence interval estimate for the true value of σ.

9. A sample of ten sheets of typing paper yielded the fact that their widths had a standard deviation of 0.01″. Use this information to determine the 95% confidence interval estimate for the value of σ.

10. The standard deviation of the weights of women in a certain nursing class was 12 lb. If there are 23 women in this class, determine the 90% confidence interval estimate for the true value of σ.

11. The treasurer of a new time-sharing condominium has received 1000 applications for a new water-side resort. She selects 15 applicants at random and discovers that the average annual family income for this sample is \$42,500 with a standard deviation of \$5000. Use this information to determine the 99% confidence interval estimate for the value of σ.

12. The consumer affairs reporter for a local newspaper bought 5 rolls of a certain kind of film at five different stores. This sample had a mean price of \$2.59 and a standard deviation of \$0.12. Use this information to determine the 90% confidence interval estimate for σ.

10.4 Comparison of Variances—Two Populations

As we have stressed before, means are insufficient for describing data. Let us look at a case where they do not even provide an adequate comparison. Two consecutive winters several years ago are remembered very differently. One is remembered for its harsh January of low temperatures and heavy snowfall. The other is remembered as being rather mild with very few extremely cold periods. Thus it came as a surprise when the National Weather Service reported that the mean temperature for the winter months was almost the same in both years. What made these similar sets of mean winter temperatures seem so different? Winter number 1 had very severe January temperatures, preceded by a mild December and followed by blooming daffodils at the end of February; the extremes averaged out. However, winter number 2 had fairly constant temperatures throughout. The two winters differed in the *variability* of their temperatures, and their *variances* need to be compared.

Comparing variances requires the introduction of a new test statistic, the ***F* ratio.** Two population variances are compared by forming a ratio of their corresponding sample variances; H_O: $\sigma_1^2 = \sigma_2^2$ uses a test statistic of

$$F = \frac{s_1^2}{s_2^2}.$$

This F statistic is described by the F distribution, which is a probability distribution.

> ***F distribution*** ■ Given two normally distributed populations with variances σ_1^2 and σ_2^2, the random variable $F = s_1^2/s_2^2$ is described by the F distribution, and probabilities for F can be found by using areas under the F curve.

The F distribution is actually a family of curves (just as t and χ^2 are families of curves), dependent on degrees of freedom. Now there are two samples and two degrees of freedom: $n_1 - 1$ for the numerator; $n_2 - 1$ for the denominator. We write this as

$$\text{df} = (n_1 - 1, n_2 - 1)$$

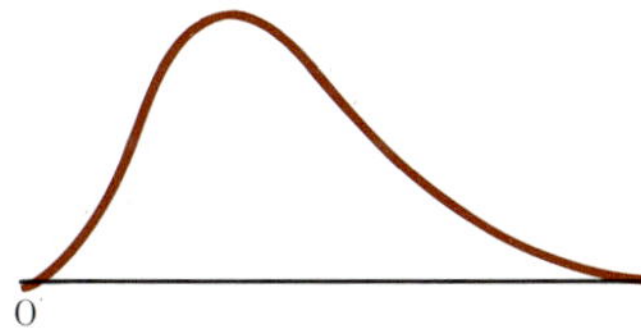

Figure 10.15

Since s_1 and s_2 can never be negative, the F curve will start at 0 and be skewed to the right (Fig. 10.15). As always with probability distributions, the total area under the curve is 1.

Before we examine the new table required to find critical values for F, let us examine H_0 and the simple calculation for F. The null hypothesis does not establish an order for σ_1^2 and σ_2^2. We wrote it as $\sigma_1^2 = \sigma_2^2$ but could have conveyed the same information with $\sigma_2^2 = \sigma_1^2$. Order is irrelevant here, as opposed to the way we dealt with the comparison of two means using z and t. Consequently, it makes no difference which sample variance goes in the numerator of F. If H_0 $(\sigma_1^2 = \sigma_2^2)$ is true, under ideal circumstances $F = s_1^2/s_2^2 = 1$; thus in order to perform a hypothesis test, we must see how far the F calculation deviates from 1 (Fig. 10.16).

Figure 10.16

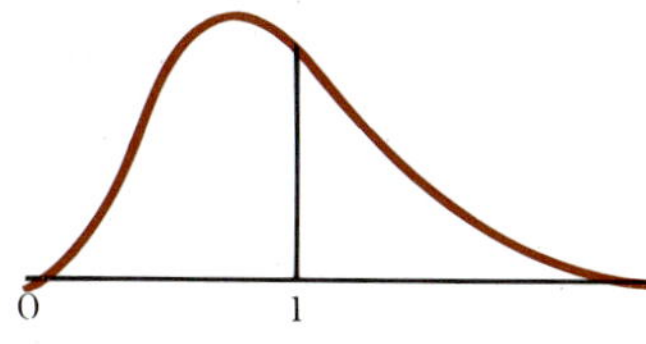

If we look at Table D.5, we can see from the diagram that it provides right-tail critical values and from the table itself that these values are all numbers larger than 1. Thus we can simplify our work by recognizing that use of the larger s as the numerator of F will always result in a value larger than 1. Doing this, we may not always get a value out in the right-tail rejection region, but we will always get a test statistic in that direction. If we decide to put the larger s^2 in the numerator, we may ignore left-tail F values; we only need to consider from H_A whether to split α before making it the area in the right tail.*

* If you want to find a left-tail critical value, you can do so by first finding the right-tail value and then taking its reciprocal:

$$F_L = \frac{1}{F_R}.$$

Now let us examine Table D.5 in detail. It consists of three pairs of pages, one pair for each of $\alpha = 0.01$, 0.025, and 0.05. We will use the notation $F_{0.01}$, for example, when we want to refer to that part of the table with 0.01 as the area in the right tail. Across the top we find the degrees of freedom for the numerator, and down the left side we find df for the denominator. We read the desired critical value where the appropriate column and row intersect.

EXAMPLE 10.11 Find the critical values for

a) $F_{0.01}$ with df = (9, 11). **b)** $F_{0.025}$ with df = (14, 19).
c) $F_{0.05}$ with df = (50, 30). **d)** $F_{0.05}$ with df = (50, 50).

Solution

a) We use the $F_{0.01}$ pages of Table D.5. We find 9 across the top and 11 down the side. Following the 9 column down until we reach the 11 row, we get 4.63 as the desired critical value.

b) We go to the $F_{0.025}$ pages. Looking for a 14 column, we find only 12 and 15. Therefore we choose the closer degrees of freedom, 15. The 19 row and the 15 column give us 2.62 as the critical value.

c) We look at the $F_{0.05}$ pages. We do not find a 50 column; 50 is equidistant between 40 and 60, so we must choose either 2.01 or 1.94 (on the 30 row) as the critical value.

	40	60
	·	·
	·	·
	·	·
30 · · ·	2.01	1.94

The more cautious approach is to choose the larger critical value—in this case, 2.01. Any test statistic that is greater than 2.01 will also be greater than 1.94 and thus result in a conclusion to reject H_0. However, the opposite is not true.

d) If df for the numerator and df for the denominator are equidistant from table values,

	40	60
	·	·
	·	·
	·	·
40 · · ·	1.88	1.80
60 · · ·	1.74	1.67

We again choose the largest critical value—in this case, 1.88—from among the possibilities. □

REMEMBER

When faced with df not on the table:

1. Choose the closer df.
2. If table degrees of freedom are equidistant from the sample df, choose the larger critical value.

Let us take another look at the winter temperature case presented at the beginning of this section.

EXAMPLE 10.12 Two consecutive winters produced the following data on mean temperature, in °F. Let n = the number of days used in calculations.

	$\bar{X}$	s^2	n
Winter 1	31.6	182.5	60
Winter 2	33.7	98.6	50

Is there greater variability in temperature for Winter 1 at the 0.05 level of significance?

Solution We assume that there is no difference in variability unless sample data can prove otherwise.

$$\text{H}_0\text{: } \sigma_1^2 = \sigma_2^2.$$

The question and earlier explanations lead us to favor Winter number 1 as having the greater variability in temperature, so

$$\text{H}_\text{A}\text{: } \sigma_1^2 > \sigma_2^2.$$

A one-tail test is indicated by H_A and $\alpha = 0.05$, so we use the $F_{0.05}$ pages of Table D.5 for Fig. 10.17. We cannot find the critical value until we set up the test statistic and decide which sample is to be used for the numerator and which for the denominator. Putting the larger variance in the numerator, we get

Figure 10.17

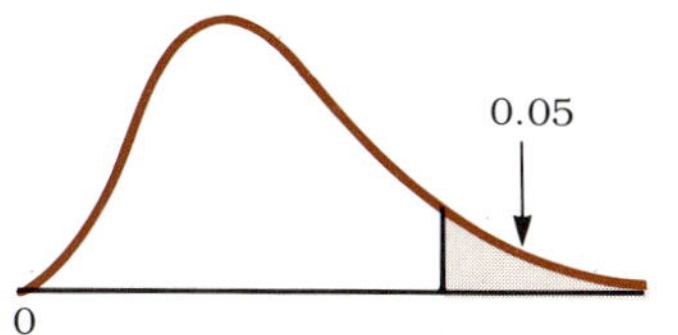

$$F^* = \frac{182.5}{98.6} = \frac{s_1^2}{s_2^2}.$$

Thus df for the numerator is $n_1 - 1$, and df for the denominator is $n_2 - 1$, or df = (59, 49). Note that the sample results were given as s^2 and we did not need to square. The degrees of freedom for the numerator are not on the table but are closer to 60 in the table and the degrees of freedom for the denominator are closer to 40; and thus the critical value is 1.64.

$$F^* = \frac{182.5}{98.6} = 1.85 > 1.64.$$

We can reject H_0.

Summary

$H_0: \sigma_1^2 = \sigma_2^2$.

$H_A: \sigma_1^2 > \sigma_2^2$.

Test region: Reject H_0 when $F^* > 1.64$.

Test statistic: $F^* = \dfrac{182.5}{98.6} = 1.85$.

Conclusion: Reject H_0. There was greater variability of temperature during the first winter. □

EXAMPLE 10.13. Two power saws are being compared to determine whether B is more reliable, that is, has a smaller variation from one piece of wood to the next. Forty-six pieces of lumber are chosen at random; 21 are run through saw A and 25 through saw B. The pieces worked on by A yielded a standard deviation of 9.3 cm and those worked on by B a standard deviation of 7.9 cm. Test the hypothesis that $\sigma_A = \sigma_B$ for $\alpha = 0.01$.

Solution We are told that $H_0: \sigma_A = \sigma_B$. We can test either variances or standard deviations. We decide to look for a smaller standard deviation for saw B, so $H_A: \sigma_A > \sigma_B$. Since $\alpha = 0.01$ we use the $F_{0.01}$ pages of Table D.5, set up F^*, and determine df.

$$F^* = \frac{s_A^2}{s_B^2} = \frac{9.3^2}{7.9^2} \qquad \begin{matrix} n_A - 1 = 20. \\ n_B - 1 = 24. \end{matrix}$$

With df = (20, 24), we get 2.33 as the critical value (Fig. 10.18).

$$F^* = \frac{9.3^2}{7.9^2} = 1.39 < 2.33.$$

We cannot reject H_0.

Figure 10.18

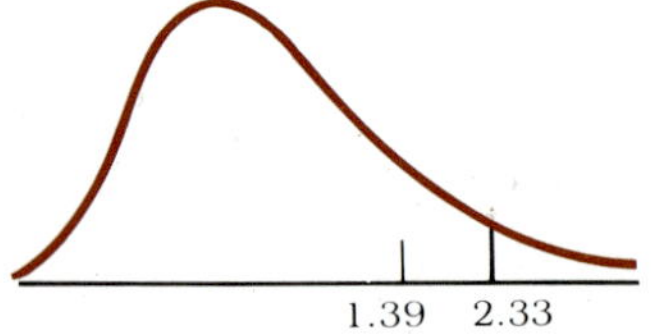

Summary

$H_0: \sigma_A = \sigma_B$.

$H_A: \sigma_A > \sigma_B$.

Test region: Reject H_0 when $F^* > 2.33$.

Test statistic:

$$F^* = \frac{9.3^2}{7.9^2} = 1.39.$$

Conclusion: We cannot reject H_0. Saw B does not have significantly less variance than saw A. □

In Section 9.2 we referred to the testing of means using small samples with *unequal* population variances and noted the Fisher-Behrens formula for degrees of freedom. Now we have the tool, the F test, to determine whether two variances are equal. However, using the F test with $H_A: \sigma_1^2 \neq \sigma_2^2$ requires normality in both population number 1 and population number 2. If there is any doubt about whether either or both are normal, this test should not be used; and large samples should be used with a z, rather than a t, test statistic to compare means. Because of the sensitivity to population normality in the F test, we will not consider its use for testing means with small samples.

EXERCISES/Section 10.4

1. Find the critical values using the following information.

a) $F_{0.01}$, df = (6, 7)
b) $F_{0.025}$, df = (20, 14)
c) $F_{0.05}$, df = (19, 25)
d) $F_{0.025}$, df = (24, 32)
e) $F_{0.05}$, df = (35, 35)

2. Find F values from Table D.5 using the following conditions.

a) $F_{0.01}$, df = (25, 25)
b) $F_{0.05}$, df = (34, 36)
c) $F_{0.025}$, df = (50, 90)
d) $F_{0.01}$, df = (70, 50)
e) $F_{0.025}$, df = (90, 90)

3. To examine seasonal variation in volume of stock traded, in millions of shares, the following data were gathered. Is there greater variation in the fall, at the 0.05 level of significance? Let n = the number of days for which the data were obtained.

	n	$\bar{X}$	s
Fall	25	45.0	5.2
Spring	30	38.2	3.9

4. Commercial growers need varieties of produce that have very little variation in maturation time in order to make harvesting efficient. The variety of corn planted last year had too many immature ears when harvested. This year's planting of a new variety provided the following information for comparison. Is the new variety more consistent in maturation time, at the 0.01 level of significance? Let n = the number of ears of corn tested and X = the maturation time in days.

	n	$\bar{X}$	s^2
Old variety	36	75	37.2
New variety	25	90	15.4

5. A manufacturer of handheld power tools advertises that his equipment is reliable and offers a warranty that is longer than most in the industry. His studies show that the mean life of his tools is beyond the length of the warranty. A new plant is opened and replacements under the warranty begin to increase. If the average life of the power tools has not changed, has the standard deviation, measuring the variability in lifetime, increased? If so, more equipment would be breaking down sooner (and also more would be lasting longer). At the 0.05 level of significance, is the new plant producing tools with more variability in length of time till breakdown? Let n = the number of tools tested and $\overline{X}$ = the mean lifetime in years.

	n	$\overline{X}$	s
Old plant	40	4.2	1.2
New plant	40	4.1	2.1

6. Is the variation of heights of 13-year-old boys greater than that of 13-year-old girls, at the 0.05 level of significance?

	n	s^2
Boys	10	12.96
Girls	15	5.76

7. Is there greater variation in the number of hours of TV viewing on Saturdays than on Sundays? Are Sunday viewers more faithful, at the 0.025 level of significance?

	n	s
Saturday	9	1.44
Sunday	8	0.61

8. Is there greater variation in the diameter, in mm, of wire produced on Mondays than on Wednesdays? Use $\alpha = 0.05$.

	n	s
Monday	10	7.28
Wednesday	15	4.45

Study Notes

KEY TERMS

χ^2 distribution ■ If random samples of size n are taken from a normally distributed population with variance σ^2, then the random variable $\chi^2 = (n - 1)s^2/\sigma^2$ is used to study σ^2 or σ with probabilities from a χ^2 curve.

F distribution ■ Given two normally distributed populations with variances σ_1^2 and σ_2^2, the random variable $F = s_1^2/s_2^2$ is described by the F distribution, and probabilities for F can be found by using areas under the F curve.

OUTLINE

I. Proportions—one population
   A. Use a normal distribution if $np > 5$ and $nq > 5$.
   B. $\mu_{\hat{p}} = p, \quad \sigma_{\hat{p}} = \sqrt{pq/n}$.
   C. Hypothesis testing:
      1. $z = \dfrac{\hat{p} - p}{\sqrt{pq/n}}$.
      2. Use p as given in H_0 whenever possible; $\hat{p}$ is used only once in calculating z.
   D. Confidence interval estimation:
      1. $\hat{p} \pm z\sqrt{\hat{p}\hat{q}/n}$.
      2. Use $\hat{p}$ because there is no p.
   E. Determination of sample size:
      1. $n = \dfrac{z^2pq}{e^2}$.
      2. If no information is available, let $p = q = 0.5$
II. Comparison of proportions—two populations
   A. Use a normal distribution.
   B. $\mu_{\hat{p}_1-\hat{p}_2} = p_1 - p_2$,

      $\sigma_{\hat{p}_1-\hat{p}_2} = \sqrt{(p_1q_1/n_1) + (p_2q_2/n_2)}$

C. Hypothesis testing:
   1. $H_0: p_1 - p_2 = 0$.
   2. $$z = \frac{\hat{p}_1 - \hat{p}_2}{\sqrt{p'q'\left(\frac{1}{n_1} + \frac{1}{n_2}\right)}},$$
      where p' is a pooled proportion and $q' = 1 - p'$.

D. Confidence interval estimation:
$$\hat{p}_1 - \hat{p}_2 \pm z\sqrt{(\hat{p}_1\hat{q}_1/n_1) + (\hat{p}_2\hat{q}_2/n_2)}.$$

III. Standard deviation and variance—one population
   A. Use the χ^2 distribution:
      1. There are no negatives.
      2. df $= n - 1$.
   B. Hypothesis testing:
      1. $$\chi^2 = \frac{(n-1)s^2}{\sigma^2}.$$
      2. Read sample results carefully; square s but not s^2.
   C. Confidence interval estimation:
      1. To estimate σ^2,
$$\frac{(n-1)s^2}{\chi^2_{\alpha/2}} < \sigma^2 < \frac{(n-1)s^2}{\chi^2_{1-\alpha/2}}.$$
      2. To estimate σ,
$$\sqrt{\frac{(n-1)s^2}{\chi^2_{\alpha/2}}} < \sigma < \frac{(n-1)s^2}{\chi^2_{1-\alpha/2}}.$$

IV. Comparison of variances—two populations
   A. Use the F distribution:
      1. $F = s_1^2/s_2^2$.
      2. df $= (n_1 - 1, n_2 - 1)$.
      3. Put the larger sample variance in the numerator.
      4. When faced with df not in the F table,
         a. Choose the closer df.
         b. If table degrees of freedom are equidistant from the sample df, choose the larger critical value.
   B. $H_0: \sigma_1^2 = \sigma_2^2$.

REVIEW PROBLEMS

1. Of 30 guests invited to a meet-the-candidates party, only 7 showed up. Estimate, with 90% confidence, the true proportion of those likely to attend such a gathering.

2. Of the 55 cars sold by a local auto dealer during a recent month, 10 were equipped with a stereo-tape player. Estimate, with 95% confidence, the true proportion of car buyers who choose to have their cars so equipped.

3. Of 50 homeowners in a certain real estate development, 12 subscribe to a lawn-care service. Estimate, with 99% confidence, the true proportion of homeowners who subscribe to such a service.

4. An organization of broadcasters claims that TV is the primary source of news for 85% of all Americans. A student in communications thinks that this figure is too high and surveys 150 people chosen at random. When asked to identify their primary source of news, 120 named television. Is this sufficient evidence to reject the broadcasters' claim, at the 5% level of significance?

5. "Imports Capture 30% of American Market," proclaimed a headline in a local paper recently. Thinking that this figure is too high, a reporter for another newspaper surveyed 200 randomly chosen new-car buyers. Of these, 48 had bought an import. Can the headline be rejected, at the 1% level of significance?

6. A publisher of diet books claims that, at any one time, about ⅓ of the population is on a diet of one kind or another. A disbelieving nutritionist surveys 50 randomly selected people. Of these, 22 are on a diet. Does this justify rejection of the publisher's claim, at the 10% level of significance?

7. The producer of an educational television show aimed at preschoolers claims that children who have been exposed to her program achieve more when they get to elementary school than children who do not watch her show. A first-grade teacher decided to test this claim by examining the test results of children who did watch the show and of children who did not watch the show. The result was that, of 15 children who had watched the show, 11

performed satisfactorily or better; of 18 children who had not watched the show, 12 performed at or above a satisfactory level. Does this evidence justify the producer's claim, at the 5% level of significance?

8. A commonly held belief is that walking is good for you. To test this a researcher examined the health records of 125 people, 48 of whom included a vigorous daily walk among their activities and the rest of whom did not. Of the walkers, only 9 had any history of heart trouble. Of the nonwalkers, 30 had a history of heart trouble. Using this criterion, does the evidence justify the commonly held belief, at the 1% level of significance?

9. An organization of American vintners claims that American wines are as good as European wines. To help establish this claim, they formed a panel of expert wine tasters. The panel was to take 25 sips of European wine and 15 sips of American wine. Fifteen of the sips of European wines were correctly identified as European, and 10 sips of the American wines were correctly identified as American. Is there any difference in the tasters' ability to identify correctly the wines, at the 10% level of significance?

10. With 95% confidence, estimate the difference in percent of the population registered Republican between two suburbs, one east of a certain large city and the other west of the same city.

	n	p
East suburb	300	15
West suburb	500	20

11. With 99% confidence, estimate the difference in percent of car buyers who buy American cars between those who are aged 40 or older and those who are younger than 40.

	n	p
40 or older	40	76
Under 40	50	32

12. With 90% confidence, estimate the difference between males and females in percent of students passing math at the freshman level.

	n	p
Male	120	80
Female	130	85

13. Due to shipping considerations, uniformity in size is important to an orchard owner. He claims that this year's peaches vary in diameter with a standard deviation of no more than 5 mm. The shipper selects a dozen peaches at random and calculates the standard deviation to be 6.6 mm. Is this sufficient evidence to reject the grower's claim? Use $\alpha = 5\%$.

14. An office manager believes that morale in his area can be maintained if the employees know that they are all making about the same salary for doing comparable work. He wants the standard deviation of gross annual salary to be no more than \$200. He chooses the names of five employees at random and learns from the payroll office that the standard deviation for this sample is \$250. Is this statistically significant at the 1% level?

15. One test of mechanical ability is the amount of time it takes to assemble a child's bicycle. A certain manufacturer thinks his brand can be put together in 30 min with a standard deviation of no more than 5 min. He learns that, in a random sample of 25 mail-in-information postcards included with each bike, $s = 9$ min. Is this statistically significant at the ½% level?

16. A landing party of three extra-terrestrials agrees to take a human IQ test. The trio scores an average of 193 out of a 200 maximum with $s = 2$. Use this information to determine the 95% confidence interval estimate for σ.

17. A random sample of ten written driving tests had $s = 3.5$. Use this sample information to find the 90% confidence interval estimate for σ.

18. A random sample of 18 light bulbs manufactured by one company had a standard devia-

tion of life span of 12 hr. Use this sample information to find the 99% confidence interval estimate for σ.

19. A sample of 20 randomly chosen test results on a standardized achievement test had a standard deviation of 35. Use this information to determine the 90% confidence interval estimate for σ.

20. It was a popular belief in a certain town that automobiles which came off the local plant's production line on Mondays should be avoided because of greater variability in the number of defects per car. A consumer-rights organization accumulated the following data. Does this data indicate a statistically significant difference at the 1% level?

	n	s
Monday cars	15	5.3
Non-Monday cars	10	2.8

21. Reliability in germination is an important characteristic in grass seed. Of two otherwise equivalent varieties of seed, the one with less variability is preferable. Fifty-one seeds of each of two types of grass are tested, and the results are summarized in the following table. Are these results statistically different at the 5% level of significance?

	n	s
Type A seed	51	5
Type B seed	51	3

22. A bottler is considering the purchase of a new bottling machine. If the new machine does a significantly more uniform job than the old one, he will buy it. The following data were gathered to compare the performance of the two machines. Are these results statistically different at the 10% level of significance? Let n = the number of bottles tested; s is in oz.

	n	s
Old machine	21	0.50
New machine	15	0.35

Analysis of variance (ANOVA)

11

11.11 Testing for Equality—Three or More Means

In Chapter 10 we discussed the testing of the means of two populations. Now we proceed to the testing of the equality of three or more means. We will assume that all samples come from normal populations, that they are all independent and of size n, and that they have a common variance σ^2. We could test the means in pairs using techniques already learned, but that method would be incorrect. The procedure for testing whether the means of k different populations are equal requires an examination of the common variance.

$$H_0: \mu_1 = \mu_2 = \mu_3 = \ldots = \mu_k,$$

with any differences in the corresponding sample means due to chance, as opposed to

$$H_A: \text{The means are not all equal.}$$

This will be tested by looking at measures of variance. We will estimate the variance in two different ways. If the two estimates are relatively close to each other, we will say that H_0 has credibility. If these estimates are not close to each other, we will reject H_0.

At first glance, it may not be obvious that testing variances will help us analyze means, but that turns out to be exactly what is required. Let us assume a null hypothesis that states $\mu_1 = \mu_2 = \mu_3 = \mu_4$, with all differences in sample means due to chance alone. We obtain two estimates of the common variance by utilizing two ideas we used before: the central limit theorem and pooling sample variances to estimate a common population variance (as we did for the two-sample t test). From the central limit theorem,

$$\sigma_{\bar{X}} = \frac{\sigma}{\sqrt{n}} \quad \text{or} \quad \sigma_{\bar{X}}^2 = \frac{\sigma^2}{n}.$$

Thus $\sigma^2 = n\sigma_{\bar{X}}^2$. We will find $s_{\bar{X}}^2$, the variance of the sample means, to estimate $\sigma_{\bar{X}}^2$, by calculating the variance *between* sample means. We obtain this estimate by applying the equation for calculating the variance (from Chapter 4) to the sample means themselves:

$$s_{\bar{X}}^2 = \frac{\Sigma(\bar{X} - \bar{\bar{X}})^2}{k - 1} \quad \text{and} \quad \bar{\bar{X}} = \frac{\Sigma\bar{X}}{k},$$

where $\bar{X}$ is a sample mean and $\bar{\bar{X}}$ is the mean of the sample means. Thus the first estimate for σ^2 is

$$ns_{\bar{X}}^2 = \frac{n\Sigma(\bar{X} - \bar{\bar{X}})^2}{k - 1},$$

when sample sizes are equal.

Our second estimate for σ^2 comes from calculating the mean variance *within* the individual samples. When we have common sample sizes, we pool the individual variances s_1^2, s_2^2, s_3^2, s_4^2 to obtain s_i and divide by the number of variances:

$$\frac{\Sigma s_i^2}{4}.$$

In general, the pooled estimate for σ^2 for k samples is

$$\frac{\Sigma s_i^2}{k} = s_p^2.$$

We compare these two estimates for σ^2 using an F test.

$$F^* = \frac{\text{Variance between means}}{\text{Variance within samples}} = \frac{ns_{\bar{X}}^2}{s_p^2}.$$

What will this F comparison tell us about the means in H_0? If all sample means are close together, the variance between them, $s_{\bar{X}}^2$, will be small. If the sample means vary greatly from one another, then $s_{\bar{X}}^2$ will be large and thus F^* will be large; as before, when F^* is sufficiently large, we reject H_0 in favor of H_A. The denominator, which tells us about the average variance within the samples, is not based on any comparison of sample means. It depends only on sample variances and thus is not affected by differences in sample means. Any calculation that produces a large F^* must do so in its numerator, which uses the $\bar{X}_i$'s:

Large differences among the $\bar{X}_i$'s $\rightarrow$ large $s_{\bar{X}}^2$ $\rightarrow$
large F^* $\rightarrow$ rejection of H_0.

As before, the F test involves only right-tail critical values, an F larger than 1, and rejection of H_0 for large values of F^*. Let us summarize our assumptions before we do some calculations.

1. The populations are assumed to be normal with equal variances.
2. All samples are independent random samples of the same size, n.
3. Data are classified on the basis of one variable.
4. There are k samples from k populations. Thus we have $\bar{X}_1, \bar{X}_2, \bar{X}_3, \ldots, \bar{X}_k$ and $s_1^2, s_2^2, s_3^2, \ldots, s_k^2$.
5. H_0 assumes that all differences among sample means are due to chance and that all means from the populations involved are actually equal.
6. H_A claims that all population means are not equal.
7. The hypothesis test uses the F statistic—comparing the variance between samples to the variance within sam-

ples—to estimate the common population variance:

$$F^* = \frac{ns_{\bar{X}}^2}{s_p^2}.$$

8. The F test uses only right-tail critical regions.

A comment needs to be made about these assumptions. The analysis-of-variance procedure is still effective if the populations are not all precisely normal, so long as they are not too far removed from normality. If samples are all equal in size, the procedure works reasonably well even if the population variances are not all equal. This procedure is not sensitive to minor deviations from the normality and equal-variance assumptions.

We move to a more precise level by employing the equations we discussed and following through to complete the hypothesis test. Let us consider the following data and examine its display in Fig. 11.1 in order to demonstrate the required calculations and to test $H_0: \mu_1 = \mu_2 = \mu_3 = \mu_4$.

Figure 11.1

Sample 1: 7, 8, 14	$\bar{X}_1 = 8.33$	$s_1^2 = 14.33$
Sample 2: 13, 14, 15	$\bar{X}_2 = 14$	$s_2^2 = 1$
Sample 3: 6, 7, 8	$\bar{X}_3 = 7$	$s_3^2 = 1$
Sample 4: 8, 9, 11	$\bar{X}_4 = 9.33$	$s_4^2 = 2.33$

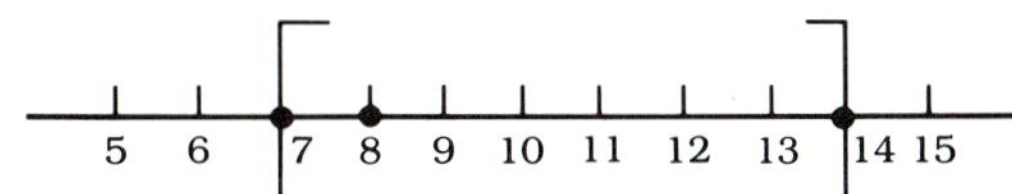

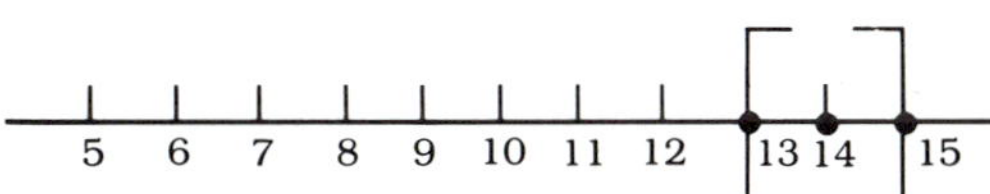

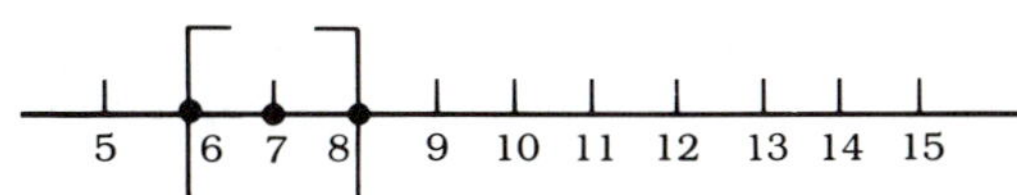

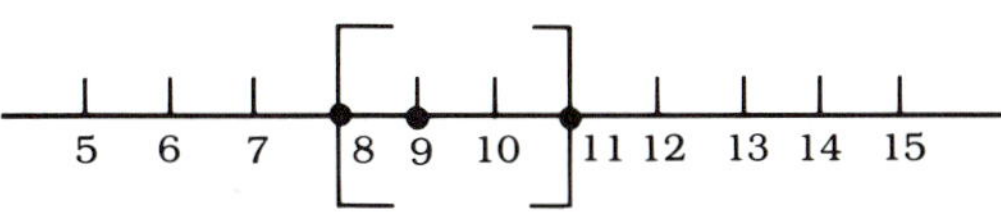

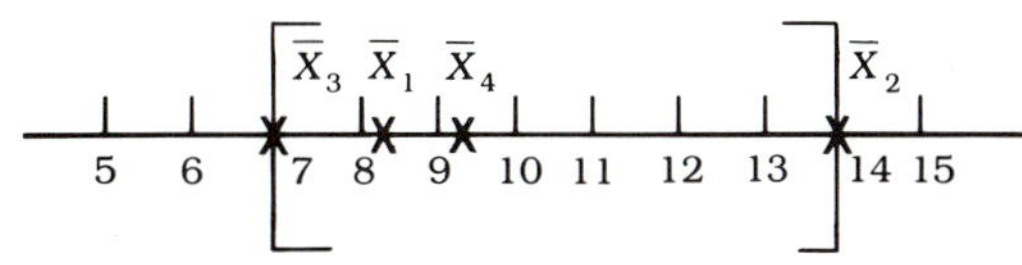

We need to find the variance among means. The mean of the sample means is

$$\bar{\bar{X}} = \frac{\Sigma \bar{X}}{k}.$$

Substituting, we get

$$\bar{\bar{X}} = \frac{8.33 + 14 + 7 + 9.33}{4} = 9.67.$$

We must calculate $s_{\bar{X}}^2$:

$$s_{\bar{X}}^2 = \frac{\Sigma(\bar{X} - \bar{\bar{X}})^2}{k - 1}$$

$$= \frac{(8.33 - 9.67)^2 + (14 - 9.67)^2 + (7 - 9.67)^2 + (9.33 - 9.67)^2}{4 - 1}$$

$$= 9.25.$$

Thus

$$ns_{\bar{X}}^2 = 3 \cdot 9.25 = 27.75,$$

$$s_p^2 = \frac{14.33 + 1 + 1 + 2.33}{4} = 4.67,$$

and, finally,

$$F^* = \frac{3 \cdot 9.25}{4.67} = 5.94.$$

We have not discussed finding the number of degrees or freedom associated with the numerator and the denominator of F^* and must now do so in order to find the critical value. Let us assume an α of 0.05, which is placed in the right-tail critical region. The *numerator* of F^* comes from the variance among k samples:

$$\text{df} = k - 1.$$

The *denominator* comes from the mean of the individual standard deviations and its number of degrees of freedom is the sum of the degrees of freedom of the samples:

$$\text{df} = k(n - 1).$$

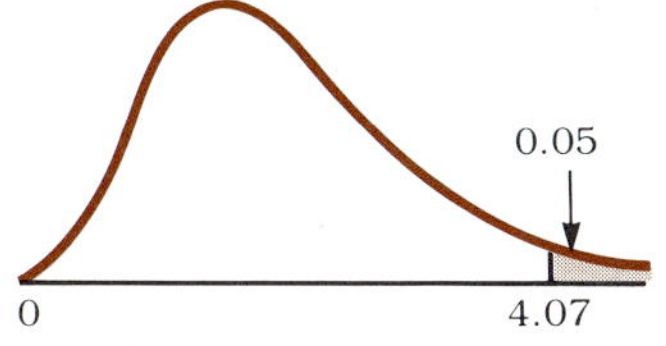

Figure 11.2

Thus the degrees of freedom are

$$df_n = k - 1 = 4 - 1 = 3;$$

$$df_d = k(n - 1) = 4(3 - 1) = 8.$$

Using the $F_{0.05}$ pages in Table D.5, we obtain $F = 4.07$ (Fig. 11.2). Our calculated $F^* = 5.94 > 4.07$; thus we reject H_0 and say that all the means are not equal.

Summary

H_0: $\mu_1 = \mu_2 = \mu_3 = \mu_4$.

H_A: The means are not all equal.

Test region: Reject H_0 when $F^* > 4.07$.

Test statistic: $F^* = \dfrac{3 \cdot 9.25}{4.67} = 5.94.$

Conclusion: Reject H_0.

The variation between sample means in the numerator of F^*, which measured the spread of the population means, was sufficiently large to cause us to conclude that at least one mean was significantly different from the rest. If the sample means had been close together, the measure of their spread would have been small, F^* would have been close to 0, and F^* would not have been near the right-tail rejection region. A large variation between means is necessary for us to conclude that they are not all equal.

In analyzing different estimates of a common variance, we used a procedure called **analysis of variance** or **ANOVA.** It is used extensively to test for the existence of differences between population means and for differences associated with other kinds of treatments or factors in experimental situations. We can easily translate the terminology used so far into the terminology required for testing the effects of various types of procedures. *Variation between means* becomes *variation between different treatments* (or levels of treatment) or *variation between factors* being tested. Variation within samples is attributed to random-chance error, and it is assumed that these errors cancel each other if they are truly the result of random chance. The word **replicate** is used to indicate a single piece of data within one level of treatment, so that each treatment group consists of n replicates. These treatment groups (samples) are formed by producing the replicates in random order as independent observations. We will look at some examples in the next section as we further examine ANOVA calculations.

EXERCISES/Section 11.1

1. Higher education is increasingly available to people of all ages through expanded services and offerings at local public colleges. A dean of students decided to determine whether age is a factor that influences success in college. He gathered the following data on 15 randomly selected students, 5 each from three different ages. At the 0.05 level of significance, are the means for the three ages all equal?

Age	Grade-Point Average				
18	2.01	2.87	2.75	3.06	2.94
27	2.87	2.99	3.08	3.15	3.24
36	2.98	2.87	3.64	3.50	3.41

2. In an attempt to justify budget requests for more officers in certain precincts, a police commissioner collected the following data on the incidence of crime in five of the city's precincts. He chose six randomly selected weeks of the previous year and recorded the crime rate for selected categories of crimes for each precinct. The weeks to be examined were chosen independently for each precinct. At the 0.05 level of significance, are the means for the different precincts all equal?

Precinct	Crime Rate					
1	73.2	68.1	52.3	65.3	70.6	61.2
2	42.1	48.9	51.3	38.6	49.1	44.6
3	49.8	52.3	64.8	61.3	68.9	59.1
4	60.2	63.1	67.4	69.5	61.7	57.8
5	72.3	79.4	62.6	73.4	75.6	71.2

3. A supervisor decided to examine the quality of production of workers on various days of the week to see whether any differences were related to specific days. Each Monday, Wednesday, and Friday he randomly selected four employees and calculated net production (gross production − defectives) of items over a two-month period and gathered the following data. At the 0.05 level of significance, do all these days have equal production rates?

Day	Mean Net Production			
M	72	84	68	78
W	83	85	90	86
F	74	79	80	71

4. Is there any difference among income levels of these four groups? Use $\alpha = 0.05$. Income of sample members is in thousands of dollars.

Group	Income of Sample Members					
A	11.2	16.3	15.7	18.2	16.7	19.2
B	20.1	19.7	18.2	19.6	17.2	23.6
C	17.6	18.2	16.3	15.4	20.7	18.2
D	11.4	18.7	21.6	27.8	24.5	10.1

11.2 The ANOVA Table: Form and Calculations

In this section we introduce the equations that are commonly used for calculations in ANOVA and then set up the ANOVA table, which allows for quick calculations of F^*. The methods used in Section 11.1 are limited to the examination of one variable. In this section we develop and use equations that allow for the analy-

NEWS & VIEWS 11.1

Doctor withdraws caffeine charge

A Japanese pathologist who earlier said caffeine can cause cancer reversed his finding yesterday.

Dr. Shozo Takayama said he erred during his 1974–76 experiments with rats fed various amounts of caffeine in water.

He said he has repeated the experiment and found no significant difference in the cancer rates among three groups of rats fed water and varying amounts of caffeine.

He said he found that a pneumonia epidemic among the rats altered the outcome of the original experiment.

Source: *Baltimore Sun*, January 19, 1982.

sis of other types, as well. News & Views 11.1 gives an example of research requiring an ANOVA approach.

The procedure requires an examination of the variance of all the pieces of data. This analysis presupposes that variance within any level of treatment or sample is basically the same for each treatment or sample, but that the variation between samples is large if the means for the samples are different.

Recall that variance for the entire data set can be calculated as

$$s^2 = \frac{n\Sigma X^2 - (\Sigma X)^2}{n(n-1)} = \frac{\Sigma(X - \bar{\bar{X}})^2}{n-1},$$

where $\bar{\bar{X}}$ is the mean of all the data combined, or $\bar{\bar{X}} = \Sigma X/N$, and N is the total number of pieces of data, or $N = nk$. If we multiply both fractions by $n - 1$ and cancel,

$$\cancel{(n-1)}\left[\frac{n\Sigma X^2 - (\Sigma X)^2}{n\cancel{(n-1)}}\right] = \frac{\Sigma(X - \bar{\bar{X}})^2}{\cancel{(n-1)}}\cancel{(n-1)},$$

we get

$$\Sigma X^2 - \frac{(\Sigma X)^2}{n} = \Sigma(X - \bar{\bar{X}})^2.$$

The quantity $\Sigma(X - \bar{\bar{X}})^2$ is called the *sum of squares* for the total set of data, or SST. It is calculated by using all data values and the overall mean $\bar{\bar{X}}$, or it can be calculated without first determining $\bar{\bar{X}}$ by using

$$\Sigma X^2 - \frac{(\Sigma X)^2}{N}.$$

We can think of the total sum of squares as consisting of two components: the variation due to the level or group, measured by the *sum of squares between groups*, SSB; and the variation due to experimental chance error caused by taking different replicates, as measured by the *sum of squares within groups*, SSW.

$$\text{SST} = \text{SSB} + \text{SSW}.$$

The *mean square*, MS, provides a measure of the average variation attributable to each of the sources (between and within). All of this can be summarized in an ANOVA table:

Source of Variation	df	SS	MS
Between groups	$k - 1$		
Within groups	$k(n - 1)$		
Total	$kn - 1$		

The number of degrees of freedom associated with the variation between groups is one less than the number of groups, or $k - 1$. The number of degrees of freedom associated with variation within groups is the sum of the degrees of freedom $n - 1$ for each of k groups, or $k(n - 1)$. The total degrees of freedom is the sum of the previous two, or

$$k - 1 + k(n - 1) = k - 1 + kn - k = kn - 1.$$

We find SSB by first calculating the sum of the data values for each group or sample. We will call these sums T for sample totals. Then

$$\text{SSB} = \frac{\Sigma T^2}{n} - \frac{(\Sigma T)^2}{N},$$

where n = sample size, and

N = total number of data values from all samples.

We calculate SSW by using the individual data values X and the totals:

$$\text{SSW} = \Sigma X^2 - \frac{\Sigma T^2}{n}.$$

We are now ready to calculate values for the last column of the ANOVA table, labeled MS. We calculate MS by dividing the SS value by the corresponding degrees of freedom:

$$\boxed{\text{MS} = \frac{\text{SS}}{\text{df}}.}$$

After we fill in the ANOVA table, the required hypothesis test is easy, as

$$F^* = \frac{\text{MSB}}{\text{MSW}}, \qquad \text{df} = [k - 1, k(n - 1)].$$

Let us return to the discussion of the data shown in Fig. 11.1 and rework the problem by setting up an ANOVA table. The original data were

Sample 1: 7, 8, 14.

Sample 2: 13, 14, 15.

Sample 3: 6, 7, 8.

Sample 4: 8, 9, 11.

To find the two sums of squares, we need the totals for the samples.

Sample	T	T^2
1	29	841
2	42	1764
3	21	441
4	28	784
	120	3830

$$\begin{aligned} \text{SSB} &= \frac{\Sigma T^2}{n} - \frac{(\Sigma T)^2}{N} \\ &= \frac{3830}{3} - \frac{(120)^2}{12} = 76.67. \end{aligned}$$

$$\begin{aligned} \text{SSW} &= \Sigma X^2 - \frac{\Sigma T^2}{n} \\ &= 1314 - \frac{3830}{3} = 37.33. \end{aligned}$$

Now we can fill in the ANOVA table:

Source of Variation	df	SS	MS
Between	$4 - 1 = 3$	76.67	$\frac{76.67}{3} = 27.11$
Within	$4 \cdot 2 = 8$	37.33	$\frac{37.33}{8} = 4.67$
Total	11	114.00	

It is always a good idea to use the equation for SST to check your arithmetic for SSB + SSW = SST.

$$SST = \Sigma X^2 - \frac{(\Sigma X)^2}{N} \quad \text{or} \quad \Sigma X^2 - \frac{(\Sigma T)^2}{N}.$$

$$SST = 1314 - \frac{(120)^2}{12} = 114.$$

We can easily calculate F^* from the right-hand column:

$$F^* = \frac{MSB}{MSW} = \frac{27.11}{4.67} = 5.81^*$$

Let us consider an example.

EXAMPLE 11.1 A professor of agriculture is trying to determine which of four fertilizers will produce the best yield for tomatoes. He grew the same variety of tomatoes on 16 different, randomly selected plots. The following table gives the yield, in bushels, for each plot. At the 0.05 level of significance, are the mean yields for the four fertilizers the same?

Fertilizer	Yield			
1	10.0	12.3	14.0	11.9
2	7.6	8.9	11.2	10.6
3	7.2	10.1	14.6	12.3
4	12.2	14.3	17.6	13.9

* This value does not agree exactly with the F^* calculated in Section 11.1 because of rounding.

Solution The null hypothesis states that the mean yields for the fertilizers are equal, or

$$H_0: \mu_1 = \mu_2 = \mu_3 = \mu_4.$$

The alternative hypothesis contradicts this.

$$H_A: \text{The means are not all equal.}$$

We set up the ANOVA table to determine the degrees of freedom in order to find the critical value and then to calculate F^*. The totals for each fertilizer are:

$$\begin{aligned} T_1 &= 10 + 12.3 + 14 + 11.9 &= 48.2 \\ T_2 &= 7.6 + 8.9 + 11.2 + 10.6 &= 38.3 \\ T_3 &= 7.2 + 10.1 + 14.6 + 12.3 &= 44.2 \\ T_4 &= 12.2 + 14.3 + 17.6 + 13.9 &= \underline{58.0} \\ & & 188.7 \end{aligned}$$

$$\text{SSB} = \frac{\Sigma T^2}{n} - \frac{(\Sigma T)^2}{N} = \frac{9107.77}{4} - \frac{(188.7)^2}{16} = 51.46.$$

$$\text{SSW} = \Sigma X^2 - \frac{\Sigma T^2}{n} = 2338.27 - \frac{9107.77}{4} = 61.33.$$

Source of Variation	df	SS	MS
Between	$4 - 1 = 3$	51.46	$\frac{51.46}{3} = 17.15$
Within	$4(4 - 1) = 12$	61.33	$\frac{61.33}{12} = 5.11$
Total	15	112.79	

Figure 11.3

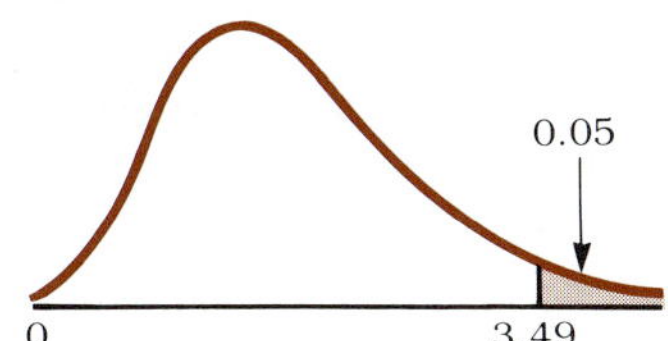

Test region: The degrees of freedom are (3, 12); we use the $F_{0.05}$ pages in Table D.5 to get $F = 3.49$ (Fig. 11.3). Thus we will reject H_0 when $F^* > 3.49$.

$$F^* = \frac{17.15}{5.11} = 3.36.$$

We cannot reject H_0. □

We have been comparing internal group fluctuations (random error) to variations between groups. If MSB is significantly larger than MSW, the means for the samples are different and the *factor being tested* makes a significant difference. Further analytical methods help us to decide which means are different and thus which treatment or procedure is most effective. If MSB is not significantly larger than MSW, we conclude that the differences are due to random error (chance alone), and we cannot reject H_0.

We have not discussed methods for handling samples of different size. The procedures are basically the same as those we have used; the only variations occur in the equations for SSB and SSW. Thus

$$\text{SSB} = \left[\frac{T_1^2}{n_1} + \frac{T_2^2}{n_2} + \cdots + \frac{T_k^2}{n_k}\right] - \frac{(\Sigma T)^2}{N}.$$

For SSB we square the totals for each sample and divide by sample size and then subtract the square of the sum of all values divided by the total number of values,

$$\text{SSW} = (\Sigma X)^2 - \left[\frac{T_1^2}{n_1} + \cdots + \frac{T_k^2}{n_k}\right].$$

and, as before,

$$\text{SST} = \Sigma X^2 - \frac{(\Sigma T)^2}{N}.$$

This process can also be extended to situations involving more than one factor or treatment. If some data are lost, destroyed, or altered (thus producing samples of different sizes) violation of the normality assumption or of the equal variance assumption is more serious but can be handled by more extensive analysis beyond the scope of this book.

EXERCISES/Section 11.2

1. An advertising agency is trying to determine the effectiveness of a series of ad campaigns. It runs the campaigns for four randomly selected weeks in three different media and determines through interviews that the following revenues, in thousands of dollars, are attributable to each source. Is there any significant difference among the media, at the 0.01 level?

Medium	Revenue			
Yellow Pages	1.4	2.1	3.6	1.2
Television	4.6	4.9	5.9	5.2
Newspaper	3.9	4.8	5.1	5.0

2. A drug company conducted a clinical trial to determine the effectiveness of a new family of drugs. Two drugs and a placebo were administered to three different treatment groups at five randomly selected locations. The following mean reaction times, in minutes, were noted. Is there any significant difference among treatment groups, at the 0.01 level?

Treatment	Mean Reaction Time				
Drug 1	21.3	25.0	24.2	27.9	31.8
Drug 2	19.2	21.5	20.6	24.7	22.6
Placebo	27.9	28.1	26.3	34.7	30.1

3. The life of batteries is the subject of much advertising. To determine the relative usefulness of different brands of batteries, a researcher inserted four types of batteries into five flashlights each and turned them on, recording the length of time, in hours, before the lights failed. Are the mean lives the same for all types of batteries tested, at the 0.05 level of significance?

Battery	Life				
Carbon					
Brand A	24.2	31.7	28.9	26.7	29.1
Brand B	31.7	33.6	32.8	37.1	34.5
Alkaline					
Brand C	41.3	45.6	61.2	71.3	52.7
Brand D	31.9	38.2	46.2	48.1	50.0

4. The gasoline mileage for three cars from each of four different compact models were compared. Are the mean miles per gallon for the compact equal, at the 0.05 level of significance?

Compact	Mileage		
A	24.6	31.2	30.6
B	25.9	34.6	38.7
C	28.2	39.2	41.2
D	20.4	29.1	28.7

Study Notes

KEY TERMS

ANOVA table ■ The table that summarizes the calculations involved in an analysis of variance procedure.

mean square

$$\text{MS} = \frac{\text{SS}}{\text{df}}$$

A measure of the average variation attributable to each of the sources *between* and *within*.

replicate ■ A piece of data within one level of treatment.

sum of squares ■

SSB—a measure of the variation in data due to the level of treatment or group.

SSW—a measure of the variation due to experimental chance error.

SST—a measure of the total variation in data and equal to SSB + SSW.

OUTLINE

I. Assumptions
   A. Populations concerned are normal.
   B. All populations have a common variance σ^2 that is to be estimated in two ways.
   C. All samples are independent random samples of size n.

II. Hypotheses
   A. H_0: $\mu_1 = \mu_2 = \mu_3 = \ldots = \mu_k$, with any differences among sample means due to chance.
   B. H_A: Not all means are equal.

III. F test for variances
   A. $F^* = \dfrac{ns_{\bar{x}}^2}{s_p^2}$, where $s_{\bar{x}}^2 = \dfrac{\Sigma(\bar{X} - \bar{\bar{X}})^2}{k-1}$.
   df = $[k-1, k(n-1)]$.
   B. Only the right-tail critical region is used.

IV. Formulas
   A. SST = SSB + SSW.
      1. $\text{SST} = \Sigma X^2 - \dfrac{(\Sigma X)^2}{N} = \Sigma X^2 - \dfrac{(\Sigma T)^2}{N}$.

2. $\text{SSB} = \dfrac{\Sigma T^2}{n} - \dfrac{(\Sigma T)^2}{N}.$

3. $\text{SSW} = \Sigma X^2 - \dfrac{\Sigma T^2}{n}.$

4. $\text{MS} = \dfrac{\text{SS}}{\text{df}}.$

B. ANOVA table (at right)

Source of Variation	df	SS	MS
Between groups	$k - 1$		
Within groups	$k(n - 1)$		
Total	$kn - 1$		

C. $F^* = \dfrac{\text{MSB}}{\text{MSW}}$ $\text{df} = [k - 1, k(n - 1)].$

REVIEW PROBLEMS

1. In an attempt to ascertain growth rates in livestock, an agricultural researcher fed four different feeds to groups of three animals. She noted the following weight gains, in pounds, over a ten-week period. Are the population mean weight gains for the different feeds the same, at the 0.05 level of significance?

Feed	Animal Weight		
A	59	47	49
B	36	41	37
C	64	67	63
D	59	47	62

2. An education researcher desired to test the effectiveness of three different modes of instruction. He chose introductory psychology as the course common to many students and examined the results of a departmental final for four sections each that had been taught using the different methods. Are the mean final exam grades the same for all modes of instruction, at the 0.01 level of significance?

Mode	Mean Grade			
Self-study	73	71	69	70
Large-lecture presentations	74	75	69	72
Small-class presentations	76	74	79	82

3. A weight-loss clinic tested two different diet regimens, with and without exercise, on groups of four patients to look for significant differences. They collected the following data on weight loss, in pounds, during the first week. Are the mean weight losses equal, at a 0.05 level of significance?

Treatment	Mean Weight Loss			
Diet A—with exercise	9.0	7.2	11.4	4.0
Diet A—no exercise	6.0	5.1	3.0	2.0
Diet B—with exercise	7.0	6.0	4.8	8.1
Diet B—no exercise	4.0	3.2	9.0	2.6

4. The mileage performance of three brands of unleaded gasoline was tested on 12 different cars. Each car was given a tankful of each gasoline and run until it ran out of gas. The following table gives the distances obtained (in miles). Are the mean distances for the gasolines equal, at the 0.05 level of significance?

Gasoline	Distance			
Brand A	424	298	361	415
Brand B	390	291	352	384
Brand C	285	221	275	280

Linear correlation and regression analysis

12

12.1 Linear Correlation Analysis

Thus far we have used statistical inference to answer questions about hypotheses and to make estimates about parameters. Techniques of statistical inference can also be used to answer another type of question: Is there any relationship between two variables? If the variables are quantitative, we can answer that question by using correlation analysis. If the variables are qualitative, we will learn how to proceed in Chapter 13.

Is there a relationship between height and weight, between grade-point average in high school and quality-point average in college, between age and salary, or between a husband and wife's years of college? To answer any of these questions, we need to gather **bivariate data,** or two pieces of information from each individual in a sample. If we call height X and weight Y, the values of these variables are presented as (X, Y), an **ordered pair,** with X and Y called the *coordinates* of the data point. Each member of the sample produces an ordered pair of data, the values of which must be recorded in the same order: X first and Y second. This pairing order must be preserved for the analysis.

Does it make any difference which variable is labeled X and which Y? In algebra, the X variable is called the *independent* variable and generally is the variable used to calculate values for Y. We are using gathered data values for Y at this point, but in Section 12.2 we will predict values for Y based on values of X. Thus, whenever prediction is desirable, Y should always be the variable for which values are to be predicted; X is the variable used to do the predicting. Thus grade-point average in high school would be labeled X and quality-point average in college, Y; it makes no sense to try to predict the opposite. If prediction is not pertinent, as in height and weight, it makes no difference which variable is labeled X.

As in algebra the first procedure to follow with ordered pairs is to graph them. In statistics this graph is called a **scatter diagram.**

> ***scatter diagram.*** ■ A graph of bivariate data used to analyze relationships.

We begin by constructing a horizontal axis on which we put the values for X and a vertical axis on which we put values for Y, as in Fig. 12.1. A dot is placed at the intersection of X and Y values, say at (3, 2).

Each ordered pair produces a dot, and thus there should be as many dots as there are sample members. The axes usually, but do

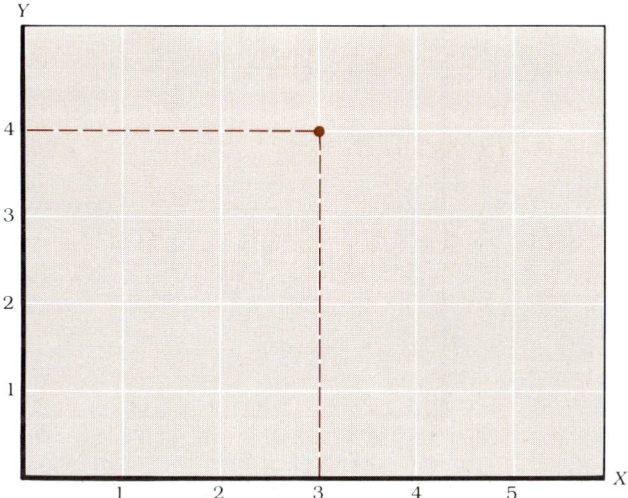

Figure 12.1

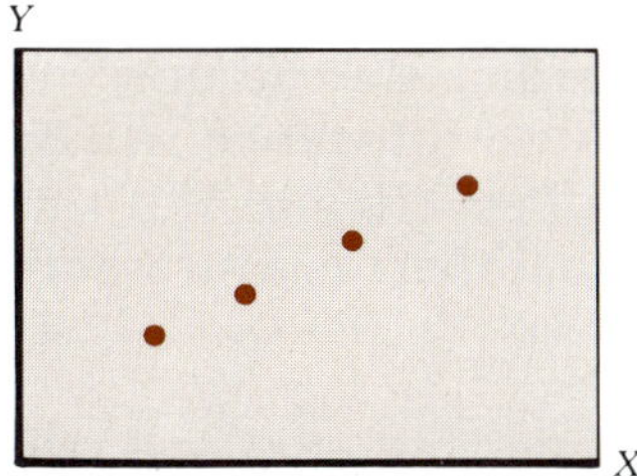

Figure 12.2

not need to, intersect at 0, but intervals along each axis must be evenly spaced. We will be looking for patterns; if intervals are distorted along either axis, any pattern will be distorted. Many patterns are possible, some of which are shown in Fig. 12.2. We will consider only those that form a straight line because they occur often and are the simplest to study. The study of curvilinear patterns often requires techniques from numerical analysis that are beyond the scope of this text. Let us examine three examples, beginning with scatter diagrams.

EXAMPLE 12.1 The Statistical Abstract of the United States provides the following information. The numbers given are in thousands.

Year	Number of Golfers	Number of Courses with 18 Holes or More
1960	4.4	2.7
1965	7.8	3.8
1970	9.7	4.8
1975	12.0	6.3
1980	13.0	6.9

Is there a linear relationship between the number of golfers and the number of courses?

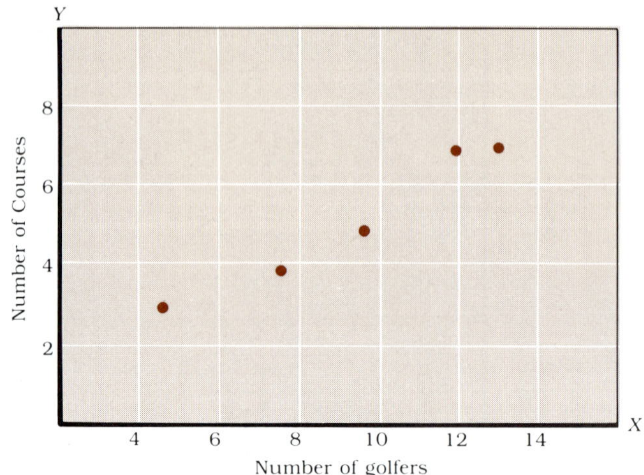

Figure 12.3

Solution We begin to answer that question by constructing a scatter diagram. In order to determine which variable is X, we ask ourselves whether prediction is a reasonable activity in this problem. We might only want to examine trends, but we might also want to use the number of golfers to predict the number of courses, or vice versa. Therefore it makes no difference which variable is labeled X. Let X = the number of golfers and Y = the number of courses in Fig. 12.3. It appears that there is a straight-line pattern, rising from left to right. □

EXAMPLE 12.2 Ten couples compared notes and obtained the following data on the number of years of college attended.

Husband	Wife
4	4
1	0
7	2
6	4
5	4
4	0
7	6
2	5
2	4
4	6

Is there a relationship between the level of a husband's education and that of his wife?

Solution Let X = the number of years of college attended by a husband and Y = the number of years of college attended by a wife. Now there appears to be no pattern (Fig. 12.4). □

EXAMPLE 12.3 When infants are born, they are rated with an apgar score by the obstetrician about five minutes after birth. The following information was gathered by five hospitals by grouping babies with the same apgar scores, recording the number of deaths that occurred, and converting deaths to neonatal mortality rates per thousand live births.

Apgar Score	Neonatal Mortality Rate
0	500
1	475
2	380
3	350
5	150

Is there a relationship between apgar score and neonatal mortality rate?

Solution If any predicting is to be done, it will involve the use of apgar score to predict mortality rate, so let X = apgar score and

Figure 12.4

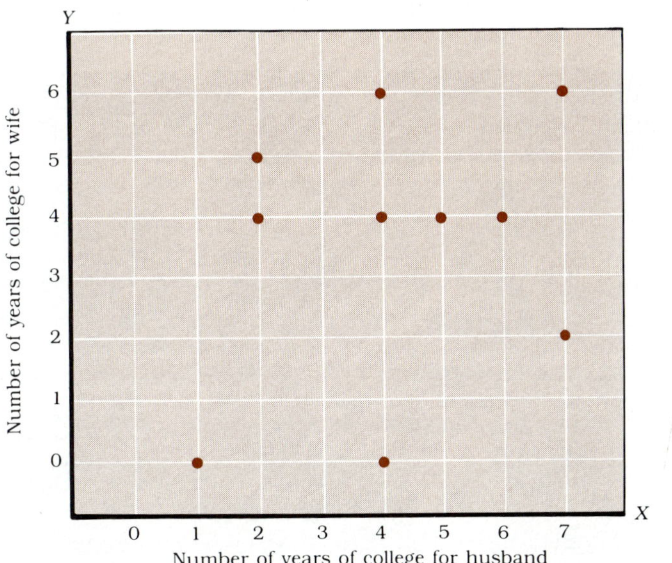

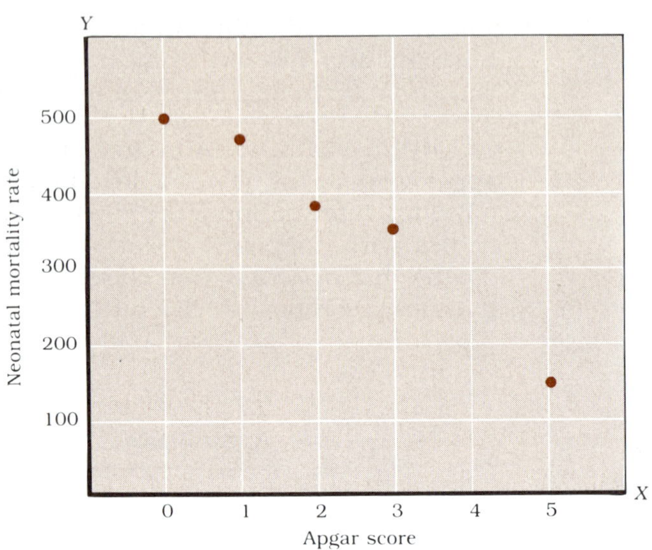

Figure 12.5

Y = neonatal mortality rate in Fig. 12.5. Again we have a straight-line pattern, but now the pattern falls from left to right. □

Let us examine what we have thus far. Example 12.1 gave us a straight-line pattern that rises from left to right. If we examine the original data carefully, we see that as the values of one variable increased so did the values of the other: Small values went with small values; large values went with large values. This type of relationship is called a **positive correlation**. The News & Views 12.1 article gives a further example of a positive correlation.

NEWS & VIEWS 12.1

Blood pressure

It may be worth going to college if only from a health standpoint.

According to a national survey recently published in the American Journal of Epidemiology, high blood pressure is 40% less prevalent in college graduates than it is in people who have had less than 10 years of schooling.

Why should this be so? Probably because the educated are more affluent, can visit doctors more frequently for a checkup, and are more knowledgeable about the dangers of high blood pressure and how to treat it.

Source: Llyod Shearer in *Parade*, May 7, 1978, p. 19.

NEWS & VIEWS 12.2

Doctors may be harmful

Is modern medical treatment more likely to kill than cure? Statistics gathered after "doctor strikes" indicate that today's physicians may not be living up to the first part of the Hippocratic oath, admonishing them to do no harm.

As a protest to soaring rates for malpractice insurance, doctors in Los Angeles went on strike in 1976. The result with no doctors around? An 18 percent drop in the death rate. That same year, according to Dr. Robert S. Mendelsohn, doctors in Bogota, Colombia, refused to provide any services except for emergency care. The result was a 35 percent drop in the death rate. When Israeli doctors drastically reduced their daily patient contact in 1973, the Jerusalem Burial Society reported that the death rate was cut in half. The only similar drop had been 20 years earlier at the time of the last doctors' strike.

Source; Reprinted from "Significa" by Irving Wallace, David Wallechinsky, and Amy Wallace; *Parade* Magazine; October 4, 1981; p. 27.

Example 12.2 showed us a case where no pattern and thus no correlation exists. Example 12.3 provided a straight-line pattern that falls from left to right; when we examine these data, we find that as the values of one variable increased, the values of the other decreased, or small values of one went with large values of the other. This relationship is called a **negative correlation.** Again we can look to newspaper and magazine articles for examples of a negative correlation, as in News & Views 12.2 and 12.3.

In the three examples presented, two sets of data have obvious patterns and one has no pattern. Does the configuration in Fig. 12.6 represent a linear pattern? In this case, as in most real

Figure 12.6

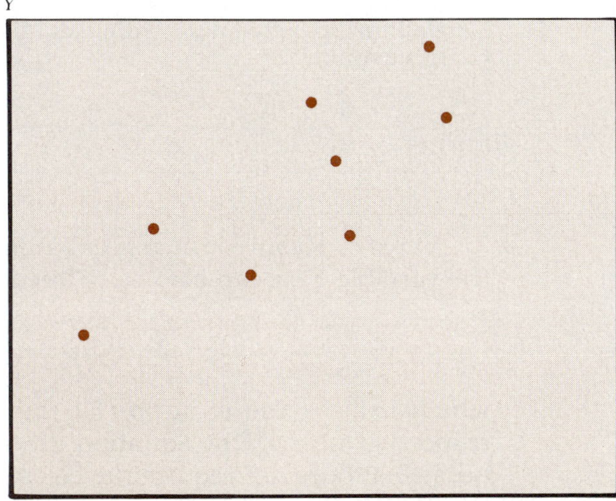

NEWS & VIEWS 12.3

Live short for a long life

It is often noted that the average American woman lives about eight years longer than the average American man. Scientists have tried to explain those extra years as resulting from female hormones, lack of stress and other factors, but one element hasn't been examined: height. What if women, who average 5′4″ in America, live longer because they are 5 inches shorter than the average male?

Not a popular idea, since we admire stature over shortness. Should we stop looking down on short people? Thomas Samaras thinks so. His theory is that tallness itself promotes a shorter lifespan.

After studying groups of athletes, U.S. Presidents, successful businessmen and even giants, Samaras concluded that average age at death declined with increasing height for all of these groups. Shorter men outlived taller ones by 6%–20%. Five Presidents under 5′8″ averaged 80.4 years at death; five over 6′1″ averaged only 66.8 years. Giants taller than 7′7″ averaged only 39.8 years at death.

Is there an optimum size for human beings? Perhaps we should study this question before the average gain in height for our race as a whole—one inch every 30 years—turns us into 21st-century dinosaurs.

Source: Reprinted from "Significa" by Irving Wallace, David Wallechinsky, and Amy Wallace; *Parade* Magazine; December 27, 1981; p. 13.

situations, the perception of the pattern may call for some judgment on our part. Rather than allow this judgment to be purely intuitive, we use a measure called the **correlation coefficient** (sometimes called the Pearson product-moment coefficient of correlation), which is designated by the letter r. The correlation coefficient r measures how values of Y change linearly with respect to values of X.

correlation coefficient. ■ A measure of the strength of the linear relationship between two variables, X and Y, and calculated using the equation

$$r = \frac{n\Sigma XY - \Sigma X \Sigma Y}{\sqrt{n\Sigma X^2 - (\Sigma X)^2}\,\sqrt{n\Sigma Y^2 - (\Sigma Y)^2}}.$$

Since r measures the relationship between the variable X and the variable Y, it can also be defined by the equation

$$r = \frac{\Sigma(X - \bar{X})(Y - \bar{Y})}{\sqrt{\Sigma(X - \bar{X})^2 \Sigma(Y - \bar{Y})^2}},$$

which utilizes the *variation* of the X and Y values from their respective means. The equation given in the box is easier to use because it does not require the calculation of means and involves fewer subtractions.

The correlation coefficient is a fraction with values from -1 to $+1$, inclusive, or

$$-1 \le r \le +1.$$

If all the points in a scatter diagram fit exactly on a straight nonhorizontal line, then r will be either -1 or $+1$; the correlation is said to be *perfect.* If $r = 0$, there is no correlation. The solution of most real-world problems does not lead to such obvious conclusions for interpreting r.

After practicing calculations, we will test hypotheses using r to draw conclusions about the strengths of relationships. However, before we calculate r for Examples 12.1–12.3, let us present some words of caution concerning use of the equation for r, since it is fairly complicated, and point out several places where errors in calculation may be spotted quickly.

1. In all the arithmetic we will do, the quantity inside a square-root sign must be positive. Thus the denominator must have $\sqrt{+} \cdot \sqrt{+}$, which is ultimately positive.
2. In the numerator, the same operation is not repeated; that is, the numbers substituted for the Σ expressions are not the same because $\Sigma XY \neq \Sigma X \Sigma Y$. Similarly, in the denominator (except for multiplying by the value of n twice), substitutions for the Σ notations are not the same numbers. Thus if you find yourself using a number twice (except for n) in the denominator, you have made a mistake.
3. The numerator is the source of the $(+)$ or $(-)$ sign for r, since the denominator is always positive. If a scatter diagram indicates a negative correlation, be sure that r is negative also: Check the numerator to make sure that it is negative.
4. Remember that n is the sample size and is represented in this case by the number of points on the scatter diagram or the number of rows of data. It is *not* the number of numbers because each member of the sample produces two pieces of data, or numbers.
5. Always multiply before subtracting.
6. If your calculations result in a value of r larger than 1 or smaller than -1, that is, $|r| > 1$, then something is wrong with the calculations.

EXAMPLE 12.1 (cont.) To continue our work with Example 12.1, let us make a table of values and extend by calculating it X^2, Y^2, and XY, which we need for substitution into the equation.

Solution

	X	Y	X^2	Y^2	XY
	4.4	2.7	19.36	7.29	11.88
	7.8	3.8	60.84	14.44	29.64
	9.7	4.8	94.09	23.04	46.56
	12.0	6.3	144.00	39.69	75.60
	13.0	6.9	169.00	47.61	89.70
Σ	46.9	24.5	487.29	132.07	253.38

$$r = \frac{n\Sigma XY - \Sigma X \Sigma Y}{\sqrt{n\Sigma X^2 - (\Sigma X)^2}\sqrt{n\Sigma Y^2 - (\Sigma Y)^2}}$$

$$= \frac{5 \cdot 253.38 - 46.9 \cdot 24.5}{\sqrt{5 \cdot 487.29 - (469)^2}\sqrt{5 \cdot 132.07 - (24.5)^2}}$$

$$= \frac{117.85}{\sqrt{236.84}\sqrt{60.1}} = \frac{117.85}{119.31} = +0.988.$$

The positive result agrees with the scatter diagram and is very close to +1. This indicates a strong correlation. □

EXAMPLE 12.2 (cont.) We follow the same procedure for Example 12.2 and make the following calculations.

Solution

	X	Y	X^2	Y^2	XY
	4	4	16	16	16
	1	0	1	0	0
	7	2	49	4	14
	6	4	36	16	24
	5	4	25	16	20
	4	0	16	0	0
	7	6	49	36	42
	2	5	4	25	10
	2	4	4	16	8
	4	6	16	36	24
Σ	42	35	216	165	158

$$r = \frac{n\Sigma XY - \Sigma X \Sigma Y}{\sqrt{n\Sigma X^2 - (\Sigma X)^2}\sqrt{n\Sigma Y^2 - (\Sigma Y)^2}}$$

$$= \frac{10 \cdot 158 - 42 \cdot 35}{\sqrt{10 \cdot 216 - (42)^2}\sqrt{10 \cdot 165 - (35)^2}}$$

$$= \frac{110}{\sqrt{396}\sqrt{425}} = \frac{110}{410.24} = +0.268.$$

The value of r is somewhat close to 0. After we make the calculations for Example 12.3, we will test this value to determine whether a relationship exists. □

EXAMPLE 12.3 (cont.) The calculations for Example 12.3 are:

Solution

	X	Y	X^2	Y^2	XY
	0	500	0	250,000	0
	1	475	1	225,625	475
	2	380	4	144,400	760
	3	350	9	122,500	1050
	5	150	25	22,500	750
Σ	11	1855	39	765,025	3035

$$r = \frac{n\Sigma XY - \Sigma X \Sigma Y}{\sqrt{n\Sigma X^2 - (\Sigma X)^2}\sqrt{n\Sigma Y^2 - (\Sigma Y)^2}}$$

$$= \frac{5 \cdot 3035 - 11 \cdot 1855}{\sqrt{5 \cdot 39 - (11)^2}\sqrt{5 \cdot 765{,}025 - (1855)^2}}$$

$$= \frac{-5230}{\sqrt{74}\sqrt{384{,}100}} = \frac{-5230}{5331.36} = -0.981.$$

The minus sign of the calculation agrees with the negative correlation indicated on the scatter diagram. The value of r is almost -1, which indicates a strong correlation. □

As we have already indicated, interpretation of scatter diagrams and calculated values for r cannot be arbitrary. Thus we test these values for r as we tested other sample results in previous chapters. The parameter in question is the correlation coefficient that measures the relationship between two *populations* from which the samples came; we denote it by a Greek letter, ρ (rho). The null hypothesis assumes that there is no relationship.

$$H_0: \rho = 0.$$

The alternative hypothesis comes from the assumption that there is a relationship.

$$H_A: \rho \neq 0.$$

We will use a two-tail test and obtain critical values from Table D.6, with

$$\boxed{\text{df} = n - 2.}$$

The value of α at the top of Table D.6 is the area in both tails. The right-tail critical value is given in the body of the table; the left-tail critical value is the negative of the right-tail value. Now we are ready to test the values of r we calculated for Examples 12.1–12.3 at the 0.05 level of significance.

EXAMPLE 12.1 (cont.)

Summary

H_0: $\rho = 0$.

H_A: $\rho \neq 0$.

Test region: Since $n = 5$, df = 3.

Reject H_0 when $r^* > 0.878$ or $r^* < -0.878$.

Test statistic: As previously calculated, $r^* = +0.988$.

Conclusion: Reject H_0. The number of golfers is linearly related to the number of courses. □

EXAMPLE 12.2 (cont.)

Summary

H_0: $\rho = 0$.

H_A: $\rho \neq 0$.

Test region: Since $n = 10$, df = 8.

Reject H_0 when $r^* > 0.632$ or $r^* < -0.632$.

Test statistic: As previously calculated, $r^* = +0.268$.

Conclusion: We cannot reject H_0 and cannot conclude that a linear relationship exists, based on the data. □

EXAMPLE 12.3 (cont.)

Summary

H_0: $\rho = 0$.

H_A: $\rho \neq 0$.

Test region: $n = 5$ and df = 3.

Reject H_0 when $r^* > 0.878$ or $r^* < -0.878$.

Test statistic: As previously calculated, $r^* = -0.981$.

Conclusion: Reject H_0. Apgar score is related to neonatal mortality rate. □

The square of r, r^2, is called the **coefficient of determination.** It is the proportion of the variation in the values of Y that is attributable to the linear relationship of Y with X.

> ***coefficient of determination.*** ▪ The proportion of the variation in Y attributable to the linear relationship of Y with X, denoted r^2.

In Example 12.1, we calculated $r^* = 0.988$. Since $r^2 = 0.976$, we can say that more than 97% of the variation in the number of golf courses is due to a linear relationship with the number of golfers. For Example 12.3, $r^* = -0.981$ and $r^2 = 0.962$. Thus more than 96% of the variation in mortality rate is due to a linear relationship with apgar scores. We examine these relationships in detail in Section 12.2.

Before we leave linear correlation, we need to make some comments about what correlation does *not* do. *Correlation analysis does not establish cause and effect,* only the existence of a relationship. For example, is it reasonable to say that a low apgar score *causes* a high mortality rate? Of course not. Both are caused by the state of infants' health. Which is stimulus and which is response: Do more golfers cause more courses to be built, or does the availability of more courses stimulate more golfing? Perhaps something else, such as socioeconomic status or the state of the economy causes changes in both.

Wrongly attributed cause and effect occurs often enough in correlation analysis that we will look at some further correlations for which cause and effect obviously do not apply. Next time someone says, ". . . causes . . . ," recall these examples and think critically about what the speaker has said.

1. There is almost a perfect positive correlation between the number of Swedish children born and the number of stork nests in Swedish churches. The more stork nests there are, the more children are born. Do storks bring children?
2. Small children observe that the bigger you are, the older you are; that the older you are, the smarter you are. Does increased age always cause increased size? Does increased size cause increased IQ?
3. Most auto accidents occur at low speeds. Accident rate is negatively correlated with speed. Do low speeds cause accidents? Is it safer to drive fast?

4. Most auto accidents occur within 40 miles of the driver's home. Distance from home is negatively correlated with accident rate. Do distances near home cause accidents? Is it safer to drive far from home?
5. A study has indicated that the larger a minister's salary the more money is spent on liquor in a neighborhood. Salary and liquor expenditure are positively correlated. Do highly paid preachers drive their congregations to drink?
6. More people die of tuberculosis in Arizona, with its dry climate, than in any other state. The level of humidity correlates negatively with the incidence of tuberculosis. Does dry air cause TB?
7. Someone with nothing better to do once determined that children who have big feet spell better than those who have small feet. Foot size positively correlated with spelling ability. Do big feet cause good spelling ability?
8. The death rate in Alaska, with its severe climate, is one of the lowest. The death rate in Miami with its mild climate is one of the highest. Death rate and severity of climate are negatively correlated. Does climate affect death rate?

The explanations for these correlations are given at the end of the chapter.

How are cause and effect established if correlation analysis is insufficient? The determination of cause and effect is a more difficult task than correlation analysis. It requires the use of controlled experiments in which the effects of confounding variables are held constant; that is, the only change from one group of sample members to another, or from one part of the experiment to another, is the application of the treatment under study. That treatment is then viewed as having *caused* any observed effect. It is difficult to say that cigarette smoking is the cause of lung cancer, since not all smokers get lung cancer and not all victims of lung cancer smoke. Thus the Surgeon General's warning in Fig. 12.7 is not phrased in those terms. Be wary of cause-and-effect conclusions when studies involve complex issues and many factors.

Figure 12.7

Warning: The Surgeon General Has Determined That Cigarette Smoking Is Dangerous To Your Health

EXERCISES/Section 12.1

1. Seven sample numbers have the following X and Y values.

X	4	5	6	8	9	10	12
Y	1	2	4	5	8	8	9

   a) Draw a scatter diagram for these data.
   b) Calculate the correlation coefficient r.
   c) Are X and Y related, at the 0.01 level of significance?

2. The following data give farm productivity in billions of workhours of labor versus the number of persons supplied with food per farmworker from 1910–1970.

Year	Workhours	Persons Supplied
1910	22.5	7.1
1920	24.0	8.3
1930	22.9	9.8
1940	20.5	16.7
1950	15.1	15.5
1960	9.8	25.8
1970	6.5	47.1

   a) Draw a scatter diagram to display the variables workhours and persons supplied.
   b) Calculate the correlation coefficient r.
   c) Are workhours of labor related to number of persons supplied, at the 0.01 level of significance?

3. Is there a correlation between the average tax rate, expressed as a percent of personal income for a single person, and the number of movie theaters, in thousands, in a country?

	Average Tax Rate (%)	Number of Movie Theaters
Canada	19	1.1
United Kingdom	25	1.5
Austria	13	0.6
France	8	5.8
West Germany	19	3.1
Italy	5	5.9
Japan	10	2.5
Norway	25	0.4
Spain	3	5.2
Sweden	36	1.2

   a) Draw a scatter diagram for these data.
   b) Calculate the correlation coefficient r.
   c) Is r significant at the 0.05 level?

4. A heart-disease study was conducted in Los Angeles over a ten-year period. As part of the study, the following data on heights, in inches, and weights, in pounds, were collected.

Weight	Height	Weight	Height
149	68	144	66
182	68	130	67
185	70	162	69
161	68	175	68
175	67	155	66

   a) Plot a scatter diagram for these data.
   b) Calculate the correlation coefficient r.
   c) At the 0.05 level of significance, is there a relationship between weight and height?

5. Is there a relationship between the annual number of days of rain a country receives and the price, in dollars, of a loaf of white bread?

	Number of Days	Price of Bread
United States	121	0.82
Canada	166	0.65
Austria	96	1.14
Japan	115	0.84
Switzerland	134.6	0.70
Netherlands	206	0.56

a) Plot a scatter diagram for these data.
b) Calculate r, the correlation coefficient.
c) Is there a relationship between annual number of days of rain and the price of white bread, for $\alpha = 0.05$?

6. Did the number of hours spent studying affect the grade received by students on an exam?

Hours	Grade
2.3	72
1.5	69
6.2	88
7.1	97
5.2	91
0.5	42
3.6	81

a) Plot a scatter diagram for these data.
b) Calculate the correlation coefficient r.
c) At the 0.01 level of significance, is there a relationship between length of time spent studying and exam grade?

7. Weather forecasting is an important art, which derives much from science and makes extensive use of statistical methods. Mean temperatures, in °F, and precipitation, in inches, for the month of July for 11 cities are:

City	Temperature	Precipitation
Chicago	73.9	3.33
Honolulu	77.7	1.31
New Orleans	82.4	6.37
Mexico City	62.4	4.10
Rio de Janeiro	67.5	1.60
Berlin	64.6	3.10
Dublin	60.1	2.60
Paris	65.5	2.20
Manila	80.6	17.40
Hong Kong	81.7	11.40
Delhi	86.4	7.60

a) Plot a scatter diagram for these data.
b) Calculate r, the correlation coefficient.
c) For $\alpha = 0.02$, is there a relationship between temperature and amount of precipitation?

8. The following data relate frustration level to performance in racquetball for ten players. Length of time spent waiting for a court, in hours, and number of wins are shown for each player.

Player	Time Spent Waiting	Wins
1	2.0	3
2	5.5	1
3	1.0	4
4	0.5	3
5	3.0	3
6	2.5	2
7	1.5	3
8	0.0	5
9	1.0	4
10	0.5	4

a) Plot a scatter diagram for these data.
b) Calculate r, the correlation coefficient.
c) For $\alpha = 0.02$, is there a relationship between time spent waiting and number of wins?

9. Recognition of the incidence of child abuse in our society is increasing. The following data show the ages of children, in years, involved in incidents of suspected child abuse and the number of children in each child's family.

Age of Child	Children in Family
1	1
3	5
4	2
14	1
11	4
6	3
5	1
3	2
1	4
4	3
10	1
9	2

a) Plot a scatter diagram for these data.
b) Calculate the correlation coefficient r.
c) At the 0.05 level of significance, is there a relationship between age of child and

number of children in the family in suspected child abuse cases?

10. A local brewery listed the following data on sales and television advertising expenditures, both in thousands of dollars, for a seven-month period.

Month	TV Advertising Expenditures	Sales
January	0.5	50
February	0.9	90
March	0.4	30
April	0.7	90
May	1.1	91
June	0.75	95
July	0.8	95

a) Plot a scatter diagram for these data.
b) Calculate the correlation coefficient r.
c) At the 0.01 level of significance, is there a relationship between advertising expenditures and sales?

11. Is grade-point average a good indicator of preparation for graduate school, as measured by scores on the Graduate Record Exam?

Grade-Point Average	GRE Score
2.64	970
2.71	780
2.07	1020
3.13	940
2.89	1320
2.86	1040
3.03	880
3.14	1110
2.08	760

a) Plot a scatter diagram for these data.
b) Calculate the correlation coefficient r.
c) Is there a relationship between gpa and score on the GRE?

12. The following data give stock information. Amounts are in dollars. *Note:* stock prices are quoted in eighths of dollars; you might want to convert the fractions to decimals before proceeding.

Dividend	Price per Share
3.56	24¼
2.02	16½
2.22	19⅝
4.35	29
2.47	18⅞
1.37	14
1.44	15⅝
5.40	61
8.48	83
1.50	14½
0.44	6¼
2.92	28¼

a) Plot a scatter diagram for these data.
b) Calculate the correlation coefficient r.
c) At the 0.05 level of significance, is there a relationship between dividend paid and the price of stock?

13. A group of eight new employees was rated by their supervisors to determine a dexterity rating and the number of items produced.

Rating	Production
6	123
2	64
3	76
4	100
9	162
6	140
7	141
1	52

a) Plot a scatter diagram for these data.
b) Calculate r, the correlation coefficient.
c) At the 0.05 level of significance, are dexterity and production level related?

14. The following data pertain to litter size of cats from mothers of various ages (in months).

Age	Litter Size	Age	Litter Size
12	6	12	4
17	5	19	5
14	7	15	6
19	4	18	6
17	6	14	5

a) Plot a scatter diagram for these data.
b) Calculate r, the correlation coefficient.
c) At the 0.01 level of significance, are age and litter size related?

15. NFL statistics report the following:

Leading Passers	Attempts	Completions
1	4129	2272
2	3084	1708
3	2701	1552
4	3817	2118
5	2613	1410
6	2895	1553
7	2128	1088
8	1565	872
9	2995	1693
10	2308	1170

a) Plot a scatter diagram for these data.
b) Calculate r, the correlation coefficient.
c) At the 0.01 level of significance, are passing attempts and number of completions related?

16. The following information shows the amount of international trade in photographic supplies, in millions of dollars, for five selected years.

Year	Exports	Imports
1960	63	82
1965	130	178
1970	250	356
1975	606	726
1980	1507	2654

a) Plot a scatter diagram for these data.
b) Calculate the correlation coefficient r.
c) At the 0.05 level of significance, are levels of exports and imports related?

17. Is there a relationship between number of churches, given in thousands, and number of members, given in hundreds of thousands, for 12 faiths and denominations?

Faith or Denomination	Churches	Members
Seventh Day Adventists	3.3	4.5
American Baptist Convention	6.0	15.0
Southern Baptist Convention	34.2	123.0
Buddhist Churches	0.06	1.0
Churches of Christ	18.0	24.0
Greek Orthodox	0.5	2.0
Jehovah's Witnesses	6.1	5.0
Mormon	7.5	33.2
American Lutheran	4.8	24.6
United Methodist	39.4	101.9
United Presbyterian	8.7	29.1
Roman Catholic	23.9	484.6

a) Plot a scatter diagram for these data.
b) Calculate r, the correlation coefficient.
c) For $\alpha = 0.05$, are the number of churches and size of membership related?

18. Given below is a ten-year safety record for U.S. commercial aviation. Aircraft miles flown are in billions.

Year	Aircraft Miles Flown	Number of Accidents
19X0	1.13	90
19X1	1.10	84
19X2	1.17	70
19X3	1.23	77
19X4	1.34	79
19X5	1.54	83
19X6	1.71	75
19X7	2.16	70
19X8	2.50	71
19X9	2.74	63

a) Plot a scatter diagram for these data.
b) Calculate r, the correlation coefficient.
c) At the 0.05 level of significance, is the number of accidents related to mileage?

19. NFL statistics show the number of yards gained, in thousands, for the nine leading lifetime rushers.

Rusher	Attempts	Yards Gained
1	2359	12.3
2	1929	9.7
3	1941	8.6
4	1727	7.3
5	1378	6.3
6	1286	5.9
7	1320	5.9
8	1413	5.5
9	1069	5.2

a) Plot a scatter diagram for these data.
b) Calculate r, the correlation coefficient.
c) At the 0.01 level of significance, are attempts and yards gained related?

20. Is there a relationship between age, in years, and diastolic blood pressure?

Age	DBP
31	80
61	110
61	88
44	94
58	74
52	80
40	90
49	75
34	80
37	70
63	90
28	80
40	82

a) Plot a scatter diagram for these data.
b) Calculate r, the correlation coefficient.
c) At the 0.05 level of significance, is blood pressure related to age?

Explanations for the correlation examples on pp. 281–282.

1. As populations increase from town to town so do the number of churches providing nesting sites for storks. Thus the stork and baby populations in larger towns are larger than those in smaller towns solely because they have more people.
2. The older relatives and friends of growing children are bigger and, rather than being smarter, have had more opportunity to learn. These relationships certainly do not exist among adults.
3. **and 4.** Most driving is done close to home and usually at slow speeds. The relatively greater amount of time spent driving slowly near home increases the probability of an accident.
5. People in any neighborhood who can afford to pay their clergy well usually can afford a large liquor bill.
6. People with TB are often urged to go to a dry climate. They have the disease when they arrive.
7. The study included growing children, and the children with larger feet were older.
8. Age is the factor to consider here. The population of Alaska is relatively young and that of Miami is fairly old.

12.2 Linear Regression Analysis

Prediction is very difficult, especially about the future.
Niels Bohr

We have been looking for linear patterns. Now we will be more specific and less intuitive about fitting a straight line to a series of dots on a scatter diagram. Although there are several methods that can be used, we will define a *line of best fit* by using the procedure known as the **least squares method.** The theoretical and technical details of this procedure require mathematical skills beyond those needed for the rest of this text; thus we summarize and demonstrate the method with diagrams and only the basic equations required.

Let us consider the scatter diagram with only four points in Fig. 12.8(a). Now let us add a line that passes through the general area containing the points, as in Fig. 12.8(b). Many such lines are possible, but we want to determine which line best fits the data.

We first measure the vertical distance of each point to the line, as shown in Fig. 12.9. Each point has coordinates (X, Y) and corresponds to another point directly above or below it on the line. This point on the line has the same X coordinate as the data point but a different Y coordinate. Ultimately, we will calculate

Figure 12.8

(a)

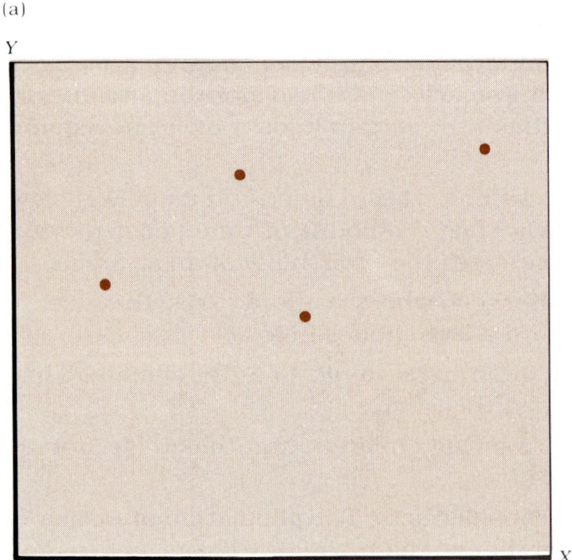

(b)

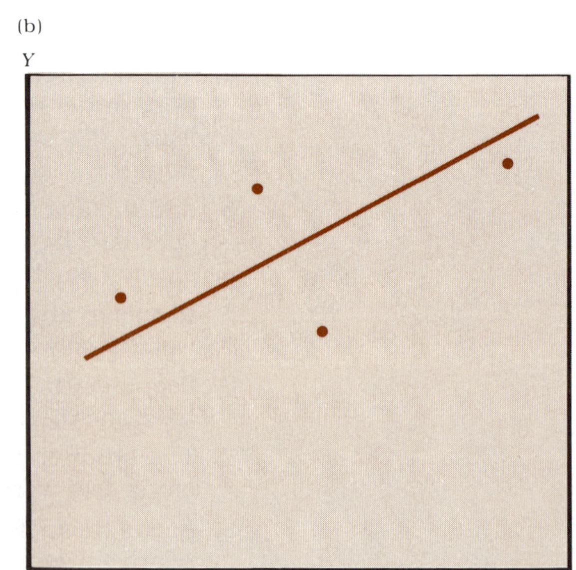

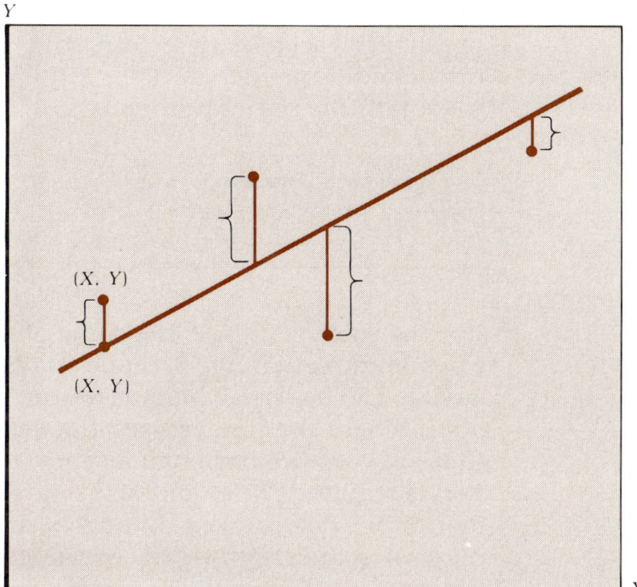

Figure 12.9

values for Y of points on the line using sample results; we distinguish the calculated Y by adding the extra symbol (ˆ) to the letter Y. Thus we will call this Y coordinate of a point on the line $\hat{Y}$ (read "Y hat"). We can find the distance from any point to the line merely by finding the difference in the Y coordinates, or $Y - \hat{Y}$. If we do this for each point, we find that some distances are positive (for points above the line) and some are negative (for points below the line). To obtain an aggregate measure of the distances from the line, we need to be sure that negatives and positives do not cancel each other. Therefore we square each distance to remove negative signs and then add them: $\Sigma(Y - \hat{Y})^2$.

In theory, we would determine such a sum for every possible line that passes in the vicinity of the points. The line that produces the smallest sum of the squared distances (*least squares*) is said to be the line of best fit. It is called the **regression line** of Y on X, and this analysis is called **linear regression analysis.**

> ***regression line.*** ■ The line of best fit through a set of data points, as determined by the least squares method.

Any line is determined by two pieces of information (or points). In this case we will use the two pieces, b_0 and b_1, in the equation that describes a line: $Y = b_0 + b_1X$. The variable X is the predicting variable, as before, and $\hat{Y}$ is now the variable to be calculated or predicted, using values of X. Substituting $\hat{Y}$ for Y, we obtain the **regression equation.**

> ***regression equation.*** ■ The linear equation of the regression line, expressed as
> $$\hat{Y} = b_0 + b_1X.$$

The coefficient b_1 is the *slope* of the regression line and indicates its direction and steepness. (We will soon see that it is related to r.) The term b_0 stands for the *Y-intercept* of the line, or the point where the line crosses the vertical axis when $X = 0$. For those of you who have had algebra, one further word of explanation is required. We cannot use two data points as sufficient information for determining b_0 and b_1 and thus the equation of the regression line. Nothing in our work so far guarantees that any of the data points actually lie on the line. The coefficient b_1 must be calculated.

$$b_1 = \frac{n\Sigma XY - (\Sigma X)(\Sigma Y)}{n\Sigma X^2 - (\Sigma X)^2}.$$

Now compare the equation for b_1 to the equation for r. The numerators are identical, and the denominator of b_1 is the same as the quantity under the first square root sign of the denominator of r. Previously we stated that a quantity under a square root sign will be positive, and so the denominator of b_1 is positive. Thus the numerator of b_1 determines the $(+)$ or $(-)$ sign of b_1, as was the case for r; the signs of r and b_1 will agree. Since we will have already calculated r before looking for a specific line, the calculation of b_1 requires only that we divide the numerator of r by the *quantity* under the first square root sign of its denominator. Do not take the square root.

Since b_1 measures the change that occurs in Y as X changes, it can be calculated using

$$b_1 = \frac{\Sigma(X - \bar{X})(Y - \bar{Y})}{\Sigma(X - \bar{X})^2}.$$

As with the preceding equation for b_1, in the box, this one is related to the definitional equation for r. However, it is not so easy to use because it requires means and many subtractions—two for each pair of data values.

One convenient property of the regression line is that it goes through the point $(\bar{X}, \bar{Y})$. Thus

$$\bar{Y} = b_0 + b_1\bar{X},$$

from which we can solve for b_0,

$$\boxed{b_0 = \bar{Y} - b_1\bar{X}.}$$

Let us return to Examples 12.1 and 12.3, in which we found strong data correlations.

EXAMPLE 12.1 (Cont.) Find the equation of the regression line that describes the relationship between the number of golfers and the number of golf courses (both given in thousands).

Year	*X* Number of Golfers	*Y* Number of Courses
1960	4.4	2.7
1965	7.8	3.8
1970	9.7	4.8
1975	12.0	6.3
1980	13.0	6.9

Solution We use the equation $\hat{Y} = b_0 + b_1X$. Referring to our calculation of r on p. 278, we see that

$$r = \frac{117.85}{\sqrt{236.84}\,\sqrt{60.1}}.$$

Thus

$$b_1 = \frac{117.85}{236.84} = 0.50.$$

We need

$$\bar{X} = \frac{46.9}{5} = 9.38;$$

$$\bar{Y} = \frac{24.5}{5} = 4.9.$$

$$b_0 = \bar{Y} - b_1\bar{X}.$$

$$b_0 = 4.9 - 0.50(9.38) = 0.21.$$

$$\hat{Y} = 0.21 + 0.50X.$$

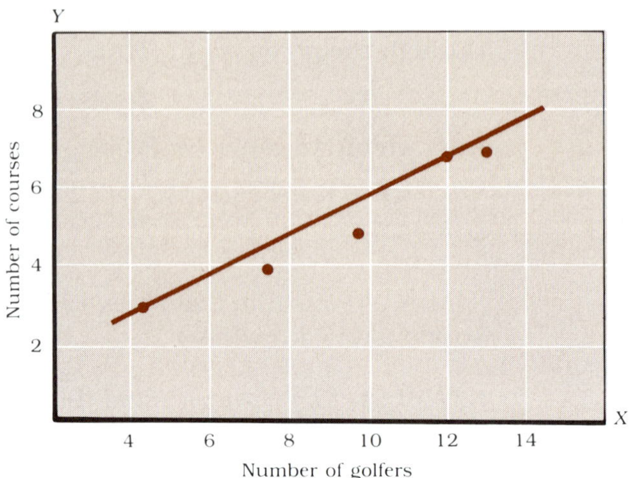

Figure 12.10

Thus the line of best fit can be plotted as shown in Fig. 12.10. □

EXAMPLE 12.3 (Cont.) Find the equation of the regression line that describes the relationship between apgar score and neonatal mortality rate (deaths per thousand babies born).

X Apgar Score	Y Neonatal Mortality Rate
0	500
1	475
2	380
3	350
5	150

Solution Again, we use the equation $\hat{Y} = b_0 + b_1X$. Referring to our earlier calculation of r on p. 279, we find that

$$r = \frac{-5230}{\sqrt{74}\,\sqrt{384{,}000}}.$$

Thus

$$b_1 = \frac{-5230}{74} = -70.68.$$

We need

$$\bar{X} = \frac{11}{5} = 2.2;$$

$$\bar{Y} = \frac{1855}{5} = 371.$$

$$b_0 = \bar{Y} - b_1\bar{X}.$$

$$b_0 = 371 - (-70.68)(2.2) = 371 + 155.50 = 526.50.$$

$$\hat{Y} = 526.50 - 70.68X.$$

Thus the line of best fit can be plotted as shown in Fig. 12.11. □

The equation of the regression line is used to predict values for the Y variable using values of the X variable. In Example 12.1, we found $\hat{Y} = 0.21 + 0.50X$, where X = the number of golfers and Y = the number of golf courses. Thus we can use the number of golfers to predict or estimate the number of courses. If $X = 10$, we substitute into the regression equation to find

$$\hat{Y} = 0.21 + 0.05(10) = 0.21 + 5 = 5.21.$$

Looking at the list of data, we see that the pair (10, 5.21) fits between 1970 and 1975, as common sense would tell us.

Figure 12.11

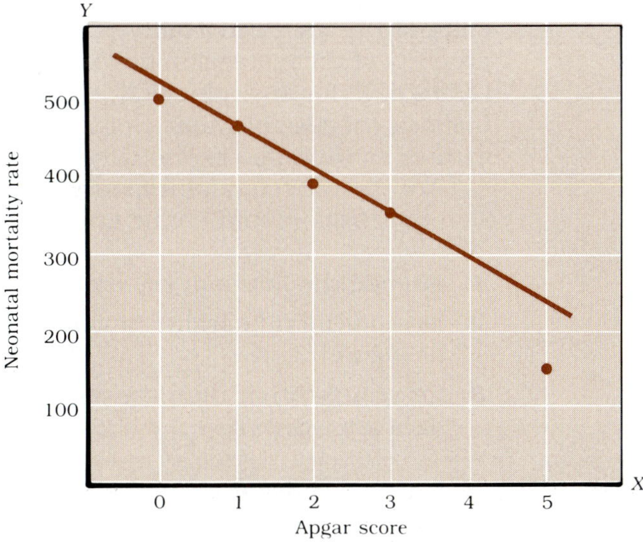

NEWS & VIEWS 12.4

Francis Galton, a cousin of Charles Darwin, first proposed the concepts of correlation and regression. He noticed that exceptionally tall fathers (on the average) had shorter sons, that is, sons who were closer to the mean height of adult males than were their fathers. This also works in reverse: Exceptionally short fathers (on the average) have taller sons. He called this phenomenon "regression toward the mean," a falling back, if you will, to the mean.

If inflation slowed after elected-official Jones came into office, or if the unemployment rate rose, did he necessarily cause the change? We are supposed to believe that, but the change may merely be the result of regression: low values increasing toward the mean or high values falling toward the mean. Many times when measurements are taken at different times or data are collected repeatedly (as with government statistics or test scores in education) extremes are not, on the average, repeated; rather they *regress* back to the mean from one set of data to another. This is a statistical phenomenon to be expected and may have little to do with politicians' or teachers' efforts.

For Example 12.3, we determined that $\hat{Y} = 526.50 - 70.68X$, where X = apgar score and Y = neonatal mortality rate. If we choose $X = 4$, we can predict that

$$\hat{Y} = 526.50 - 70.68(4) = 526.50 - 282.72 = 243.78$$

or about 244, and again the pair (4, 244) fits into the list of data. If we found babies with an apgar score of 4 at every hospital, would each hospital have a neonatal mortality rate of exactly 244? Certainly not! How would the mortality rates of all those hospitals compare with the 244 we calculated? The 244 would be an estimate of the *mean* mortality rate of the hospitals. An important property of the regression line is that it goes through the mean of values of Y: for any value of X, the regression line passes through the mean of possible values for Y. Thus $\bar{X}$ will always be associated with $\bar{Y}$, as we noted previously.

Are there circumstances in which prediction using a regression equation should *not* be attempted? There are at least three:

1. outside the limits of the data values for X;
2. beyond the time frame in which the data were collected (past or future); and
3. for groups other than the population from which the original data were drawn.

Let us discuss each of these conditions separately. The data collected presents limits on the values of X that were used to create the regression line and thus the equation $\hat{Y} = b_0 + b_1X$. Let us consider the graph of a relationship between X and Y (Fig. 12.12a). If the data collected had produced values between X_1 and X_2, we would find a regression line to approximate them as shown in Fig. 12.12(b). If the data had values between X_2 and X_3, we would have found a very different regression line (Fig. 12.12c). And finally, if data values for X had fallen between X_3 and X_4, we would use the regression line in Fig. 12.12(d). In any of these cases, a prediction outside the interval in which the values of X fall would not produce results that replicate the true graph.

Figure 12.12

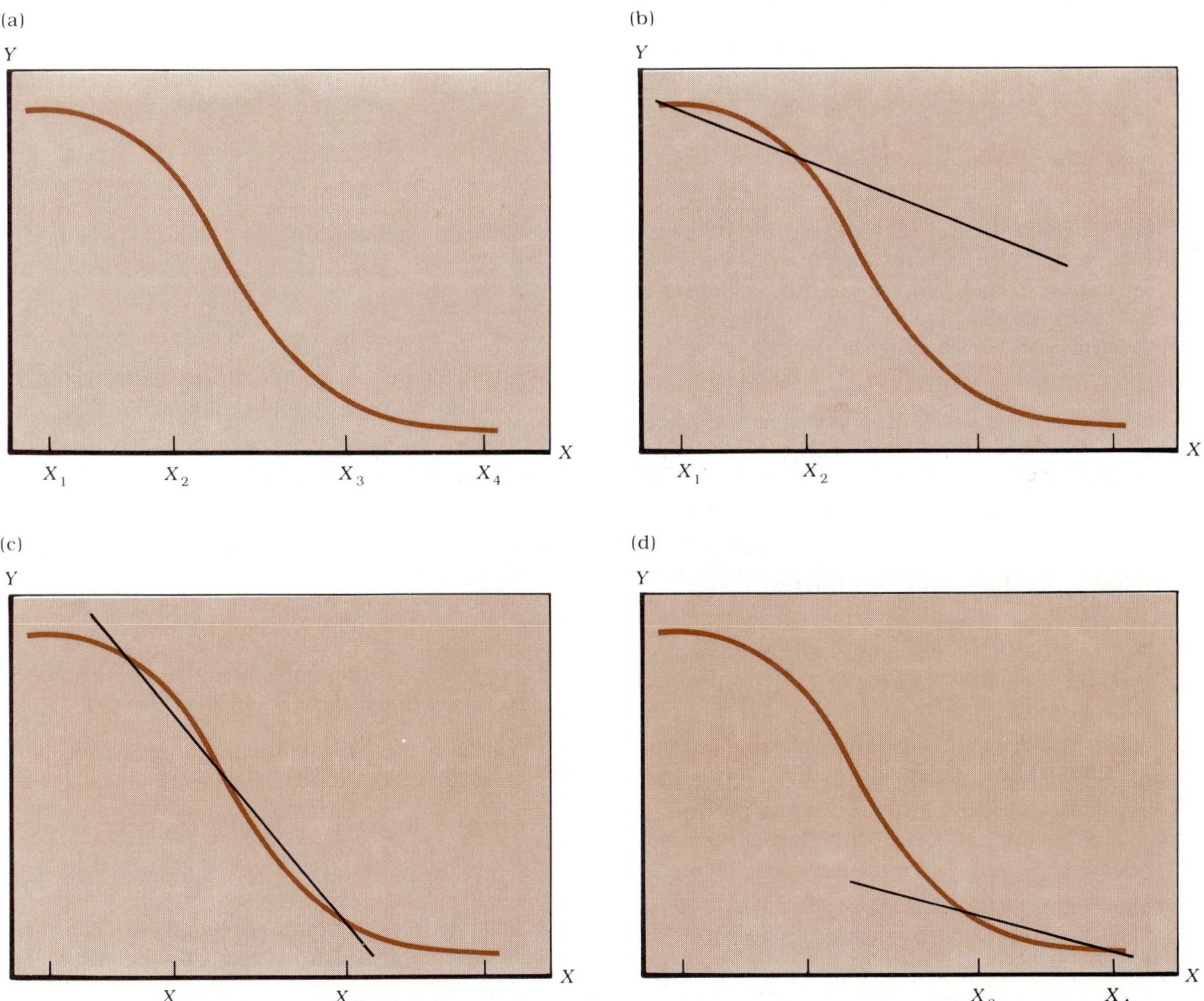

If we try to predict beyond the immediate time in which the data were gathered, problems can arise. Obviously, no one can adequately predict the future; even conclusions drawn about the past based on current data may be erroneous. Any attempt to determine average heights of people centuries ago based on current heights proves woefully inadequate, when compared to such indicators as size of armor worn in battle or heights of doors in castles. Diets have improved substantially, and people have grown larger.

A similar height example can be used to demonstrate that regression does not allow adequate estimation for other groups. Data on the heights of adult Japanese collected before World War II would be inadequate to predict height just one generation later. The generation of Japanese growing up after the war is much taller. The introduction of more red meat into their diets, competing with fish as a major source of protein, has increased height.

EXERCISES/Section 12.2

Using Exercises/Section 12.1, determine the equation of the regression line. Then use the regression equation to make the prediction called for in the following exercises, which correspond to the numbering in Exercises/Section 12.1.

1. If $X = 7$, what is $\hat{Y}$?
2. How many persons would have been supplied with food by 18 billion workhours of farm labor?
3. How many movie theaters are there, on the average, in a country with a tax rate of 20%?
4. What is the height of an individual weighing 150 lb?
5. What is the price of a loaf of white bread in a country that receives 140 days of rain annually?
6. What grade should be achieved by a student who has spent 4.5 hr studying?
7. What precipitation level would you predict for a city having a mean July temperature of 68.1°F?
8. How many wins should be expected after a player spends 4.5 hr waiting for a court?
9. How many children are there likely to be in a family with a 7-year-old involved in an incident of child abuse?
10. How many dollars of sales can be expected after a TV-advertising expenditure of $1000?
11. What GRE score would be expected for a student with a grade-point average of 2.90?
12. What price per share would be predicted for a stock that pays a dividend of $3.75?
13. Estimate production based on a dexterity rating of 5.
14. How many kittens would be expected in the litter of a mother cat that is 16 months old?
15. If 2800 attempts are made by an NFL player, how many completions can be expected?
16. In a year when exports reach $300 million, what level of imports could be expected?
17. How many members does a faith or denomination with 12,000 churches have?
18. In a year when 1.75 billion commercial-aircraft miles are flown, how many accidents can be expected?
19. If 2000 attempts are made by an NFL rusher, how many yards gained can be expected?
20. What is the diastolic blood pressure for a person of age 50?

Study Notes

KEY TERMS

bivariate data ■ Values for two variables from each individual in a sample.

coefficient of determination ■ The proportion of the variation in Y attributable to the linear relationship of Y with X, denoted r^2.

correlation coefficient ■ A measure of the strength of the linear relationship between two variables, X and Y, and calculated using the equation

$$r = \frac{n\Sigma XY - \Sigma X \Sigma Y}{\sqrt{n\Sigma X^2 - (\Sigma X)^2}\sqrt{n\Sigma Y^2 - (\Sigma Y)^2}}.$$

negative correlation ■ A relationship between two variables where an increase (or decrease) in the values of one is associated with a decrease (or increase) in the values of the other.

ordered pair ■ Two variables whose values are matched and presented in a consistent order.

positive correlation ■ A relationship between two variables where an increase (or decrease) in the values of one is associated with an increase (or decrease) in the values of the other.

regression equation ■ The linear equation of the regression line, expressed as $\hat{Y} = b_0 + b_1X$.

regression line ■ The line of best fit through a set of data points, as determined by the least squares method.

scatter diagram ■ A graph of bivariate data used to analyze relationships.

OUTLINE

I. Linear correlation analysis
   A. Label as X the variable to be used as predictor and as Y the variable to be predicted.
   B. Plot a scatter diagram.
   C.

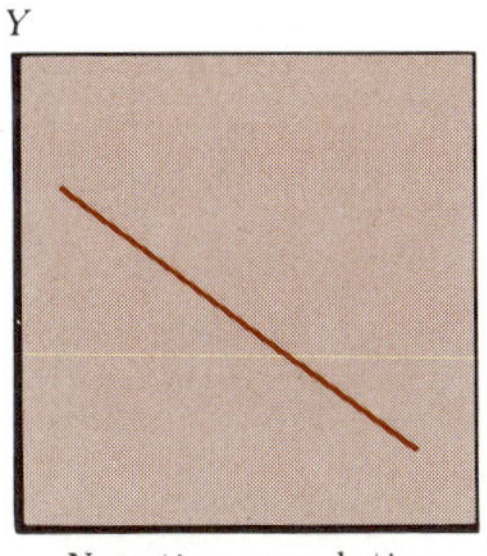

Negative correlation

Y X
Positive correlation

   D. Calculate the correlation coefficient, r:

$$r = \frac{n\Sigma XY - \Sigma X \Sigma Y}{\sqrt{n\Sigma X^2 - (\Sigma X)^2}\sqrt{n\Sigma Y^2 - (\Sigma Y)^2}};$$

$$-1 \leq r \leq +1.$$

   E. Hypothesis test for r:
      1. H_0: $\rho = 0$.
      2. H_A: $\rho \neq 0$.
      3. df = $n - 2$.
   F. The coefficient of determination explains the proportion of the variation in Y attributable to the linear relationship of Y with X.

II. Linear regression analysis
   A. Find the equation of the regression line $\hat{Y} = b_0 + b_1X$, using

$$b_1 = \frac{n\Sigma XY - \Sigma X \Sigma Y}{n\Sigma X^2 - (\Sigma X)^2},$$

   and

$$b_0 = \bar{Y} - b_1\bar{X}.$$

   B. For each value of X, the regression line goes through the mean of possible values for Y.
   C. Predict by using the regression equation: Choose a value for X, substitute it into the equation, and calculate a corresponding value for $\hat{Y}$.
   D. Do not predict
      1. outside the limits of the data values for X;
      2. beyond the time frame in which the data were collected; and
      3. for groups other than the population from which the original data were drawn.

REVIEW PROBLEMS

1. A teacher of remedial math suspects that attendance and course grade are correlated. Using her grade book, she computed the average course grade for students absent 0 days, 1 day, 2 days, 3 days, etc. The results of her work are summarized in Table 12.1 (A = 4.0, B = 3.0, C = 2.0, D = 1.0).
   a) Find r.
   b) Write the equation of the regression line.
   c) Estimate Y when $X = 2.5$.
2. A political science professor at a local state college wonders whether the precinct in which she votes is a bellweather for her state. She researches the record and discovers the percentages shown in Table 12.2 for the winning candidate for governor in the last eight gubernatorial elections. Find r. Was she correct for $\alpha = 0.05$?
3. A citizens' prison-reform action group wonders whether the number of prior convictions is correlated to length of current sentence for inmates at a certain prison. The group selected the records of 12 randomly chosen inmates and obtained the data shown in Table 12.3. Use it to find r. Length of current sentence is given in years.
4. During the course of a semester, a statistics professor accumulates a lot of data about his students. Among such data are a numerical midsemester grade and a course grade. A sample of ten of his students, chosen at random, yielded the data shown in Table 12.4.
   a) Draw a scatter diagram for these data.
   b) Calculate r.
   c) Determine the equation of the regression line.

TABLE 12.1

Days absent (X)	0	1	2	3	4	5	6
Average course grade (Y)	3.4	3.6	3.1	2.5	2.0	1.5	1.2

TABLE 12.2

Year	'52	'56	'60	'64	'68	'72	'76	'80
Winner's precinct percent	0.50	0.52	0.56	0.51	0.55	0.54	0.49	0.58
Winner's state percent	0.52	0.54	0.54	0.55	0.51	0.51	0.45	0.60

TABLE 12.3

Prisoner	1	2	3	4	5	6	7	8	9	10	11	12
Number of prior convictions	2	0	5	8	2	1	4	2	1	0	3	2
Length of current sentence	5	0.5	26	50	12	3	14	10	6	1	9	8

TABLE 12.4

Midsemester grade	99	80	61	77	24	91	63	49	95	83
Course grade	98	88	71	79	36	94	80	60	81	89

TABLE 12.5

Number of mistakes	10	12	9	5	15	3	0	7
Practice time	90	100	120	180	75	200	240	150

TABLE 12.6

Average daily temperature	25	40	12	30	15	38	33	18	20	10
Daily revenue	26	12	30	25	30	15	18	28	29	32

d) Use the equation of the regression line to estimate the course grade for a student with a midsemester grade of 85.

5. A piano teacher thinks that he has discovered a piece of music that is a good indicator of the amount of effort his pupils put into their training. He believes that there is a correlation between the number of mistakes committed by the pupil and the number of minutes per day spent practicing. A sample of eight of his students yielded the data shown in Table 12.5.

a) Draw a scatter diagram for these data.
b) Calculate r.
c) Is r significant at the 0.05 level?
d) Determine the equation of the regression line and use it to estimate the practice time of a pupil who only made one mistake.

6. The manager of a ski resort maintains records of average daily temperatures, in °F, and daily revenues, in thousands of dollars. For ten randomly selected days during the last ski season, the numbers shown in Table 12.6 were recorded.

a) Plot a scatter diagram for these data.
b) Calculate r.
c) Is r significant at the 0.05 level?
d) Determine the equation of the regression line and use it to estimate the revenue on a day having an average temperature of 28°F.

Further applications of chi-square

13

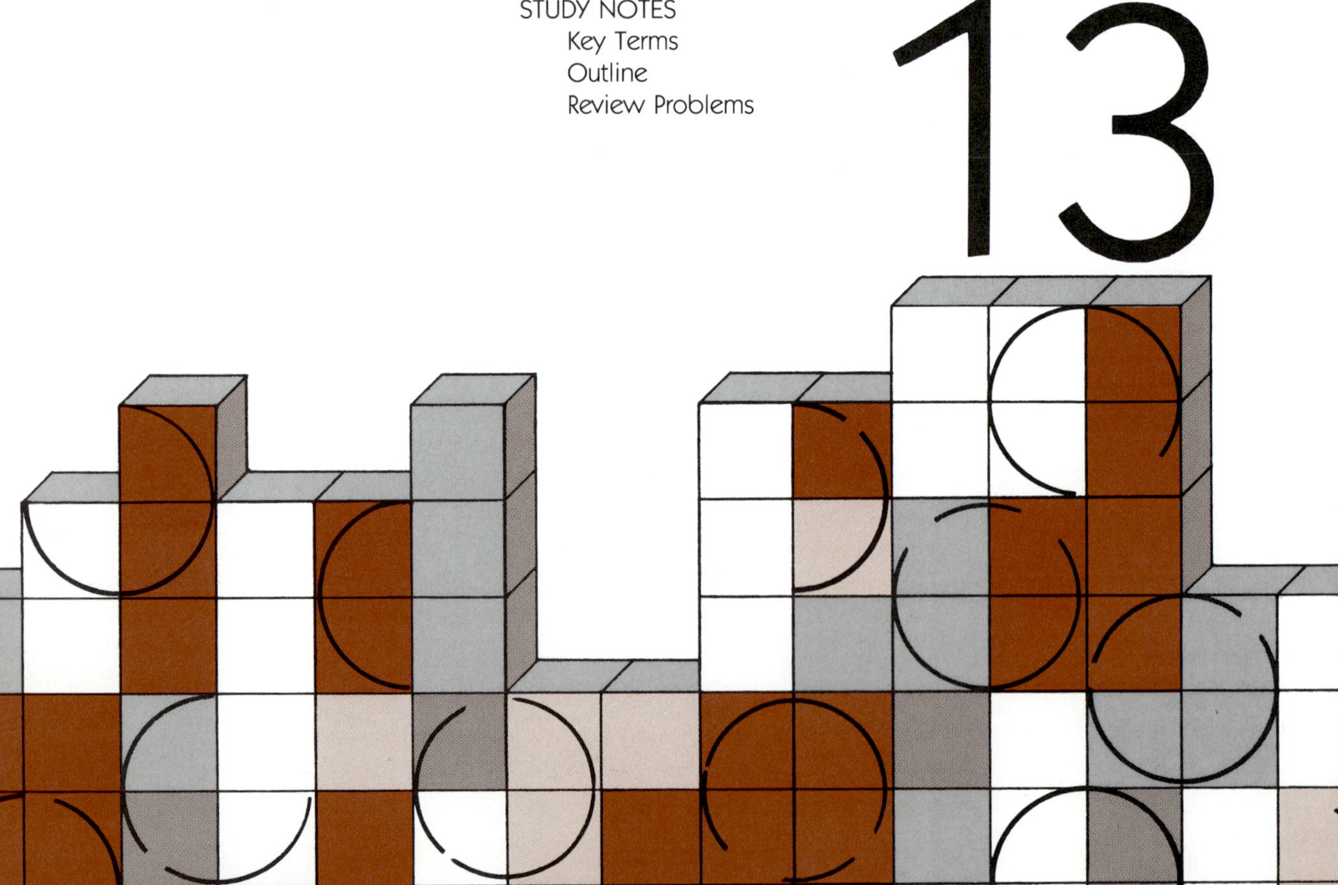

13.1 Tests for Multinomial Data—Homogeneity, Goodness of Fit

Homogeneity

In Chapter 9 we examined problems that involved binomial data. We now examine situations that require use of more than two values for a variable, or more than two categories into which we will group the members of our sample. As with binomial data, we analyze qualitative variables, that is, categorical data such as hair color, political affiliation, smoking habits, opinion, religion, and the like. The work in this chapter requires the use of probabilities associated with values of such variables; the groundwork that we lay in this section will be extended in later sections to cover other situations.

*Bi*nomial data is classified into two categories (success or failure). ***Multinomial* data** can be classified into many categories. Consider the following example.

EXAMPLE 13.1 A random sample of the final digits used by daily lottery winners resulted in the frequencies shown in the following table. Test the claim that each digit is equally likely to occur. In other words, there is no rigging of the results. Use $\alpha = 0.05$.

Final digit	0	1	2	3	4	5	6	7	8	9
Observed frequencies	7	12	8	7	15	12	9	6	14	10

Solution We are beginning a series of tests (to be continued in Chapter 14) in which H_0 does not have parameters. Rather, we will be testing statements that tell us about the probabilities associated with the values of the random variable, which in this case is the final digit of a lottery number.

H_0: The final digits are equally likely to occur.

The alternative is formed by negating this statement.

H_A: The final digits are not equally likely; that is, some are more likely than others, indicating that the game is not fair.

Since we do not know which digits to suspect, we must keep track of all of them.

We now develop a test statistic to enable us to examine these hypotheses. As before, we assume that H_0 is true and use its information. There are ten digits, 0–9. If they were equally likely, what probability would be associated with each? Answer: 1/10.

By adding the observed frequencies, we find that 100 winning numbers were examined. If the final digit of each has a probability of 1/10 of showing up, we should expect to see each digit 10 times, which is the statement of H_0 above.

Digit	0	1	2	3	4	5	6	7	8	9
Observed (O)	7	12	8	7	15	12	9	6	14	10
Expected (E)	10	10	10	10	10	10	10	10	10	10

In order to test the null hypothesis, we compare expected results with actual observations, using $O - E$. A cursory examination of the chart shows that some of these differences will be positive and some will be negative. Since we want to add them and do not want them to cancel each other, we first square each difference: $(O - E)^2$.

We also need to note that, for example, a difference of -3 (from 7–10) for the digit 0 is more significant than if that same -3 resulted, say, from a difference of 10,007–10,010. We need to put our work into the context of the size of the numbers with which we are dealing. We do this by dividing each calculation by E and then adding the results obtained for each digit. This statistic is described by the χ^2 distribution.

$$\chi^{2*} = \sum \frac{(O - E)^2}{E}$$

We calculate χ^{2*} as follows:

Digit	0	1	2	3	4	5	6	7	8	9
O	7	12	8	7	15	12	9	6	14	10
E	10	10	10	10	10	10	10	10	10	10
$O - E$	−3	2	−2	−3	5	2	−1	−4	4	0
$(O - E)^2$	9	4	4	9	25	4	1	16	16	0

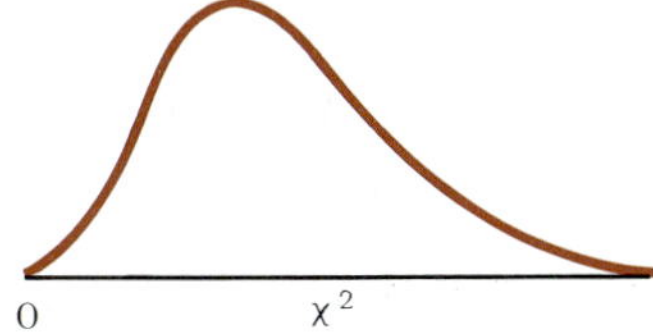

Figure 13.1

$$\chi^{2*} = \frac{9}{10} + \frac{4}{10} + \frac{4}{10} + \frac{9}{10} + \frac{25}{10} + \frac{4}{10} + \frac{1}{10} + \frac{16}{10} + \frac{16}{10} + \frac{0}{10}$$

$$= \frac{88}{10} = 8.8.$$

Before we can complete this test, we need to determine the critical value and find the rejection region in Fig. 13.1. Recall that the χ^2 distribution is never negative and starts at 0. Do we need to consider a left-tail rejection region? What would be the relationship between observed numbers and expected numbers if the calculation is to be small, that is, near 0? If the calculation is to be small, O and E must be very similar, so that $O - E$ is small. If what we observe agrees closely with what we expect, then H_0 is reinforced. Therefore a value of χ^2 in the left tail should not be a cause for rejection of H_0. Thus *all χ^2 tests comparing observed frequencies with expected ones will be one-tailed to the right.* When χ^{2*} values are large, a large discrepancy exists between what is observed and what is expected.

We use Table D.4 to find critical values. In order to read it, we need to determine the number of degrees of freedom. We are examining data classified by attribute values of the variable with k outcomes; therefore we no longer focus on n, but on the outcomes, which are usually called **cells** or classes. If the number of cells is k, then df $= k - 1$.

df $= k - 1$, where k = the number of cells.

Figure 13.2

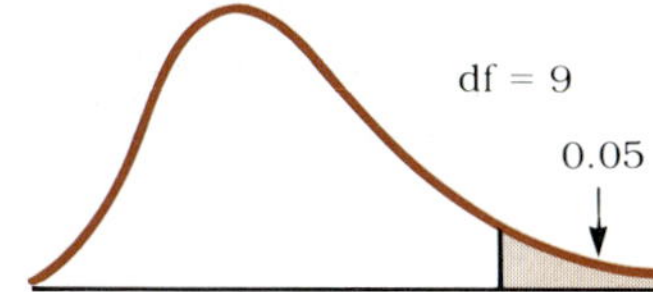

Using Fig. 13.2 and Table D.4, we find a critical value of 16.919; $\chi^{2*} = 8.8$, and thus we cannot reject H_0.

Summary

H_0: The final digits are equally likely to occur.

H_A: The final digits are not equally likely.

Test region: Reject H_0 when $\chi^{2*} > 16.919$.

Test statistic: $\chi^{2*} = 8.8$.

Conclusion: We cannot reject H_0. □

Before we look at another example, let us list the characteristics of the analysis we have just performed.

GENERALIZATIONS

1. The procedure followed in Example 13.1 is used to analyze qualitative variables. (Note the number of qualitative variables in News & Views 13.1.)
2. The experiment involves choosing a random sample of size n. The randomness allows an assumption of independence among the cells, and thus the probabilities associated with the cells will remain constant throughout the experiment.
3. Each piece of data fits into one and only one cell.
4. The null hypothesis determines the probabilities associated with each cell. It is usually expressed verbally.
5. The alternative hypothesis is formed by negating the null hypothesis.
6. The test is always right-tailed with all of α in the right tail.
7. $\Sigma O = n$.
8. For any cell, $E = np$ where p is the probability associated with the cell, as determined by H_0.
9. $\chi^{2*} = \Sigma(O - E)^2/E$, with df determined by the number of cells.

These characteristics apply throughout the rest of the chapter. However, before we move on, we need to elaborate on generalization number 3. When categorizing data, we must be sure that all the pieces fit into well-defined categories. We must provide an *other* or *miscellaneous* slot for those members of the sample that do not fit neatly into delineated categories. We must also be careful not to permit any ambiguity of interpretation. If we were studying eye color, could everyone objectively distinguish between hazel and green? Does everyone have the same definition for these two eye colors? They usually do not, so we must be careful to define them in such a way that everyone collecting data would categorize them the same way.

The χ^2 distribution is a continuous one, whereas the calculated χ^{2*} is based on discrete or attribute data. As before, when we studied the binomial distribution, np must be greater than 5 in order for χ^{2*} to be approximated by the chi-square distribution. Thus

$$E = np > 5.$$

Note that we make no such statement about O, the observed frequency. If E for any particular cell does not meet this condition, the problem may be solved by combining categories. For example,

NEWS & VIEWS 13.1

Staying single may be wiser move than remarriage, new study says

Marriage isn't always better the second time around.

A long-term study of divorced men and women finds that remarriage seldom brings an improved feeling of well-being, and that staying single may be the wiser course for some whose first marriage ended in court.

"We suspect that for every person who benefits from a remarriage, there is another whose psychological status is adversely affected by an unsuccessful remarriage," sociologists Graham B. Spanier and Frank F. Furstenberg, Jr., report in the *Journal of Marriage and the Family.*

Most studies of divorce and marriage have compared two separate groups of people—divorced vs. married, for example. Mr. Spanier and Mr. Furstenberg followed the same group of people for 2½ years to see what happened to them.

"This is one of the first studies to look at whether or not those who remarry after divorce are better off than those who do not," Mr. Spanier said in a telephone interview. "It conflicts with the undisputed public notion that the best thing that can happen to someone divorced is to get remarried."

Mr. Spanier, of the State University of New York at Stony Brook, and Mr. Furstenberg, of the University of Pennsylvania, studied the effects of divorce and remarriage on 180 people living in central Pennsylvania, primarily in Centre county.

"It is probably a fairly representative cross-section of what you find in America at large," Mr. Spanier said. "If you compare your sample with national samples, the demographic differences are not that great."

The two researchers found to their surprise that moving from the divorced state to a second marriage generally doesn't result in such things as greater satisfaction with life or higher self-esteem.

"We conclude that remarriage after divorce is not associated with enhanced well-being," the two write.

"Well-being" is a word commonly used but loosely defined by sociologists.

Mr. Spanier and Mr. Furstenberg used eight measurements to determine well-being—including satisfaction with life and health; self-esteem; thoughts of suicide; changes in drinking, smoking and gambling habits, and such physical complaints as insomnia, nervousness, headaches and indigestion.

In other words, Mr. Spanier said, their use of well-being means "basically how well people have adjusted, their overall quality of life and how comfortable they are with their social environment and psychological states."

Many studies have found that people generally feel an increased sense of well-being following their divorce. The Pennsylvania study found this continues.

"Most individuals report greater well-being three to four years after their final separation regardless of their marital status," Mr. Spanier and Mr. Furstenberg write.

But people vary considerably in just how much better they feel. And the two sociologists could find nothing that predicts who will have the greater sense of well-being after separation—not age, sex, education, whether children are involved or whether the person initiated the divorce or had it forced upon them.

"Undoubtedly, then, variation in well-being during the post-separation period must be attributed to as yet undiscovered influences," they conclude.

The researchers found that the greater the sense of well-being after divorce, the more likely a person is to remarry. They also found that people with a low sense of well-being after divorce are likely to end up with their initial feelings of failure reinforced in a second marriage.

The men and women in the study were interviewed first in the spring of 1977 and again in the summer and fall of 1979. At that time, 35 percent had remarried, 13 percent were single but living with a member of the opposite sex, and 52 percent were unmarried.

The two researchers found no significant difference between those who remarried and those who didn't in age, income, occupation, education, religion, depth of religious belief, sex or whether there were children at home.

Source: *Baltimore Sun,* October 28, 1982.

we could put all hazel- and green-eyed people in the same cell. This increases p for that cell, with the goal of increasing E for that cell.

EXAMPLE 13.2 The following information was gathered during a campaign to ban smoking in public places. A group of 150 people responded that they favored banning smoking in restaurants.

	Age		
	18–30	**31–54**	**Over 55**
Number in favor	37	59	54

At the 0.05 level of significance, does opinion vary with age?

Solution The null hypothesis states that there is no difference in the number of people in favor of banning smoking from one age group to another; that is, the probability of a randomly selected person being in favor of banning smoking is the same for each age group.

H_0: All age groups are equally likely to favor banning smoking.

H_A: All age groups are not equally likely to favor banning smoking.

Figure 13.3

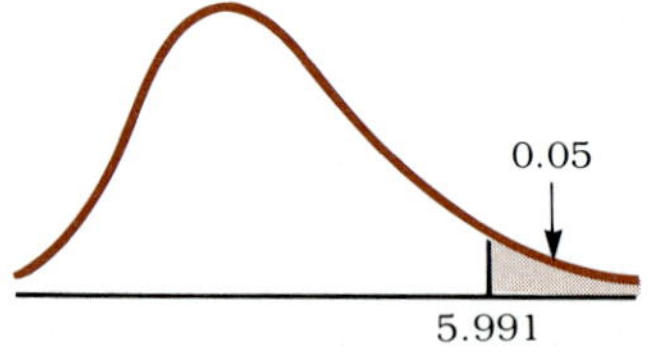

There are 3 cells, so df = 2 for Fig. 13.3. Since

$$\Sigma O = 37 + 59 + 54 = 150 = n,$$

and

$$E = np = 150 \cdot 1/3 = 50 \text{ for each cell,}$$

$$\chi^{2*} = \frac{(37 - 50)^2}{50} + \frac{(59 - 50)^2}{50} + \frac{(54 - 50)^2}{50}$$

$$= 5.32.$$

This value is less than 5.991, so we cannot reject H_0.

Summary

H_0: All age groups are equally likely to favor banning smoking in public places.

H_A: All age groups are not equally likely to favor banning smoking in public places.

Test region: Reject H_0 when $\chi^{2*} > 5.991$.

Test statistic: $\chi^{2*} = 2.34$.

Conclusion: We cannot reject H_0. All age groups are equally likely to favor banning smoking in public places. □

Goodness of Fit

The final type of problem we consider in this section is one that allows us to determine whether a set of data matches a theoretical probability distribution. We have looked at examples in which all values were equally likely, or homogeneous. We now consider other possibilities.

EXAMPLE 13.3 Three coins were tossed together 240 times, and the following results were recorded.

Number of Heads	0	1	2	3
Observed frequency	45	70	105	20

Are all of the coins fair; that is, can these results be due to chance, at the 0.05 level of significance?

Solution Normally, the distribution of probabilities associated with tossing coins follows a binomial distribution, with $n = 3$, $p = 0.5$, and $q = 0.5$. Thus

H_0: The probabilities associated with X = number of heads follow a binomial distribution with $n = 3$, $p = 0.5$, and $q = 0.5$; the coins are fair.

The alternative negates this.

H_A: The coins are not fair.

There are 4 cells, so df = 4 − 1 = 3 for Fig. 13.4. We need to find the probabilities associated with each cell in order to calculate $E = np$ for each cell. The null hypothesis gives us these probabilities as usual. Let X = the number of heads. From the binomial distribution, we get:

Figure 13.4

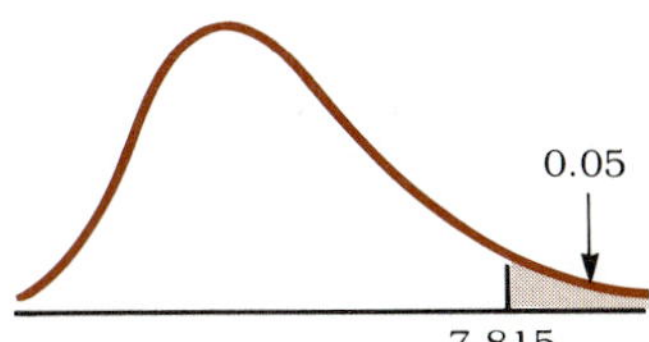

X	$P(X)$
0	$(1)(1/2)^0(1/2)^3 = 1/8$
1	$(3)(1/2)^1(1/2)^2 = 3/8$
2	$(3)(1/2)^2(1/2)^1 = 3/8$
3	$(1)(1/2)^3(1/2)^0 = 1/8$

Multiplying each of these fractions by the sample size of 240, we obtain the expected numbers for each cell.

Number of Heads	0	1	2	3
Observations	45	70	105	20
Expected values	30	90	90	30

$$\chi^{2*} = \sum \frac{(O - E)^2}{E}$$

$$= \frac{(45 - 30)^2}{30} + \frac{(70 - 90)^2}{90} + \frac{(105 - 90)^2}{90} + \frac{(20 - 30)^2}{30}$$

$$= 17.778.$$

Since $\chi^{2*} = 17.778 > 7.815$ (from Table D.4), we reject H_0.

Summary

H_0: The coins are fair and follow a binomial distribution.

H_A: The coins are not fair.

Test region: Reject H_0 when $\chi^{2*} > 7.815$.

Test statistic: $\chi^{2*} = 17.778$.

Conclusion: Reject H_0; the coins are not all fair. □

A chi-square test may also be used to determine whether a promise has been kept.

EXAMPLE 13.4 A snack-food mix is advertised as 30% sunflower seeds, 20% peanuts, 10% dried peas, 10% Japanese rice sticks, and 30% sesame sticks. After the purchase of a pound of the snack food, a customer noticed what he thought was an inordinate number of sunflower seeds and peanuts. He counted the components and found the following:

	Sunflower Seeds	Peanuts	Dried Peas	Japanese Rice Sticks	Sesame Sticks
Observed frequencies	210	140	45	45	160

Do his observations support the promise made in the advertisement, at the 0.01 level of significance?

Solution The null hypothesis is the advertised claim.

H_0: The percent distribution is as advertised.

Again we negate this to find H_A.

H_A: The percent distribution is not as claimed.

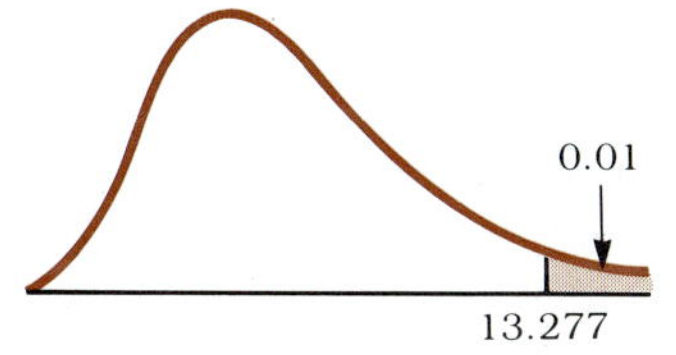

Figure 13.5

There are five cells, so df = 5 − 1 = 4 for Fig. 13.5. To find the expected numbers for each cell, we multiply the given percents by n. Recall that $\Sigma O = n = 600$.

	Sunflower Seeds	Peanuts	Dried Peas	Japanese Rice Sticks	Sesame Sticks
O	210	140	45	45	160
E	180	120	60	60	180

$$\chi^{2*} = \frac{(210-180)^2}{180} + \frac{(140-120)^2}{120} + \frac{(45-60)^2}{60} + \frac{(45-60)^2}{60} + \frac{(160-180)^2}{180}$$

$$= 18.055.$$

Since $\chi^{2*} = 18.055 > 13.277$, we reject H_0.

Summary

H_0: The percent distribution of ingredients is as advertised.

H_A: The percent distribution of ingredients is not as advertised.

Test region: Reject H_0 if $\chi^{2*} > 13.277$.

Test statistic: $\chi^{2*} = 18.055$.

Conclusion: Reject H_0; the ingredients are not mixed in the proportions claimed. □

EXERCISES/Section 13.1

1. In an attempt to examine work habits, a personnel officer collected the following data from randomly selected reports of employee tardiness (more than 15 min) by day of the week over a 3-month period. Are all days equally likely to have the same number of tardy work-

ers, at the 0.05 level of significance? Is tardiness evenly distributed throughout the week?

	M	T	W	Th	F
Number tardy	62	37	52	49	60

2. The chairperson of a psychology department wanted to determine whether all the department's instructors were equally popular. To do this, he gathered data on enrollment patterns for several semesters in which each professor taught a class at 11:00 a.m. each semester. At the 0.05 level of significance, are the four professors equally popular?

	Professor			
	A	B	C	D
Number of students	221	179	150	250

3. A major grocery chain wished to test the claim that its house-brand detergent was equally popular with two leading national brands. Of the 400 sales of detergent during the week of the study, the following choices were made. At the 0.05 significance level, are the brands equally popular?

	Brand A	Brand B	House Brand
Number selected	150	130	120

4. The deployment of nuclear missiles is concentrated in certain geographic areas of the country, and a politician was interested in learning whether sentiment for nuclear disarmament was also concentrated. He had his assistant tabulate the number of editorials in 35 newspapers from each section of the country, during a 6-month period, that spoke to this issue; he found the following results. Is concern for the nuclear disarmament issue evenly distributed throughout the country, at the 0.01 level of significance?

	East	Southwest	South	Central	Northwest
Number of editorials	28	24	19	20	19

5. Common wisdom holds that major industrial goods produced on certain days of the week are more likely to be defective. To test this assertion against a null hypothesis that the day of production made no difference in quality, a company vice-president used serial numbers to monitor the number of defectives produced by day of the week. Is imprecise manufacturing evenly distributed throughout the week, at the 0.01 level of significance?

	M	T	W	Th	F
Number of defectives	73	52	63	61	66

6. In genetics, theories are tested by examining the inherited characteristics of offspring after determining the characteristics of the parents. In the simplest case dominant, D, characteristics always show up in offspring whether pure (gene for dominant characteristic from both parents, DD) or not (one parent contributes the gene for dominant characteristic, one the gene for the recessive, R, characteristic, and the offspring is DR). This gene pair of DD or DR in an offspring is always displayed as the dominant characteristic. A recessive characteristic is manifested only when it is pure, or RR. Tables 13.1–13.3 give the gene pairs of offspring from possible pairings for parents.

TABLE 13.1

		Parent 1		
		D	R	
Parent 2	D	DD	DR	3/4 of offspring will display the dominant characteristic.
	R	DR	RR	1/4 will show the recessive trait.

TABLE 13.2

		Parent 1		
		D	D	
Parent 2	R	DR	DR	All offspring will show the dominant trait.
	R	DR	DR	

TABLE 13.3

		Parent 1		
		D	R	
Parent 2	R	DR	RR	½ of offspring will show the dominant trait. ½ will show the recessive trait.
	R	DR	RR	

Since Parent 1 in Table 13.2 and Parent 1 in Table 13.3 are indistinguishable (they both display the dominant trait), the following data were gathered to help the experimenter determine the case with which she was dealing. At the 0.05 level of significance, do these data sufficiently reflect the theory in Table 13.3 for the experimenter to continue under that assumption?

	Dominant	Recessive
Observed frequency	11	19

7. The number of tickets for speeding given by police officers at a radar trap (monitored by days) is shown in the following table. Let X = the number of tickets and f = the frequency in number of days on which X tickets were given. At the 0.01 level of significance, test this distribution to determine whether it is basically binomial with $p = 0.4$.

x	f	x	f
0	30	3	25
1	75	4	10
2	60		

8. The distribution of farms in the United States by size is as follows:

Size of farm (acres)	Percent of All Farmland
Under 500	31.7
500–999	13.9
1000–1999	11.6
2000 and over	42.8

A survey of farmers in a large state indicated the following distribution. Are these data representative of the nation, at the 0.05 level of significance?

Size of Farm (acres)	f
Under 500	160
500–999	55
1000–1999	65
2000 and over	220

9. The proportion of loans for new homes nationwide, by loan amount, guaranteed and insured by VA is:

Amount	Percent
Less than $25,000	12.2
$25,000–$29,999	23.8
$30,000–$34,999	31.2
$35,000–$39,999	19.5
$40,000 and over	13.3

A survey of VA-insured loans in a large urban area indicated the following. Is this urban area typical of the country as a whole, at the 0.05 level of significance?

Amount	Number
Less than $25,000	50
$25,000–$29,999	150
$30,000–$34,999	230
$35,000–$39,999	150
$40,000 and over	120

10. The following information shows the distribution of a sample number of families with children under 18 years of age. Let X = the number of children per family. Is this distribution binomial with $p = 0.1$? [Assume that $n = 7$ for the binomial distribution calculations, adding $P(4) + P(5) + P(6) + P(7)$ to obtain $P(4$ or more).] Use $\alpha = 0.05$.

X	*Families*
0	460
1	200
2	180
3	90
4 or more	70

13.2 Tests for Independence—Contingency Tables

Section 13.1 extended the analysis of binomial data. Similarly, this section extends to qualitative data the previous work done in correlation analysis. Correlation analysis deals with determining the existence of a relationship between two continuous variables. Now we look at situations where the variables are attributes or categories. News & Views 13.2 presents some examples of attribute variables and relationships between them.

The null hypothesis for such tests of relationships is always one of independence, as before.

H_0: The two characteristics are independent; that is, there is no relationship between them.

The alternative hypothesis negates this.

H_A: The two characteristics are not independent; that is, there is a relationship between them.

As in Section 13.1 we will use a χ^2 statistic and compare actual observations with the expected, assuming that the null hypothesis is true. The tests are right-tailed, with all of α in the right tail.

Observations will now be classified by *two* characteristics. For example, all blond-haired, blue-eyed people will be counted in the same cell. This *two-way classification of data* produces what is known as a **contingency table.**

contingency table. ■ A two-way classification of qualitative data.

NEWS & VIEWS 13.2

More schooling means less overeating, study finds

Well-educated people have the healthiest life-styles, according to a survey that found they are thinner, exercise more and smoke and drink less.

The survey of Massachusetts residents conducted by the state Department of Public Health found that 43 percent of the state's citizens are overweight, 33 percent smoke, 28 percent get no exercise and 12 percent indulge in "robust alcohol use."

It found that people with better educations have the fewest damaging health habits.

"The advantages of social position may well include behavior that enriches physical health," the researchers wrote. "Since a healthy life-style requires neither wealth nor higher education, preventive medicine and health education can perhaps produce their most profound impact within less advantaged strata of our society."

The survey of 1,091 persons was published in today's *New England Journal of Medicine.*

Among the findings were:

- Slightly more men than women describe themselves as smokers.
- People who watch five or more hours of television a day are the heaviest smokers. Almost half of them use tobacco.
- Almost 13 percent said they run or jog, and 56 percent claim to exercise at least twice a week. People who did not finish high school are the least likely to exercise.
- Robust drinkers—those who have four or more drinks at a sitting at least twice a week—tend to have a high school education or less.
- More than 90 percent of those with college educations drink at least once a week.
- People who did not complete high school are the most likely to be overweight.

Source: *Baltimore Sun*, Baltimore, April 29, 1982.

A contingency table has a number of rows (called r), one for each of the r values or categories of one variable, and a number of columns (called c), one for each of the c values of the other variable. A contingency table is sometimes called an $r \times c$ table (read "r by c") to give its dimensions. The number of degrees of freedom for such a table is

$$\text{df} = (r - 1)(c - 1).$$

Let us begin by using data collected from 1000 randomly chosen people on their hair color and corresponding eye color. Table 13.4—a 6×4 table—shows the number of people who have the same two attributes. Thus, for example, there are 100 people with both brown hair and brown eyes. We want to find, at the 0.05 level of significance, whether there is a relationship between hair color and eye color.

TABLE 13.4

	Eye Color				
Hair Color	**Brown**	**Blue**	**Green/ Hazel**	**Other**	**Totals**
Brown	100	140	200	10	450
Black	100	50	25	25	200
Blond	90	45	10	5	150
Red	5	15	5	0	25
Gray	30	40	5	0	75
Other	75	10	5	10	100
Totals	400	300	250	50	1000

Before we state the hypotheses, let us comment on the numbers to the right of and below the table. They are called **marginals.** The numbers across the bottom are called *column totals*, the numbers down the right side are called *row totals*, and the number in the bottom right-hand corner is called the *grand total* (which is equal to n).

marginals. ■ The numbers to the side of and below a contingency table created by totaling each row (row totals, T_r) and each column (column totals, T_c).

grand total. ■ Labeled n, it is found by summing either the row totals or the column totals. It may also be labeled T.

The null hypothesis is one of independence.

H_0: Hair color and eye color are independent.

The alternate hypothesis is its negation.

H_A: Hair color and eye color are not independent.

In order to calculate χ^{2*}, we need to find an E, the expected value, for each cell. Recall that $E = np$. As before, H_0 provides the necessary information to find p. Let us concentrate on the cell containing brown-haired and brown-eyed sample members. What is the probability that someone chosen randomly from the sample will have both brown hair and brown eyes?

$$P(\text{brown hair and brown eyes}) = ?$$

The null hypothesis states that hair color and eye color are independent, and thus

$$P(\text{brown hair and brown eyes}) = P(\text{brown hair}) \cdot P(\text{brown eyes}).$$

Since there are 450 brown-haired people,

$$P(\text{brown hair}) = \frac{450}{1000}.$$

Similarly,

$$P(\text{brown eyes}) = \frac{400}{1000},$$

and

$$P(\text{brown hair and brown eyes}) = \frac{450}{1000} \cdot \frac{400}{1000}.$$

Now E for that cell is

$$E = np = 1000 \cdot \frac{450}{1000} \cdot \frac{400}{1000} = \frac{450 \cdot 400}{1000}.$$

Where did these numbers come from? The numerator of E is the row total for the cell times the column total, and the denominator is the grand total.

$$E = \frac{T_r \cdot T_c}{T}$$

Table 13.5 shows the expected numbers for this problem. Each was calculated using the preceding equation. (Later we will examine a shortcut.)

TABLE 13.5

	Eye Color				
Hair Color	**Brown**	**Blue**	**Green/ Hazel**	**Other**	**Totals**
Brown	180	135	112.5	22.5	450
Black	80	60	50	10	200
Blond	60	45	37.5	7.5	150
Red	10	7.5	6.25	1.25	25
Gray	30	22.5	18.75	3.75	75
Other	40	30	25	5	100
Totals	400	300	250	50	1000

Several comments need to be made about Table 13.5 and the data:

1. The rows and columns still sum to the same totals. This must be so because we have merely resegregated the components of the same totals according to the null hypothesis.
2. Even though the original data involved counts, these numbers are estimates of expected values, averages if you will, and thus do not need to be whole numbers.
3. The most important observation to be made about Table 13.5 is that some of the expected numbers are less than 5, a situation we need to correct. We examine the table and find that the hair colors red and gray contain relatively few sample members. If we eliminate these categories and place their numbers in the *Other* category, we produce expected numbers that are now all larger than 5. The revised data are shown in Table 13.6, a new 4 × 4 table.

We recalculate the expected numbers, list them next to the observed values (so that we can ultimately calculate χ^{2*}), and circle the E's to obtain Table 13.7. We now have all the information needed to complete the test for independence. The critical region is from the χ^2 table (Table D.4) with the number of degrees of freedom

$$\text{df} = (r - 1)(c - 1) = (4 - 1)(4 - 1) = 3 \cdot 3 = 9$$

for Fig. 13.6.

Figure 13.6

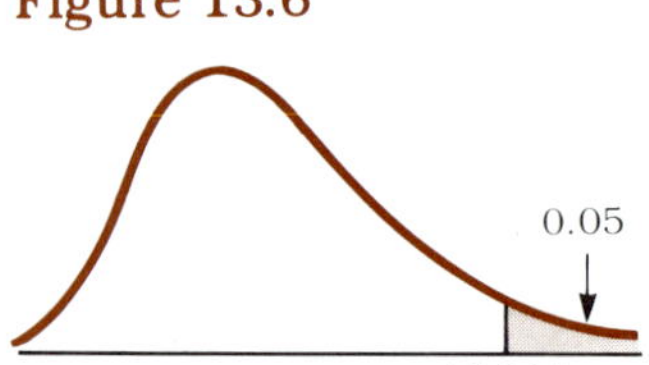

TABLE 13.6

Hair Color	Eye Color: Brown	Blue	Green/ Hazel	Other	
Brown	100	140	200	10	450
Black	100	50	25	25	200
Blond	90	45	10	5	150
Other	110	65	15	10	200
	400	300	250	50	1000

TABLE 13.7

Hair Color	Eye Color: Brown	Blue	Green/ Hazel	Other	
Brown	100 (180)	140 (135)	200 (112.5)	10 (22.5)	450
Black	100 (80)	50 (60)	25 (50)	25 (10)	200
Blond	90 (60)	45 (45)	10 (37.5)	5 (7.5)	150
Other	110 (80)	65 (60)	15 (50)	10 (10)	200
	400	300	250	50	1000

$$\chi^{2*} = \sum \frac{(O - E)^2}{E} = \frac{(100 - 180)^2}{180} + \frac{(140 - 135)^2}{135}$$

$$+ \frac{(200 - 112.5)^2}{112.5} + \frac{(10 - 22.5)^2}{22.5} + \frac{(100 - 80)^2}{80}$$

$$+ \frac{(59 - 60)^2}{60} + \frac{(25 - 50)^2}{50} + \frac{(25 - 10)^2}{10} + \frac{(90 - 60)^2}{60}$$

$$+ \frac{(45 - 45)^2}{45} + \frac{(10 - 37.5)^2}{37.5} + \frac{(5 - 7.5)^2}{7.5}$$

$$+ \frac{(110 - 80)^2}{80} + \frac{(65 - 60)^2}{60} + \frac{(15 - 50)^2}{50} + \frac{(10 - 10)^2}{10}$$

$$= 212.396.$$

Note that the denominators are the circled numbers, the E's, and are no longer all equal (as they were in Section 13.1). Since χ^{2*} is greater than 16.919, we reject H_0. We can summarize as follows.

H_0: Hair color and eye color are independent.

H_A: Hair color and eye color are not independent.

Test region: Reject H_0 when $\chi^{2*} > 16.919$.

Test statistic: $\chi^{2*} = 212.396$.

Conclusion: We reject H_0. Hair color and eye color are related.

EXAMPLE 13.5 Each person in a random sample of 1000 registered voters (classified according to family income) was asked to give an opinion of the president's foreign policy. Does the sample evidence in the following table support the hypothesis that income level is independent of opinion about the president's foreign policy? Use $\alpha = 1\%$.

Opinion	Income Level ($)			
	0–10,000	10,001–25,000	25,001–100,000	Over 100,000
Favor	100	125	150	100
Oppose	200	175	100	50

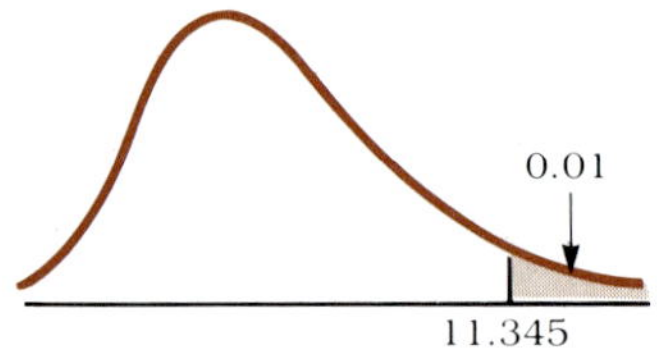

Figure 13.7

Solution

H_0: Opinion is independent of income level.

H_A: Opinion is not independent of income level.

Since there are two rows and four columns, df = (2 − 1)(4 − 1) = 1 · 3 = 3 for Fig. 13.7. In order to calculate χ^{2*}, we need to find the marginals and the expected numbers for each cell. The marginals are:

	0–10,000	10,001–25,000	25,001–100,000	Over 100,000	
Favor	100	125	150	100	475
Oppose	200	175	100	50	525
	300	300	250	150	1000

Let us begin calculating the expected numbers by using the first row.

$$E = \frac{T_r \cdot T_c}{T} = \frac{475 \cdot 300}{1000} = 142.5.$$

$$E = \frac{475 \cdot 300}{1000} = 142.5.$$

$$E = \frac{475 \cdot 250}{1000} = 118.75.$$

We have now calculated three E's, the same number as df. Since we know that the row totals and the column totals must remain unchanged, we can shorten our work by calculating the remainder of the E's by subtracting from the column or row totals.

	0–10,000	10,001–25,000	25,001–100,000	Over 100,000	
Favor	100 (142.5)	125 (142.5)	150 (118.75)	100 (71.25)	475
Oppose	200 (157.5)	175 (157.5)	100 (131.25)	50 (78.75)	525
	300	300	250	150	1000

$$\chi^{2*} = \sum \frac{(O-E)^2}{E} = \frac{(100-142.5)^2}{142.5} + \frac{(125-142.5)^2}{142.5}$$

$$+ \frac{(150-118.75)^2}{118.75} + \frac{(100-71.25)^2}{71.25} + \frac{(200-157.5)^2}{157.5}$$

$$+ \frac{(175-157.5)^2}{157.5} + \frac{(100-131.25)^2}{131.25} + \frac{(50-78.75)^2}{78.75}$$

$$= 66.00.$$

Since χ^{2*} is greater than 11.345, we reject H_0.

Summary

H_0: Opinion is independent of income level.

H_A: Opinion is not independent of income level.

Test region: Reject H_0 when $\chi^{2*} > 11.345$.

Test statistic: $\chi^{2*} = 66.00$.

Conclusion: We reject H_0. Opinion and income level are not independent. □

EXERCISES/Section 13.2

1. The data opposite were collected from a random sample of 830 people in an attempt to determine whether occupation is related to political affiliation. Is there a relationship at the 0.05 level of significance?

	Democrat	Republican	Other
Blue Collar Worker	180	120	45
White Collar Worker	90	165	50

2. The data in Table 13.8 show the socioeconomic status of parents, as reflected by occu-

TABLE 13.8

SES	Educational Level			
	HS	<2 Years College	College Grad	Graduate Degree
Professional	30	48	120	92
Clerk	226	127	62	15
Craftsperson	42	60	62	20
Laborer	32	7	6	2
Farmer	6	18	20	5

pation, and the educational level achieved by their children. Are SES of a parent and educational level achieved by a child related, at the 0.05 level of significance?

3. Mathematics is a required subject for many major fields of study. Is there a relationship between major and grade in a math course, at the 0.05 level of significance?

	A, B	C	D, F
Engineering	41	9	4
Liberal Arts	286	92	50
Business	130	70	60
Science	55	22	11

4. Does opinion concerning the budgetary deficiencies in the Social Security system vary with age? The following responses to an interviewer's question about the desirability of increasing payroll deductions to support Social Security were obtained. Is there a relationship between opinion and age, at the 0.05 level?

	Age		
	18–29	30–59	Over 60
Favor	62	114	35
Oppose	91	136	70
No opinion	18	72	2

5. Screening tests are given to job applicants to test skills required for the job and to provide information to help in the making of hiring decisions. A follow-up of successful applications produced the following information. Does the test successfully relate to later job performance, at the 0.05 level of significance?

Job Rating	Grade on Screening Test			
	A	B	C	D
1	52	32	19	6
2	34	49	25	15
3	14	21	61	72

6. Caveat emptor. Does educational level help the buyer beware? A survey of opinion on consumer legislation produced the following results. Is there a relationship between educational level and proconsumer attitude, at the 0.05 level of significance?

	Educational Level		
	HS Graduate	College Graduate	Advanced Degree
Favor	18	22	30
Oppose	24	17	19
No Opinion	18	11	11

7. As various aspects of society are affected by the computer, resistance to change is expressed. In banking, savings institutions offer a choice between passbook accounts—where transactions are recorded manually by a teller in a small passbook—or statement accounts—where transactions are recorded electronically and periodic statements mailed to the owner of the account. Is the choice of method affected by age? Use $\alpha = 0.05$.

	Age		
	18–29	30–59	Over 60
Passbook Account	9	20	60
Statement Account	41	45	14

Study Notes

KEY TERMS

cell ■ A value of the variable, an outcome of an experiment, or a location of a two-way classification of data.

contingency table ■ A two-way classification of qualitative data.

grand total ■ Labeled n, it is found by summing either the row totals or the column totals. It may also be labeled T.

marginals ■ The numbers to the side of and below a contingency table created by totaling each row (row totals, T_r) and each column (column totals, T_c).

multinomial data ■ Data that can be classified into many categories (as opposed to binomial data, which can be classified into only two categories).

CHAPTER OUTLINE

I. Characteristics and assumptions
   A. The procedure is used to analyze qualitative data.
   B. The experiment consists of choosing a random sample of size n with independence among the cells to guarantee that the probabilities associated with the cells remain constant throughout the experiment.
   C. Each data value fits into one and only one cell.
   D. H_0 determines the probabilities associated with the cells. It is expressed verbally, rather than parametrically.
   E. H_A negates H_0.
   F. All of α goes in the right tail of the χ^2 distribution.
   G. $\Sigma O = n$
   H. For any cell, $E = np$ should be greater than 5, where p is the probability associated with the cell as determined by H_0.

II. $$\chi^{2*} = \sum \frac{(O - E)^2}{E}.$$

III. Degrees of freedom
   A. For multinomial data and goodness of fit tests, df $= k - 1$, where k = the number of cells.
   B. For tests of independence using contingency tables, df $= (r - 1)(c - 1)$, where r = the number of rows and c = the number of columns.

IV. Contingency tables
   A. H_0: The two characteristics are independent.
   B. H_A: The two characteristics are not independent.
   C. $$E = \frac{T_r \cdot T_c}{T} \text{ for each cell.}$$

REVIEW PROBLEMS

1. Is air transport terrorism a uniform occurrence; that is, does it occur equally often over a number of years? Use $\alpha = 0.01$.

Year	Number of Incidents	Year	Number of Incidents
19X1	59	19X5	20
19X2	60	19X6	24
19X3	22	19X7	34
19X4	26		

2. Is there a relationship, at the 0.05 level of significance, between socioeconomic status and income level? Use the data in Table 13.9, obtained from a randomly chosen group of 700 people.

3. Is opinion on morality independent of the sex of the respondent for $\alpha = 0.05$? The following question evoked the responses shown: Will the new, more liberal morality weaken the institution of marriage?

	Opinion		
	Agree	Disagree	Don't Know
Men	61	31	8
Women	64	27	9

4. Is voter preference independent of the sex of the voter? Use $\alpha = 0.05$. When asked whether they would vote for a woman for Congress, the members of the sample gave the following preferences.

Voter	Preferred Candidate	
	Man	Woman
Man	47	9
Woman	29	15

5. Is age at which death occurs due to heart disease related to sex? Use $\alpha = 0.01$.

Age	Sex	
	Men	Women
0–24	1	1
25–44	3	1
45–54	8	4
55–64	18	10
65 and over	69	85

6. Is marital status related to age for $\alpha = 0.05$? Use the data in Table 13.10.

TABLE 13.9

SES	Income		
	Less than $20,000	$20,001–$50,000	Over $50,000
Professional	25	125	100
Clerk	75	23	2
Craftsperson	10	110	30
Laborer	70	25	5
Farmer	20	17	63

TABLE 13.10

	Age					
	18–24	25–34	35–44	45–54	55–64	65 and Over
Married	14	56	82	85	57	24
Widowed	0	8	20	80	104	174
Divorced	5	64	103	105	48	15
Single	78	85	77	84	67	54

7. Nationally, the distribution of age for the head of a household is:

Age	Percent
14–24	8.2
25–34	21.0
35–44	16.7
45–54	18.2
55–64	15.9
65 and over	20.1

In a new planned community, which claimed to provide services for all ages and socioeconomic levels, a random sample of 150 people produced the following data on age of head of household. Does the community live up to its advertising in terms of its residents' age distribution? Use $\alpha = 0.05$.

Age	f
14–24	20
25–34	40
35–44	30
45–54	25
55–64	20
65 and over	15

8. In Chapter 3 data on the weights of 90 adult males were presented. They are reproduced in Table 13.11. For these data, $\bar{X} = 162.5$, and $s = 24.5$.
   a) Find the areas under a normal distribution using $\mu = 162.5$ and $\sigma = 24.5$ for the classes in the frequency table derived from these data (p. 43). Be sure to use class boundaries.
   b) Calculate values for E for each class by multiplying the areas (probabilities) found in part (a) by 90.
   c) Using a goodness-of-fit test with $\alpha = 0.05$, determine whether these data could have come from a normally distributed population. The number of degrees of freedom for this problem is $k - 3$, where k is the number of classes. Two degrees of freedom are lost because $\bar{X}$ and s are used to estimate μ and σ.

TABLE 13.11 Weights of Adult Males (in lb)

108	144	155	167	181
111	144	156	168	182
121	145	158	169	184
123	145	159	169	185
125	146	159	170	185
130	147	160	171	186
130	148	160	171	187
131	148	161	172	188
134	149	161	174	190
138	149	162	175	192
138	150	162	175	193
140	151	162	175	195
140	152	163	176	196
141	153	164	178	198
142	154	164	178	200
142	154	166	179	212
144	154	166	180	225
144	155	167	181	247

Nonparametric statistics

14

14.1 Sign Test

Thus far, much of our work has involved tests that (1) require an assumption of a normal distribution for the population and (2) use this normality assumption and the associated parameters μ and σ to analyze data. Now we look at tests that do not have a strict requirement of a normal distribution for the population, called **distribution-free tests,** and tests that do not use the parameters μ and σ, called **nonparametric tests.**

These tests can be used under very general conditions and with fewer restrictive assumptions. They are easier to understand and require fewer calculations than parametric methods. The data need not even be numerical, as we learned in Chapter 13 for the χ^2 tests (which are nonparametric). The lack of strict normal distribution requirements, the capability of analyzing a wider range of variable types, and ease in computation are the advantages of nonparametric tests. However, these tests have disadvantages; for example, the specificity of individual data values is lost, and thus we may need a strong case in order to reject H_0. Let us deal with two such tests: the distribution-free **sign test** and the **rank-sum test.**

One-Sample Sign Test

The *sign test* provides an alternative to the one-sample t test (Chapter 8) and to the paired or matched t test (Section 9.3) to test *medians.* We use a null hypothesis concerning the median of the population.

> ***sign test.*** ■ A test of the median, rather than the mean, of a population without a normality assumption.

EXAMPLE 14.1 In an attempt to elicit sympathy for the fiscal plight of his area, a U.S. Senator stated that one-half the residents of a certain large city in his state made less than the median national income. A pollster gathered the following data on annual incomes from a sample of size 15 in that city.

\$17,100	\$13,100	\$15,000
15,600	15,200	8,000
14,200	12,000	11,500
18,700	25,600	14,100
9,000	16,100	15,900

If the median national income is \$16,000, is the senator entitled to sympathy? That is, at the 0.05 level of significance, test the median national income against an alternative of Med < \$16,000.

Solution Rather than eliciting sympathy, the senator's statement should merely remind us that, by definition, one-half of the wage earners in the country make less than the median national income; it may be that the city in his state is typical, rather than in need of financial aid. We analyze the data with a sign test.

H_0: The median is \$16,000, or Med = \$16,000.

H_A: Med < \$16,000.

The sign test requires that we merely determine whether data values are larger or smaller than the value stated in the null hypothesis by assigning each piece of data a (+) for larger than or a (−) for smaller than.

\$17,000	+	\$13,100	−	\$15,000	−
15,600	−	15,200	−	8,000	−
14,200	−	12,000	−	11,500	−
18,700	+	25,600	+	14,100	−
9,000	−	16,100	+	15,900	−

There are 4 plusses and 11 minuses. Since we are dealing with medians, the null hypothesis tells us that

$$P(\text{Med} < \$16{,}000) = P(\text{Med} > \$16{,}000) = 0.5,$$

or the plus and minus signs are randomly distributed and follow the binomial distribution with $p = 1/2$. We focus our attention on the minus signs, since H_A has a *less than* symbol; that is, $p < 1/2$. Thus we can use the binomial distribution that we studied in Chapter 6 to help us complete this analysis. For $n = 15$, $p = 1/2$, what is $P(X \geq 11)$? That is, what is the probability that, if H_0 is true, we would obtain 11 or more minus signs?

We are looking for a rejection region in the right tail of Fig. 14.1. Using the binomial formula,

$$P(X) = \binom{n}{X} p^x q^{n-x},$$

and substituting, we obtain

$$P(X \geq 11) = P(11) + P(12) + P(13) + P(14) + P(15)$$

$$= \binom{15}{11}\left(\frac{1}{2}\right)^{11}\left(\frac{1}{2}\right)^{4} + \binom{15}{12}\left(\frac{1}{2}\right)^{12}\left(\frac{1}{2}\right)^{3} + \binom{15}{13}\left(\frac{1}{2}\right)^{13}\left(\frac{1}{2}\right)^{2}$$

$$+ \binom{15}{14}\left(\frac{1}{2}\right)^{14}\left(\frac{1}{2}\right)^{1} + \binom{15}{15}\left(\frac{1}{2}\right)^{15}\left(\frac{1}{2}\right)^{0}$$

$$= 0.0417 + 0.0139 + 0.0032 + 0.0005 + 0.0000$$

$$= 0.0593.$$

We have found that the shaded area in Fig. 14.1 exceeds the predetermined level for α. Our count of 11 minuses is in the acceptance region, and we cannot reject H_0. Twelve or more minuses would have allowed rejection of H_0; 11 or less do not.

Summary

H_0: Med = \$16,000.

H_A: Med < \$16,000.

Test region: Reject H_0 if there are 12 or more minuses in the data.

Test statistic: There are 11 minuses.

Conclusion: We cannot reject H_0. The sample could have occurred by chance if H_0 is true. □

Figure 14.1

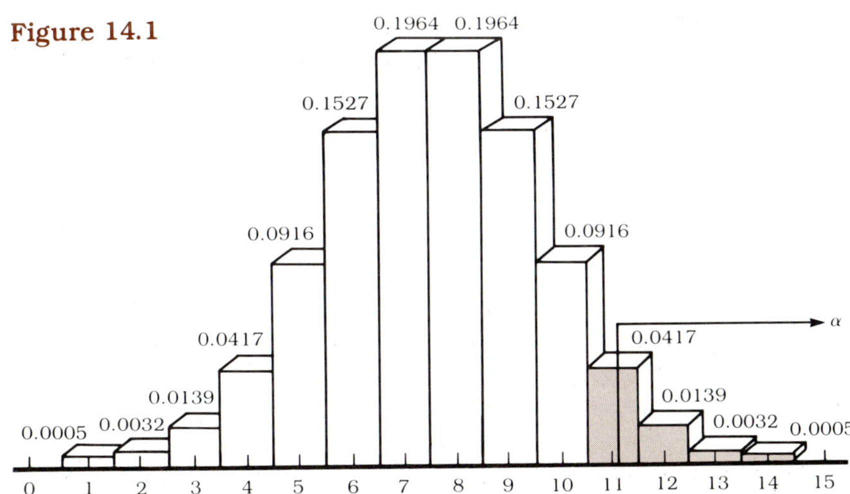

Let us review and comment on this procedure. The hypotheses require comparison to a median by labeling each data value with a (+) if it is larger than the hypothesis value or a (−) if it is smaller than that value. We then use the binomial distribution to obtain a probability associated with the number of signs and the tail of the distribution. If this probability is smaller than α, we can reject H_0; the probability that our results are due to chance alone is smaller than our predetermined, acceptable α. If this probability is larger than α, then we cannot reject H_0; the probability that our results are due to chance is too great. If we draw a histogram of the binomial distribution for $p = 0.5$ and the given n, it will graphically display whether the number of signs falls in the rejection or acceptance region, as determined by α.

One possible characteristic of the data that needs to be discussed briefly is the case of values that tie (are the same as) the hypothesis value. Since we are concentrating on the number of (+) or (−) signs, we can ignore these ties by reducing n by one for each tie.

If n is large, making calculation of the binomial probabilities tedious, we can utilize the technique discussed in Section 6.4: the normal approximation to the binomial distribution. Recall that np and nq must both be greater than 5; or since $p = q = 1/2$,

$$n\left(\frac{1}{2}\right) > 5.$$

We must transform the z score for these problems, remembering that

$$\mu = np = \frac{n}{2};$$

$$\sigma = \sqrt{npq} = \sqrt{n \cdot \frac{1}{2} \cdot \frac{1}{2}} = \frac{\sqrt{n}}{2}.$$

Thus

$$z = \frac{X - \mu}{\sigma}$$

becomes

$$z = \frac{X - (n/2)}{\sqrt{n}/2}.$$

EXAMPLE 14.2 At randomly selected times, a radar check on Meadow Drive produced the following information. Of 800 cars, 9 were being driven at the speed limit of 25 mph, 11 were being driven under the speed limit, and 780 were being driven over 25 mph. At the 0.05 level of significance, is the 25-mph speed limit unrealistically low for this heavily traveled road?

Solution In this case, we test the speed limit by using it as the median actual speed.

$$H_0\text{: Med} = 25.$$

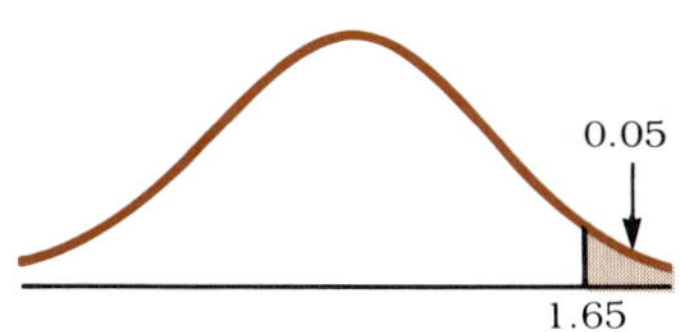

Figure 14.2

We are examining the data to find out whether 25 is too small.

$$H_A\text{: Med} > 25.$$

Since $np = 800 \cdot 1/2 = 400 > 5$, we can use the normal approximation to the binomial distribution in Fig. 14.2 and reject H_0 when $z^* > 1.65$. We need to obtain the probability of at least 780 plusses (values greater than the median) and ignore the 9 values equal to 25, which reduces n to 791. We calculate the z score to be:

$$z^* = \frac{780 - (791/2)}{\sqrt{791}/2} = 27.34.$$

Since $z^* > 1.65$, we can reject H_0.

Summary

H_0: Med = 25.

H_A: Med > 25.

Test region: Reject H_0 if $z^* > 1.65$.

Test statistic: $z^* = 27.34$.

Conclusion: Reject H_0. □

Paired-Sample Sign Test

We can use the same procedure as an alternative to the matched t test of Section 9.3, in order to test the equality of two medians. Again, we find the difference between the matched pairs of data, recording only (+) for positive changes and (−) for negative changes—and exclude ties. The null hypothesis states that no change occurs, and thus implies that H_0: $P(+) = P(-) = 0.5$. We can now proceed as we did in Example 14.1; that is compare with α the probability that the data results occurred by chance and draw a conclusion.

EXAMPLE 14.3 In order to determine the value of a remedial math program, eight students were chosen at random and given a standardized math aptitude test. After completion of the program, the same students were given an equivalent test. The grades of the students on both tests were:

Student	1	2	3	4	5	6	7	8
Pretest score	60	65	50	58	56	62	52	55
Posttest score	75	88	65	70	56	59	69	70

Did the eight students benefit from the program? That is, did their test scores significantly increase for $\alpha = 0.01$?

Solution Recording the changes that occurred at (+) or (−) we obtain:

1	2	3	4	5	6	7	8
+	+	+	+	0	−	+	+

There are 6 plusses and 1 minus. We ignore the data for person number 5 and reduce n to 7. The null hypothesis states that no change occurred; that the median difference in test scores is 0.

$$H_0\text{: Med} = 0.$$

The alternative requires that we look for an increase.

$$H_A\text{: Med} > 0.$$

Since H_A focuses our attention on a positive change, we find $P(X \geq 6)$ for $n = 7$ and $p = 0.5$.

$$P(6) = \binom{7}{6}\left(\frac{1}{2}\right)^6\left(\frac{1}{2}\right)^1 = 0.0547;$$

$$P(7) = \binom{7}{7}\left(\frac{1}{2}\right)^7\left(\frac{1}{2}\right)^0 = 0.0078.$$

From Fig. 14.3 we can see that $P(X \geq 6) > \alpha$, and we cannot reject H_0.

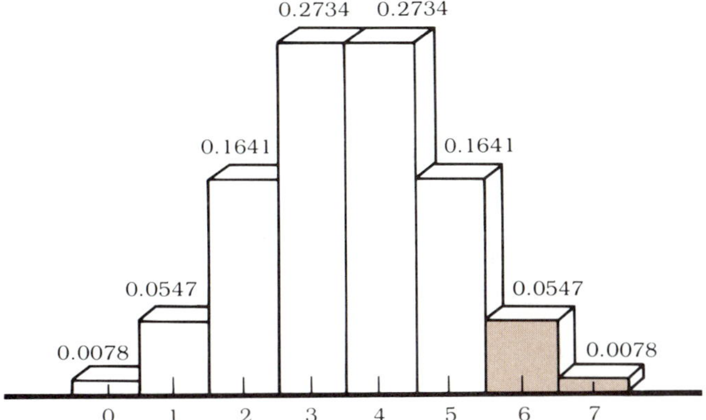

Figure 14.3

Summary

H_0: Med = 0.

H_A: Med > 0.

Test region: Reject H_0 if there are 7 plusses in the data, since $P(7) < \alpha$.

Test statistic: There are 6 plusses.

Conclusion: We cannot reject H_0. The program has not been proven effective. □

These examples demonstrate the major weakness of the sign test: It does not use all of the information in the data; each piece of data is reduced to a (+) or (−) sign, and the magnitude of any change or comparison is lost completely. In this case the changes were fairly large, but we were unable to reject H_0. This same example was used in Section 9.3 (Example 9.7), and the opposite conclusion was drawn. The t test previously used *did* allow rejection of H_0 because it utilized the size of the changes. In general, parametric tests (such as the t test) are more *powerful.* In technical terms this means that β is smaller. Recall that β is the probability of a Type II error, the probability of accepting a false null hypothesis. The **power of a test** is defined as $1 - \beta$, the probability of rejecting a false null hypothesis; thus a smaller β is desirable.

> ***power of a test.*** ■ The probability of rejecting a false hypothesis, $1 - \beta$.

For normal distributions, the sign test has a larger β than the t test and is less sensitive because it ignores the size of differences in the data. It will indicate the rejection of H_0 less often. We should not be tempted to use the sign test all the time because of its simplicity; it is better to use parametric tests if we think that the original data come from an approximately normal population. In situations where normality is certain, parametric tests are more powerful and lead to the rejection of false null hypotheses with higher probabilities.

EXERCISES/Section 14.1

1. A sociologist, acting as a consultant to a small manufacturing firm, administered a test to determine the level of job satisfaction among workers. He obtained the following scores:

   90, 63, 81, 82, 94, 59, 63, 73, 72, 84, 81, 67

   Previous experience had taught him that a median score of at least 65 indicated that a company was running smoothly. At the 0.01 level of significance, can he provide an optimistic report to the firm, as H_A, against an H_0: Med = 65?

2. Team statistics for the first half of the basketball season indicate that, of the top ten scorers in the league, three are scoring more than 27 points per game; one is scoring 27; and the rest are scoring less than 27 points. At the 0.01 level of significance, test the hypothesis that the median is less than 27 points per game.

3. To measure how frustration affects performance, the level of play of seven tennis players was followed after they were required to wait more than two hours for a court and were pitted against players of equal skill who were not required to wait. The number of wins for each player was: 1, 3, 2, 2, 2, 0, 2. At the 0.05 level of significance, test the hypothesis that the median number of wins is 2.5.

4. For each of the five months of the warm season when an amusement park does most of its business, the following number of entrance tickets was purchased: 14,000; 12,500; 15,500; 18,500; 9,500. The median for the industry nationwide is 18,000 tickets per month. Using a one-tail sign test at the 0.05 level of significance, test the hypothesis that the admissions figures for this particular park are representative.

5. Example 14.2 referred to a speed check. The null hypothesis of Med = 25 was rejected. Reexamine the data, using 30 mph as the benchmark. Of the 800 cars, 93 were driven at less than 30 mph; 40 were driven at 30 mph; and 667 were driven at more than 30 mph. At the 0.05 level of significance, is 30 mph a satisfactory speed limit or is it still unrealistically low?

6. A large fertilizer company, testing a new growth hormone, gathered the following information on the height of a variety of plant ten days after germination: 6 plants were under 10″ tall; 5 were 10″ tall; and 29 were over 10″ tall. If the median height for this plant is 10″ ten days after germination, has this hormone improved growth, at the 0.05 level of significance?

7. A sample of 90 adult women showed that 10 weighed less than a previously established median weight of 130 lb for the population as a whole, 5 weighed 130 lb, and 75 weighed more than 130 lb. Does this sample indicate that weights are increasing? Use $\alpha = 0.05$.

8. Cancer death rates for women in 22 industrialized countries showed 8 below and 14 above the worldwide median of 110 per 100,000 population. At the 0.01 significance level, is death due to cancer more prevalent in industrialized countries?

9. The following are election returns for randomly selected precincts by percent of votes cast for the Democratic candidate during two consecutive presidential elections. Use a sign test to determine whether the vote decreased substantially, at the 0.01 level of significance.

10. In an attempt to judge extrasensory perception, ten subjects were put through a series of tests while awake and then again while hypnotized. The following scores were obtained.

Subject	Awake	Hypnotized
1	23	23
2	12	23
3	21	21
4	15	24
5	19	21
6	15	27
7	17	22
8	21	19
9	18	19
10	14	23

Use a sign test to determine whether hypnotism improved the ESP of these subjects, at the 0.05 level of significance.

11. A study to determine the effect of nursing intervention on behavior modification of patients over a two-month period produced the following test results. Use a sign test to determine whether nursing intervention produced a significant change in behavior, as indicated by an increased score at the 0.05 level of significance.

Subject	Initial	Final
1	31	32
2	40	39
3	41	42
4	118	134
5	57	73

12. The following data come from a study that followed veterans over a ten-year period. The diastolic blood pressures obtained at the beginning and again at the end of the period are shown. Use a sign test to determine whether blood pressure changes significantly with age. Use $\alpha = 0.01$.

TABLE 14.1

Precinct	1	2	3	4	5	6	7	8	9	10	11	12
19X4	64	72	38	29	51	32	97	63	13	44	91	83
19X8	62	69	39	28	55	28	75	57	22	43	86	79

TABLE 14.2

Patient	1	2	3	4	5	6	7	8	9
Beginning DBP	80	110	88	94	74	80	80	80	90
Ending DBP	84	58	70	88	90	68	58	76	96

14.2 Rank-Sum Test

Let us consider the testing of the difference between means for two populations using the **rank-sum test.*** It requires that all data be ranked as though they were in one sample, but each sample list is still kept separate. We use the magnitude of the data points indirectly since they are ranked, and thus larger values are ranked higher. Because we use more information, the rank-sum test is more powerful than the sign test where only plusses or minuses were used without taking into account magnitude.

> ***rank-sum test.*** ■ A test of the difference between the means of two populations by ranking all data values.

We begin by treating the data as though we had one sample and put the combined data in order, thus ranking the data values while keeping the data lists separate. Then the rankings of the data for each sample are summed, producing $R_1 = \Sigma$ (Ranks of data in Sample 1) and $R_2 = \Sigma$ (Ranks of data in Sample 2). If both samples are larger than 10, the R variable is approximately normally distributed with

$$\mu_R = \frac{n_1(n_1 + n_2 + 1)}{2}$$

and

$$\sigma_R = \sqrt{\frac{n_1 n_2(n_1 + n_2 + 1)}{12}}.$$

Thus we can use the Standard Normal Distribution Table (Table D.2) to complete a hypothesis test. We test a null hypothesis which states that the two samples come from identical populations, or that the population means are equal. If so, then R_1 and R_2 should be about the same magnitude. The alternative hypothesis states that they are not. Let us proceed, using an example.

EXAMPLE 14.4 The New York Stock Exchange list of most active stocks for 198X included the following randomly selected oil-related and non-oil-related stocks with the closing prices for the year as listed. Test the hypothesis that the median closing price

* This is equivalent to the Mann-Whitney test and is sometimes called the Wilcoxon rank-sum test.

Oil-Related Stocks		Non-Oil-Related Stocks	
Exxon	29¾	IBM	96¼
Mobil	25⅛	AT&T	59⅜
Texaco	31⅛	Sony	15¼
ARCO	42	Sears	30⅛
Std. Oil of Cal.	32	Tandy	50¾
Std. Oil of Ind.	39¾	K Mart	22
Superior Oil	28¾	Mattell	16⅝
GM	62⅜	Citicorp	32½
Chrysler	49⅝	Digital	99½
Haliburton	35⅜	Apple	22⅛

for all oil-related stocks is not significantly different from that of all non-oil-related stocks, for $\alpha = 0.05$.

Solution

$$H_0: \mu_O - \mu_{N-O} = 0.$$

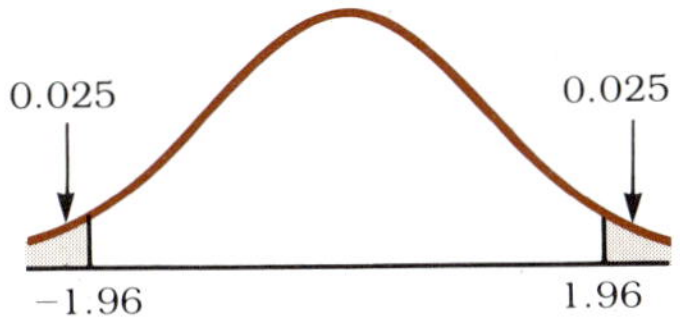

Figure 14.4

We negate this with an H_A using a two-tail alternative because the problem is not prejudiced in favor of a one-tail H_A.

$$H_A: \mu_O - \mu_{N-O} \neq 0.$$

Our test region is two-tailed in Fig. 14.4. We must now rank the data 1–20, since 20 stocks are under consideration.

Stock	Rank	Stock	Rank
Exxon 29¾	14	IBM 96¼	2
Mobil 25⅛	16	AT&T 59⅜	4
Texaco 31⅛	12	Sony 15¼	20
Atl Rich 42	7	Sears 30⅛	13
SO of Cal 32	11	Tandy 50¾	5
SO of Ind 39¾	8	KMart 22	18
Sup Oil 28¾	15	Mattell 16⅝	19
GM 62⅜	3	Citicp 32½	10
Chrysler 49⅝	6	Digital 99½	1
Halibtn 35⅜	9	Apple 22⅛	17
$R_O =$	101	$R_{N-O} =$	109

We choose one of the R's to test, using a normal distribution. Let us choose R_O. The test statistic is a z score:

$$z^* = \frac{R_O - \mu_R}{\sigma_R},$$

where

$$\mu_R = \frac{n_O(n_O + n_{N-O} + 1)}{2}$$

and

$$\sigma_R = \sqrt{\frac{n_O n_{N-O}(n_O + n_{N-O} + 1)}{12}}.$$

Next we calculate the needed parametric values:

$$\mu_R = \frac{(10)(10 + 10 + 1)}{2} = \frac{210}{2} = 105;$$

$$\sigma_R = \sqrt{\frac{10 \cdot 10(10 + 10 + 1)}{12}} = \sqrt{\frac{2100}{12}} = 13.23.$$

Thus

$$z^* = \frac{101 - 105}{13.23} = -0.30,$$

and we cannot reject H_0.

Summary

H_0: $\mu_O - \mu_{N-O} = 0$.

H_A: $\mu_O - \mu_{N-O} \neq 0$.

Test region: Reject H_0 if $z^* < -1.96$ or $z^* > 1.96$.

Test statistic: $z^* = -0.30$.

Conclusion: We cannot reject H_0. There is no significant difference between the median closing price of oil-related stocks and non-oil-related stocks. □

Several assumptions required for the rank-sum test are listed in the box on the page opposite. Let us first review them and then conclude with Example 14.5.

EXAMPLE 14.5 In an attempt to judge extrasensory perception, a psychologist put 24 subjects through a series of tests, 12 while awake and 12 while hypnotized. The following scores were obtained.

ASSUMPTIONS

1. The null hypothesis examines whether the samples come from identical populations.
2. We do not need to know the underlying distributions of the data, but we must be able to assume that the variables are continuous and that the samples are independent.
3. We rank all data as if we had one sample but keep the samples separate in order to obtain the sum of the rankings for each, R_1 and R_2. If two pieces of data have the same value (that is, a tie occurs), assign each a rank halfway between the two ranks to be assigned to the data points; for example, if two data points are equal and are to be ranked 5 and 6, assign each data point a rank of 5.5 and give the next value a rank of 7. If more than two values tie, give each the mean of the ranks to be assigned to the values and proceed to the next rank.
4. If n_1 and $n_2 \geq 10$, the R variable is approximately normally distributed. It is not necessary that $n_1 = n_2$.
5. The test statistic is a z score:

$$z^* = \frac{R_1 - \mu_R}{\sigma_R},$$

where

$$\mu_R = \frac{n_1(n_1 + n_2 + 1)}{2};$$

$$\sigma_R = \sqrt{\frac{n_1 n_2(n_1 + n_2 + 1)}{12}};$$

n_1 = sample size for the sample used to calculate the R in z^*; and

n_2 = sample size for the other sample.

Awake	Hypnotized
23	23
12	23
21	21
15	24
19	21
15	27
17	22
21	19
18	19
14	23
16	25
20	24

At the 0.05 level of significance, did hypnotism improve the mean scores of the subjects?

Solution The null hypothesis assumes that there is no change.

$$H_0: \mu_h - \mu_a = 0.$$

We are asked to look for an improvement, so

$$H_A: \mu_h - \mu_a > 0.$$

Thus the test region is one-tailed in Fig. 14.5. We must rank the data and take the ties into account.

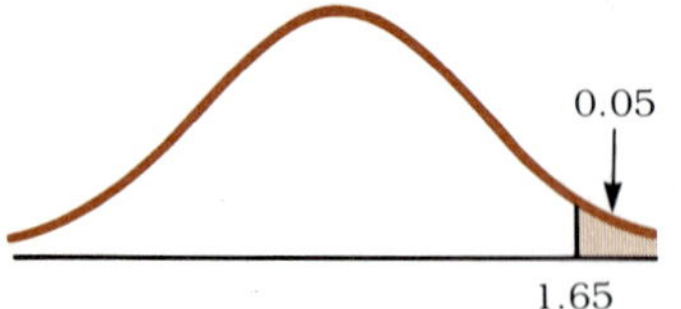

Figure 14.5

Awake	Rank	Hypnotized	Rank
23	6.5	23	6.5
12	24	23	6.5
21	11.5	21	11.5
15	21.5	24	3.5
19	16	21	11.5
15	21.5	27	1
17	19	22	9
21	11.5	19	16
18	18	19	16
14	23	23	6.5
16	20	25	2
20	14	24	3.5
	$R_A = 206.5$		$R_H = 93.5$

We need to work with only one, either R_a or R_h.* Let us work with R_a. Thus

$$\mu_R = \frac{(12)(12 + 12 + 1)}{2} = 150;$$

$$\sigma_R = \sqrt{\frac{12 \cdot 12 \cdot (12 + 12 + 1)}{12}} = 17.32$$

$$z^* = \frac{206.5 - 150}{17.32} = 3.26.$$

We can reject H_0.

Summary

H_0: $\mu_h - \mu_a = 0$.

H_A: $\mu_h - \mu_a > 0$.

Test region: Reject H_0 when $z^* > 1.65$.

Test statistic: $z^* = 3.26$.

Conclusion: Reject H_0. Hypnotism did make a difference in the mean test scores. □

We have examined two examples of nonparametric or distribution-free statistical tests. There are many others, but these should give you a sufficient understanding of this type of procedure to allow you to utilize other methods on your own.

---

* The two values of R are related because $R_a + R_h$ must equal the sum of the rankings. For samples of size n_1 and n_2, then $R_1 + R_2$ must equal the sum of the first $n_1 + n_2$ positive integers.

$$R_1 + R_2 = \frac{(n_1 + n_2)(n_1 + n_2 + 1)}{2}$$

EXERCISES/Section 14.2

1. Is the mean grade-point average the same for participants in team sports as for the participants in sports which stress individual achievement at the 0.05 level of significance?

Team Sports	Individual Sports
2.00	2.64
2.40	3.49
1.00	3.00
3.40	4.00
3.48	1.91
3.03	3.38
2.64	2.93
2.58	2.94
2.91	2.97
2.63	3.41
2.17	3.52
	3.43

2. Fuel consumption, in miles per gallon, for 28 different vehicles is given for two different speeds. Does increased speed decrease mean fuel consumption, at the 0.01 level of significance?

55 mph	60 mph
22.5	21.7
32.5	28.7
25.3	23.7
30.6	27.9
22.5	21.1
25.0	22.9
24.3	22.6
17.8	16.2
28.0	26.4
27.4	24.1
27.0	24.6
25.7	23.8
15.2	13.8
14.5	13.2

3. Perform a rank-sum test to determine whether age affects mean diastolic blood pressure. Use $\alpha = 0.05$.

DBP at Age 50	DBP at Age 60
70	60
70	66
90	72
80	68
80	84
90	108
80	64
80	82
70	68
100	66
85	90
82	72
82	88
	76
	58

4. One method of ascertaining the size of a wildlife population is by the capture-recapture method, using tagging to identify animals captured on the first outing. The following numbers of a species were observed at 28 different locations on outings one month apart. Use the rank-sum test to determine whether there is any difference in the mean number for the two consecutive sightings, at the 0.05 level of significance.

First Sighting	Second Sighting
6	7
8	6
9	10
4	4
3	2
8	11
15	12
21	19
10	12
15	16
6	8
7	7
8	6
	3
	4

5. Use a rank-sum test to determine whether there is any difference in the mean level of wear (in mm measured at the heel) for two different brands of shoe soles that were tested on 26 individuals over a two-month period. Use $\alpha = 0.05$.

Brand *X*	**Brand *Y***	**Brand *X***	**Brand *Y***
1.0	1.1	1.3	1.29
1.2	1.21	1.7	1.61
1.3	1.22	1.22	1.42
0.9	1.0	1.11	1.35
0.7	1.4	0.98	0.97
0.85	0.93	1.0	
1.14	1.13	0.74	

6. Use a rank-sum test to determine whether mean cancer death rates (per 100,000 population) differ between men and women in 22 countries. Use $\alpha = 0.01$.

Men	**Women**
199	126
108	130
116	124
178	113
176	124
131	135
168	127
137	124
163	140
159	108
137	
147	

Study Notes

KEY TERMS

distribution-free test ■ A test which does not assume a normal distribution.

nonparametric test ■ A test that does not use parameters such as μ and σ.

power of a test ■ The probability of rejecting a false hypothesis, $1 - \beta$.

rank-sum test ■ A test of the difference between means of two populations by ranking all data values.

sign test ■ A test of the median, rather than the mean, of a population without a normality assumption.

OUTLINE

I. Sign test
  A. Alternative to the one-sample *t* test and to the matched *t* test.
  B. Test for medians of continuous data.
  C. H_0: Med = ________.
  D. H_A: Med ($\neq$, $<$, $>$) ________.
  E. Assign data values (+) or (−), depending on their relationship to the value claimed in H_0, ignore ties, and reduce n accordingly.
  F. If n is small:
    1. Use the binomial distribution with $p = ½$ to determine the probability of obtaining the value of X or less, as determined in E.
    2. If the result in (1) is $<\alpha$, reject H_0.
  G. If $n/2 > 5$, use the normal approximation to the binomial:

$$z = \frac{X - (n/2)}{\sqrt{n}/2}.$$

  H. Alternative to the matched *t* test:
    1. Assign (+) for increases and (−) for decreases; ignore ties, as before.
    2. H_0: Med = 0.
    3. H_A: Med ($\neq$, $<$, $>$) 0.

II. Rank-sum test
  A. Tests the difference between means.
  B. Data must be continuous and from similarly shaped distributions.

C. Rank all data as if from one sample and then form $R_1 = \Sigma$ (Ranks from Sample 1) and $R_2 = \Sigma$ (Ranks from Sample 2).

D. R is normally distributed and

1. $z^* = \dfrac{R_1 - \mu_R}{\sigma_R}$;

2. $\mu_R = \dfrac{n_1(n_1 + n_2 + 1)}{2}$;

3. $\sigma_R = \sqrt{\dfrac{n_1 n_2(n_1 + n_2 + 1)}{12}}$.

REVIEW PROBLEMS

1. Ten items were selected to represent a typical grocery purchase. During the past year, two rose in price by less than 10%; one rose in price by 10%; and seven rose in price by more than 10%. Test the hypothesis that there was a median rise in cost of 10% or more, at the 0.05 level of significance.

2. An insurance broker hired a marketing research firm to survey the potential for sales of life insurance in the suburban community where he lived. The firm reported that, of the ten families it questioned, eight owned less than $90,000 in life insurance and two had policies for more than $90,000. Using the industry median of $90,000 of coverage for comparison, should the insurance agent expect many sales in this area? Are families in this community underinsured? Use $\alpha = 0.05$ to test the hypothesis that families living there are properly insured.

3. On a 25-man roster for a local baseball team, 21 players have batting averages under .270, and 4 players have batting averages over .270. If the median batting average for the league is .270, is this team batting significantly lower than the league, at the 0.05 level?

4. Since income is not normally distributed, it is not reasonable to test mean income. We can overcome this problem by examining the median, which nationally is $17,500 per family per year. A sample of 30 families in a small town showed that 22 have an annual income of less than $17,500, and 8 have an annual income of more than $17,500. Do the sample data indicate that residents of the town are significantly poorer than those of the country as a whole, at the 0.01 level of significance?

5. "Things Go Better with Music" is the advertising slogan of a local easy-listening-music radio station. It justifies this claim on the basis of the data in Table 14.3, which shows production levels per hour in a certain factory. Do these figures justify the slogan's claim, at the 0.05 level of significance?

6. Fluoride is supposed to reduce the incidence of tooth decay. To test this hypothesis, a producer of toothpaste chose one-dozen willing participants at random, had each use a fluoridated toothpaste for a year, and had each use a nonfluoridated toothpaste for a year. Table 14.4 records the number of cavities per person per year. Use a sign test to determine

TABLE 14.3

Employee number	1	2	3	4	5	6	7	8	9	10
With music	28	32	30	29	27	29	31	33	35	30
Without music	25	31	30	27	26	31	25	30	34	25

TABLE 14.4

Nonfluoridated	4	6	2	3	0	4	2	3	4	1	2	5
Fluoridated	3	2	1	2	1	3	0	1	1	1	2	3

whether these figures support the hypothesis, for a 10% level of significance.

7. A consumer survey of prices at two neighborhood grocery stores produced the following comparative price lists for name-brand items. Use a rank-sum test to determine whether there is a difference in the mean prices, at the 0.05 level of significance.

Item	Store A	Store B
1	$0.57	$0.59
2	0.39	0.40
3	0.32	0.33
4	0.57	0.69
5	0.85	0.95
6	0.59	0.59
7	0.33	0.33
8	0.81	0.81
9	0.29	0.26
10	0.45	0.45
11	0.40	0.41
12		0.41
13		0.89

8. Heights, in inches, of preschoolers of comparable age are shown. Is there any difference in mean height for girls and boys? Use a rank-sum test with $\alpha = 0.05$.

Boys	Girls	Boys	Girls
30.2	29.4	45.9	45.9
34.6	33.8	39.2	40.1
37.8	37.5	41.6	42.1
40.8	40.7	45.1	45.2
43.4	43.4	38.9	39.0

9. Use a rank-sum test to determine whether there is a difference in mean IQ scores for two different socioeconomic groups, at the 0.05 level of significance.

Group 1	Group 2
100	100
110	93
104	101
95	102
81	120
99	82
103	71
104	95
96	100
101	99
134	123
120	112
93	95
134	
125	

Appendixes

A. OTHER TYPES OF GRAPHS
B. SUMMATION NOTATION
C. FACTORIAL NOTATION
D. TABLES
E. ANSWERS TO ODD-NUMBERED PROBLEMS

APPENDIX A
Other types of graphs

A brief explanation and examples of the major types of graphs are presented in this appendix. The data for these graphs came from *Historical Statistics of the United States* and the *Statistical Abstract*, which is published by the Department of Commerce, Bureau of the Census. Some of the numbers were rounded for ease in demonstrating the characteristics of graphs and for simplifying arithmetic. Graphs at best can never be as precise as a list of numbers but, if well done, they can show clearly relationships and patterns of data—and that is what we should look for when using them. We start with line graphs, continue with pie (or circle) charts, and end with bar graphs and a variation on the bar graph, the picture graph.

A.1 Line Graphs

Line graphs are constructed using a vertical axis and a horizontal axis. The quantities identified in the title of the graph are usually displayed on the vertical axis, and years or categories on the horizontal axis. A scale is selected for each axis that groups quantities into appropriate sizes (for comprehension and amount of space available). The intervals for quantities along each axis are spaced equally. Consider the graph in Fig. A.1. The title tells us that rates are being considered, so dollars go on the vertical axis at equally spaced intervals. Dots are placed at the proper height for the corresponding year (horizontal distance) and then connected. Note that since $0 is possible, the vertical scale starts with 0.

Figure A.2 is a line graph that shows information about the labor force. Again, the intervals are equally spaced on the vertical axis and on the horizontal axis; the vertical scale starts with 0.

We can create different impressions by using different scales on line graphs. Let us do the same graph over and push the years closer together, as in Fig. A.3. Now we see a much steeper trend. It looks as though the percentage of women in the labor force is increasing more rapidly than before, but all we have done is change the scale. By condens-

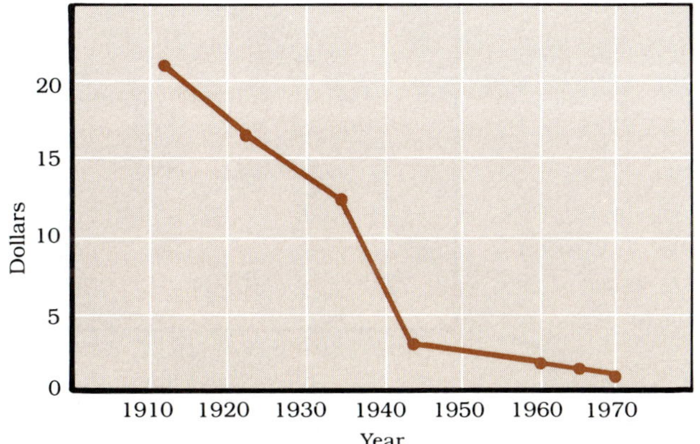

Figure A.1
Telephone toll rates between New York City and San Francisco (station-to-station, daytime, 3-minute call)

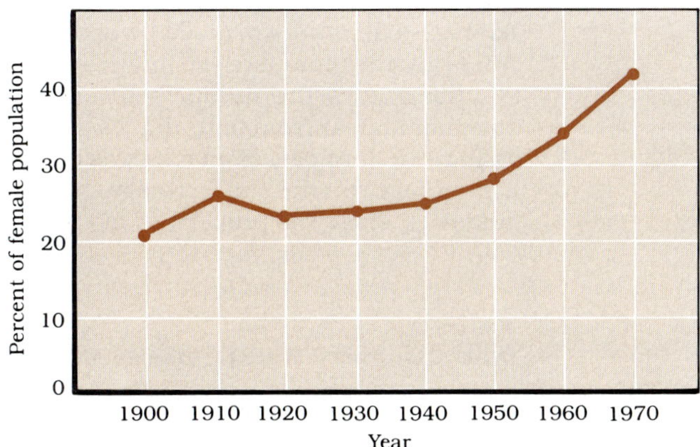

Figure A.2
Female labor force

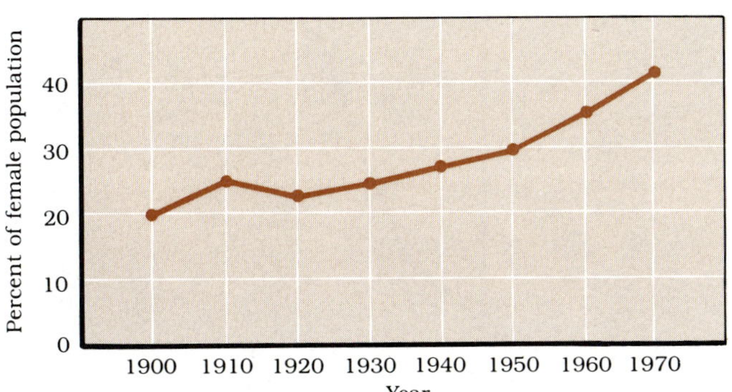

Figure A.3
Female labor force

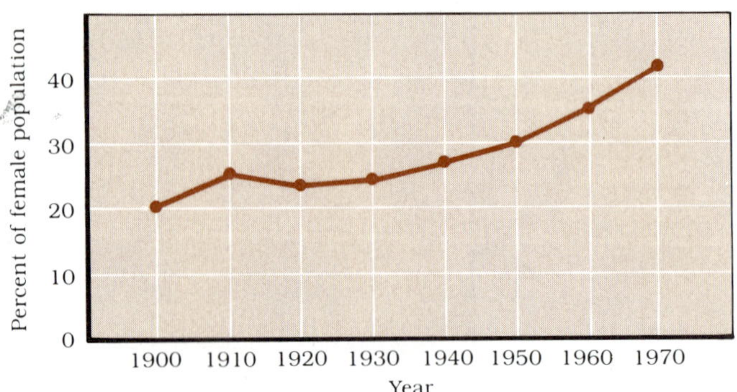

Figure A.4
Female labor force

ing the vertical scale we can really flatten out the graph (Fig. A.4). The equal spacing of intervals still holds in each case.

Figure A.5 is a comparative line graph. Two sets of data are plotted on the same axes for purposes of comparison. For simplicity, the graphmaker chose a scale of billions of dollars, rather than just dollars, so that small numbers could be used in labeling the vertical axis. Figure A.6 is another comparative line graph.

Sometimes the numbers involved in graphing are relatively close together and far from 0. In that case a symbol such as ⌇ or —∿— is used to indicate that part of an axis was eliminated in order to avoid excessive empty space on the paper; this is called a break in the scale. Figures A.18 and A.19 show the proper use of the vertical break symbol.

Figure A.5
Outlays for federal space program: 1960 to 1976

Source: Chart prepared by U.S. Bureau of the Census. Data from U.S. Office of Management and Budget.

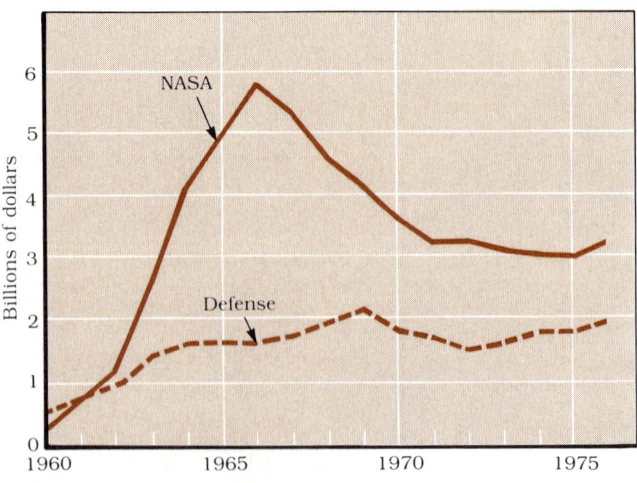

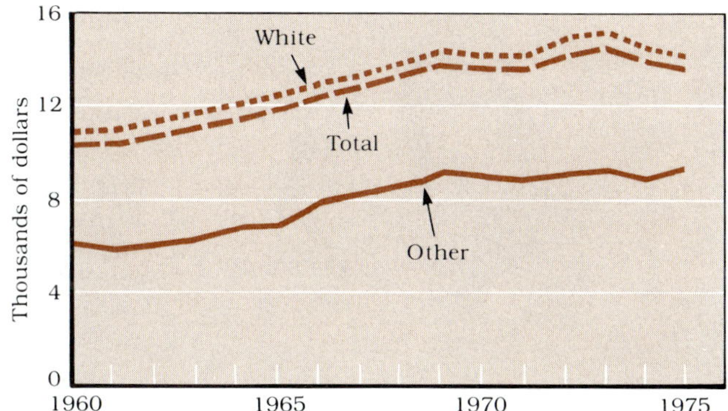

Figure A.6
Median annual money income in constant (1975) dollars of families, by race: 1960 to 1975

Source: U.S. Bureau of the Census.

A.2 Pie Charts

For pie charts, we use a circle to represent the total and divide it into proportional wedges to show parts of the whole. Each part represents a percentage of the total, and the size of each wedge is found by multiplying those percentages by the total number of degrees (360°) in a circle. This idea can be applied to any shape; for example, a dollar bill is often cut up to show revenue or expenditures in proportion to the quantities being displayed.

Figure A.7 is a pie chart that shows the amount of fuel consumed, in quadrillion BTUs, for various purposes. It also shows the percentage that each figure is of the whole. The units of consumption add up to 71; the 19.0 for industrial usage, for example, is 27% of this total. We multiply

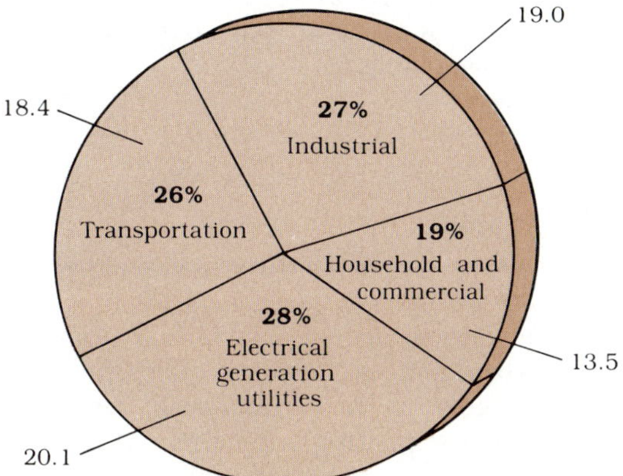

Figure A.7
Consumption of fuel resources (in quadrillion BTU)

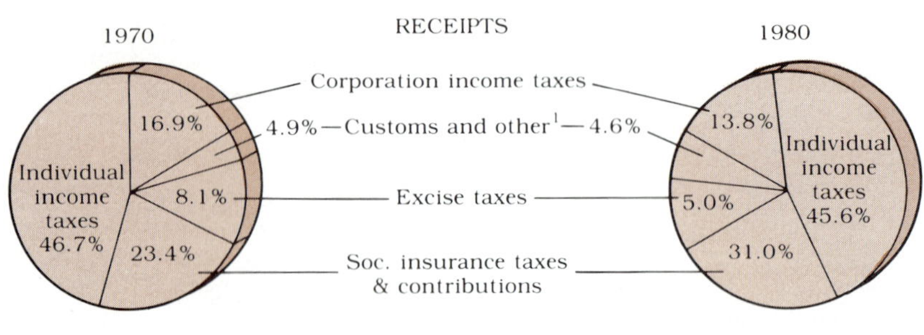

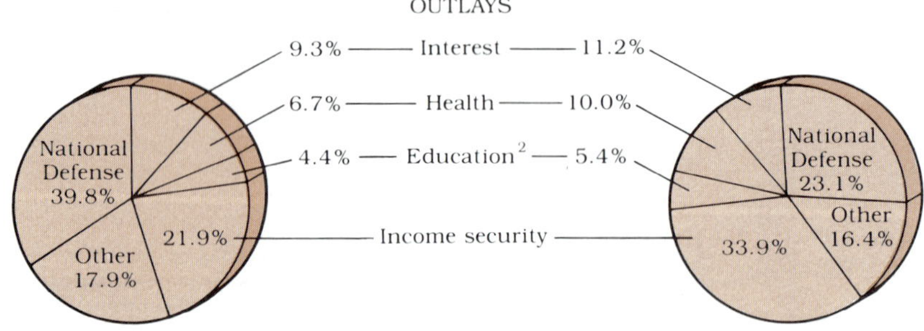

Figure A.8
Federal budget—percent distribution by function: 1970 and 1980

[1] Other includes gift taxes and other receipts.

[2] Includes training, employment, and social services.

Source: Chart prepared by U.S. Bureau of the Census.

27% by 360° to determine the number of degrees that are needed to represent the industrial-usage portion of the whole.

$$27\% \cdot 360° = 0.27 \cdot 360° = 97°.$$

The process is repeated for each segment until the chart is completed.

Figure A.8 gives some further examples of pie charts. Note the amount of labeling that should be done in order to make the graphs easily readable.

A.3 Bar Graphs

Bar graphs are similar to line graphs in that they have vertical and horizontal axes. However, quantities can be placed on either axis. If they are listed on the vertical scale, we draw a bar up to the proper height instead of putting a dot there. Again this axis must start with 0, each axis must have equally spaced intervals, all bars should be of equal

Figure A.9
Homicides 1900–1970 (per 100,000 population)

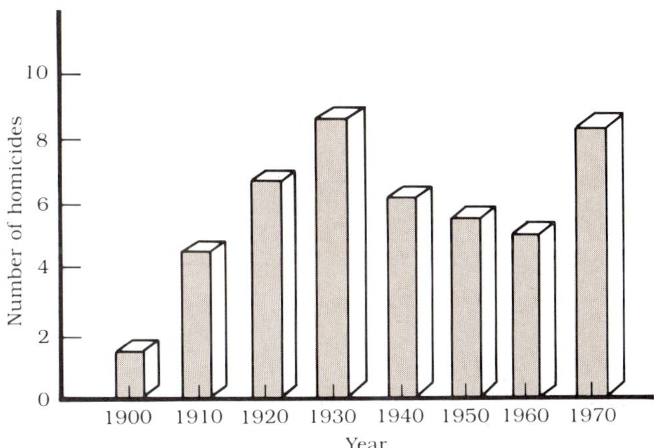

width. The break-in-scale should never be used with a bar graph because the impression created by a bar graph depends on the length of the bars. A break in scale would shorten the bars and distort the impression. Figure A.9 is a vertical bar graph.

If the horizontal scale is used to display quantities, it should start with 0 and be divided into equal spaces; the vertical would then display categories or years. Figures A.10 and A.11 are horizontal bar graphs.

Figure A.10
Marital status of women in the civilian labor force in 1900

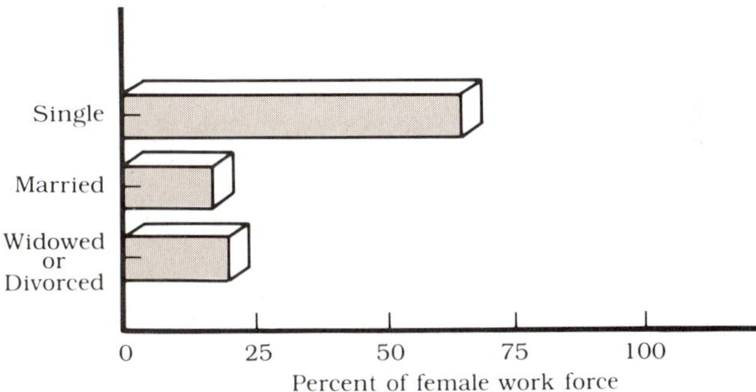

Figure A.11
Marital status of women in the civilian labor force in 1970

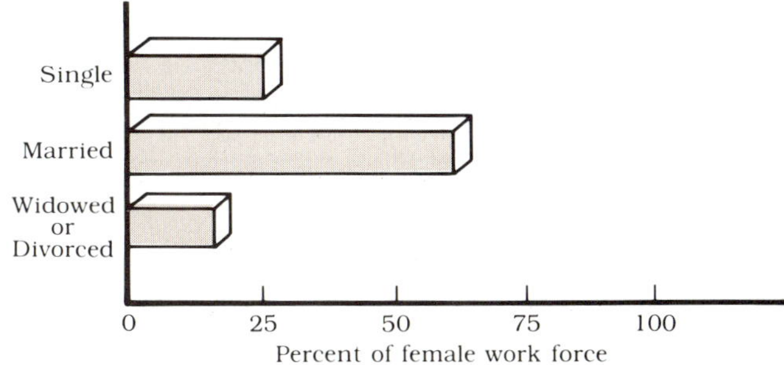

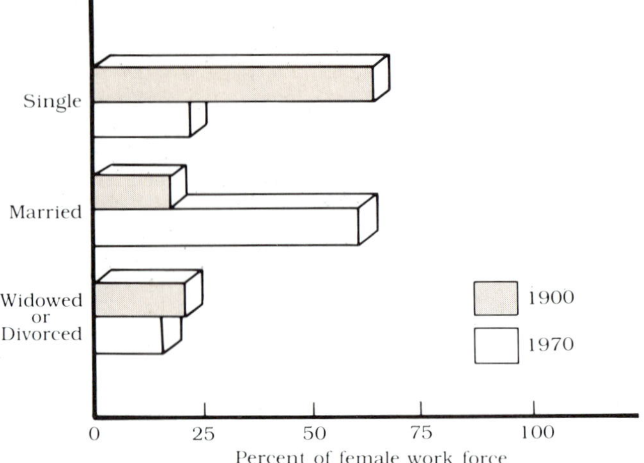

Figure A.12
Marital status of women in the civilian labor force

Since these graphs contain similar information, we can put them together to make a *comparative* bar graph (Fig. A.12). However, we need the key to tell us which bars refer to 1900 and which to 1970.

Figures A.13–A.16 are some further examples of comparative bar graphs. Note that each uses a 0 starting point for the axis representing quantities, and that all intervals (and bar widths) along each axis are equal.

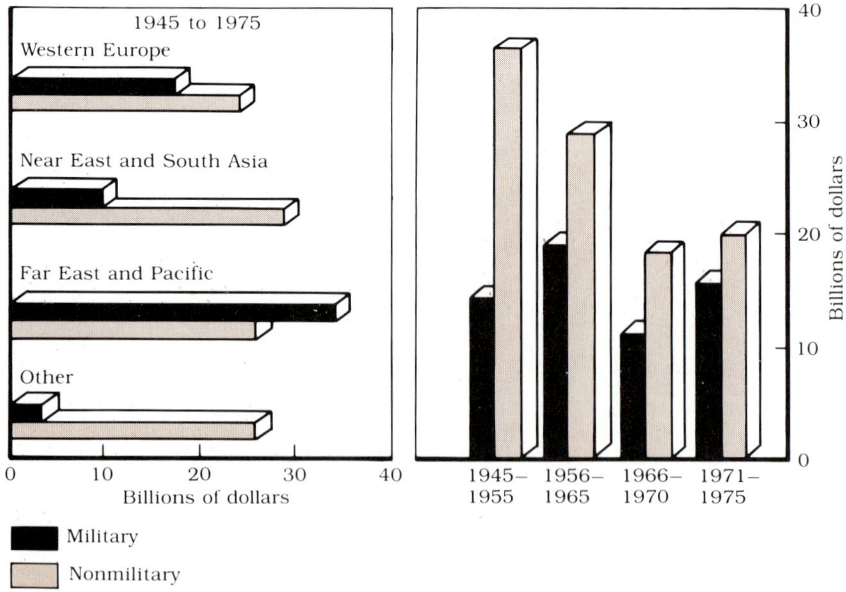

Figure A.13
U.S. foreign aid since World War II

Source: Chart prepared by U.S. Bureau of the Census. Data from U.S. Bureau of Economic Analysis and Board of Governors of the Federal Reserve System.

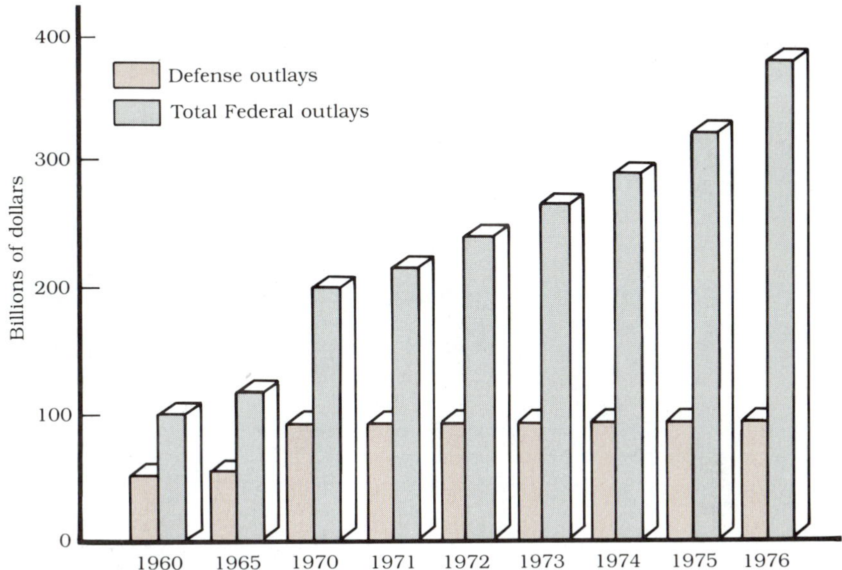

Figure A.14
National defense and total budget outlays: 1960 to 1976

Source: Chart prepared by U.S. Bureau of the Census. Data from U.S. Office of Management and Budget.

Figure A.15
Worldwide military expenditures: 1970 to 1978

Source: Chart prepared by U.S. Bureau of the Census.

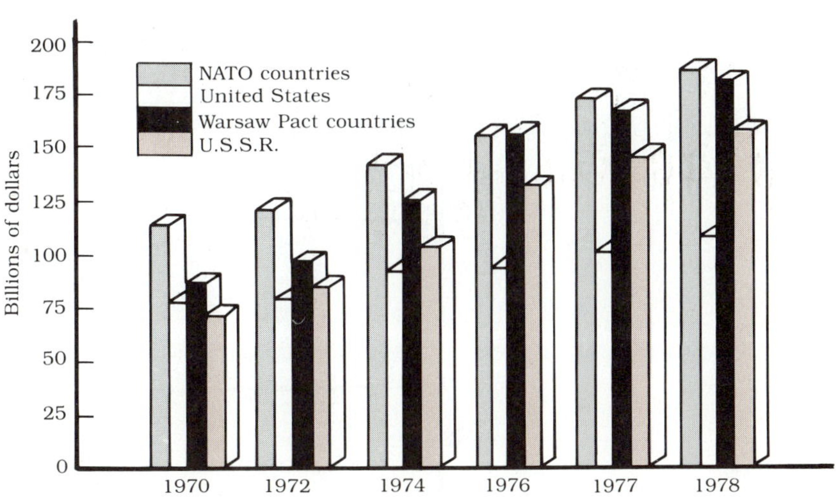

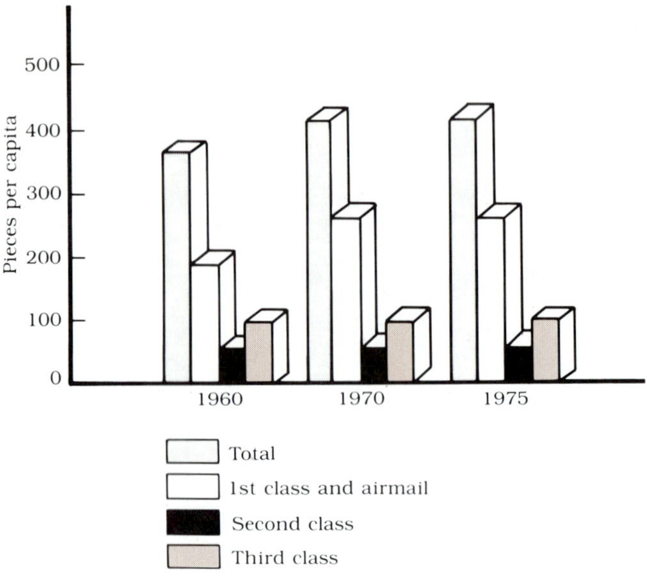

Figure A.16
Pieces of mail handled: 1960, 1970, and 1975

Source: Chart prepared by U.S. Bureau of the Census. Data from U.S. Postal Service.

Sometimes bars and lines can be combined effectively to provide more information than either could show alone, as in Fig. A.17. Figure A.18 is another combined line graph—that shows trends—and bar graph—that puts these trends in perspective.

Figure A.17
Birth and death rates: 1950 to 1974

Source: Chart prepared by U.S. Bureau of the Census. Data from U.S. National Center for Health Statistics.

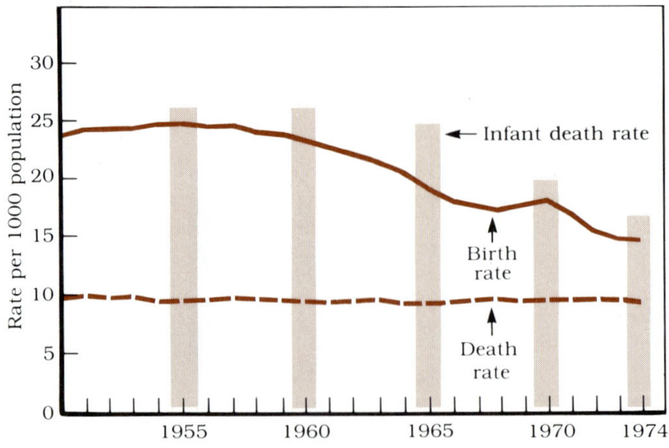

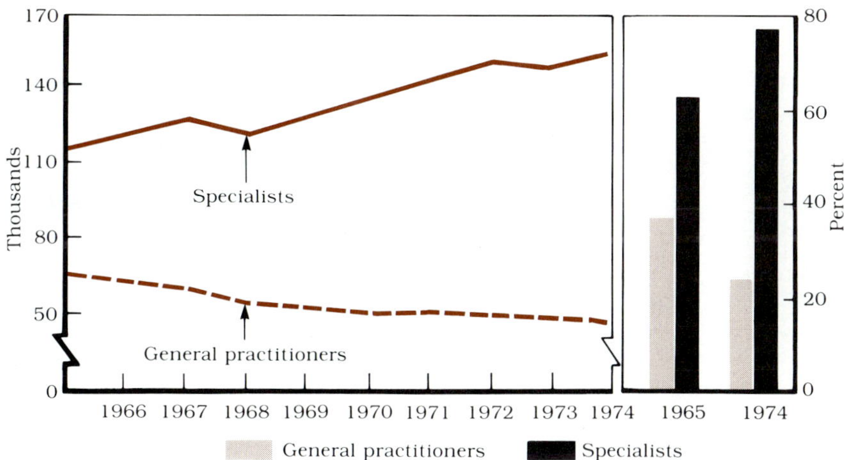

Figure A.18
Office-based private physicians, by type of practice: 1965 to 1974

Source: Chart prepared by U.S. Bureau of the Census. Data from U.S. National Center for Health Statistics and American Medical Association.

Finally, Fig. A.19 employs two different vertical scales—one on the left for the lines and one on the right for the bars—to combine economic information and show the effects of inflation. The left-hand vertical scale has its own 0 point, with a break in scale. Note that this break allows both the line and bar graphs to be displayed together without causing distortion.

Figure A.19
Gross national product (GNP) in current and constant 1972 dollars: 1960 to 1975

Source: Chart prepared by U.S. Bureau of the Census. Data from U.S. Bureau of Economic Analysis.

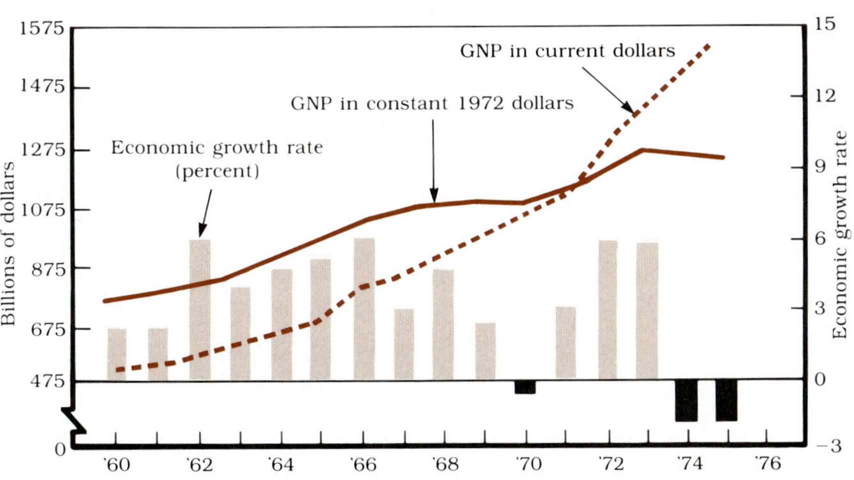

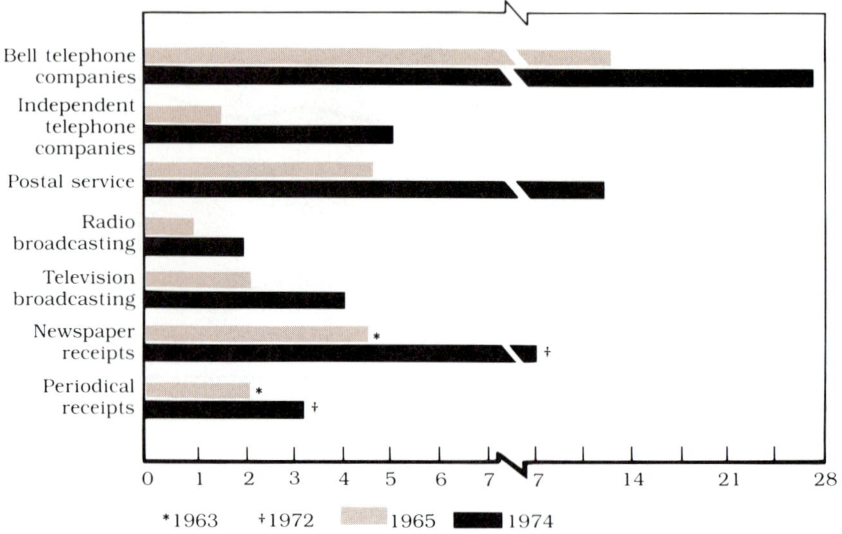

Figure A.20
Operating revenues of selected media: 1965 and 1974

Source: Chart prepared by U.S. Bureau of the Census.

We have been emphasizing the desirability of starting at 0 and spacing intervals equally along an axis. Let us look at some examples where these criteria are not followed. In Fig. A.20, a ⁓ is used to interrupt the bars and change the scale. If the original scale had been used for the entire graph, some of the bars would have been very long. This change allows everything to fit neatly on the paper—but unless you are alert, you may miss the break in scale. Without the change in scale, the first dark bar would be more than five times as long as the second dark bar; with the change, it is only a little more than twice as long, and an incorrect impression may be given.

Let us consider the information given in Fig. A.1 concerning telephone rates between New York City and San Francisco:

1945	$2.50
1965	$2.00

The rate was cut by 20%, which we can show on a graph (Fig. A.21). It looks like a much bigger change than 20%, doesn't it? However, if we graph the same information properly, using a 0 point (Fig. A.22), a truer impression is created.

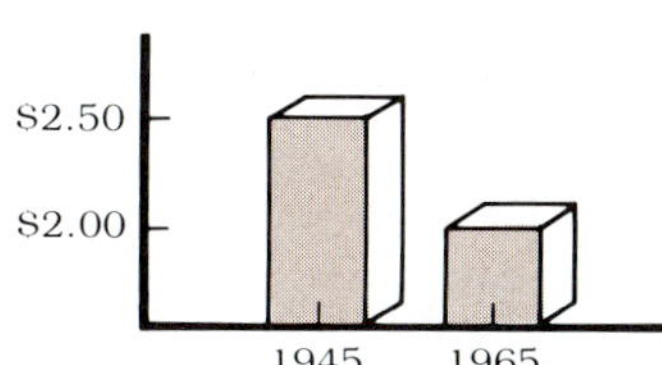

Figure A.21
Telephone rates between New York City and San Francisco

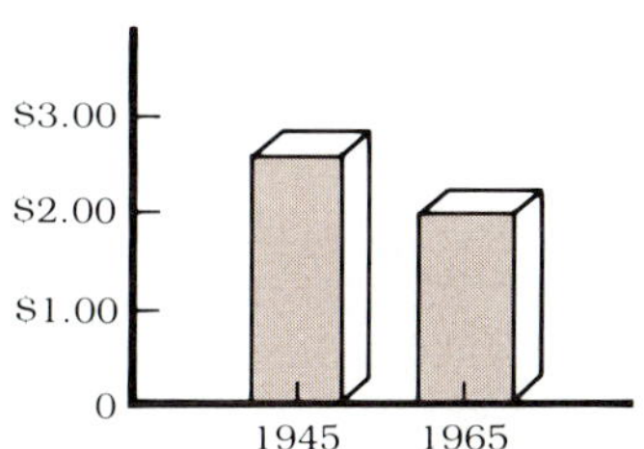

Figure A.22
Telephone rates between New York City and San Francisco

Picture Graphs

What can happen if we decide to get fancy with our graphs? Consider the amounts, in millions of pounds, of U.S. wool production in two different years:

1940	372
1970	162

We note that production was cut by more than one-half. Instead of uninteresting bars, we decide to draw pictures of sheep to show this change, as in Fig. A.23. The vertical scale tells us that one picture is more than twice as high as the other, but the graph gives the impression of a much greater difference. The eye sees the change in *area*, not the change in height. Here the horizontal scale is distorted to keep the shapes of the sheep proportional, while showing the proper heights. The horizontal intervals for the sheep are not equal, and so the impression created is a false one.

Figure A.23
U.S. wool production

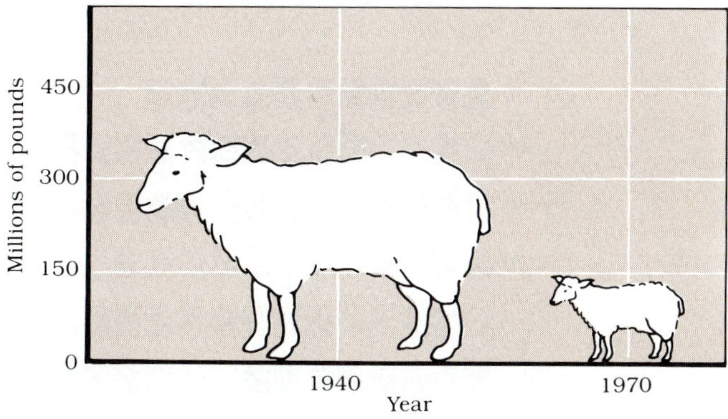

Picture graphs can be a useful variation of the bar graph, but we must be careful of possible distortions. A better way to utilize pictures is to use units that are all the same size, as in Fig. A.24, rather than one picture that has been blown up or shrunk in size.

Figure A.24
Changes in farming: 1940 to 1975

Source: Chart prepared by U.S. Bureau of the Census. Data from U.S. Department of Agriculture, Statistical Reporting Service and Economic Research Service.

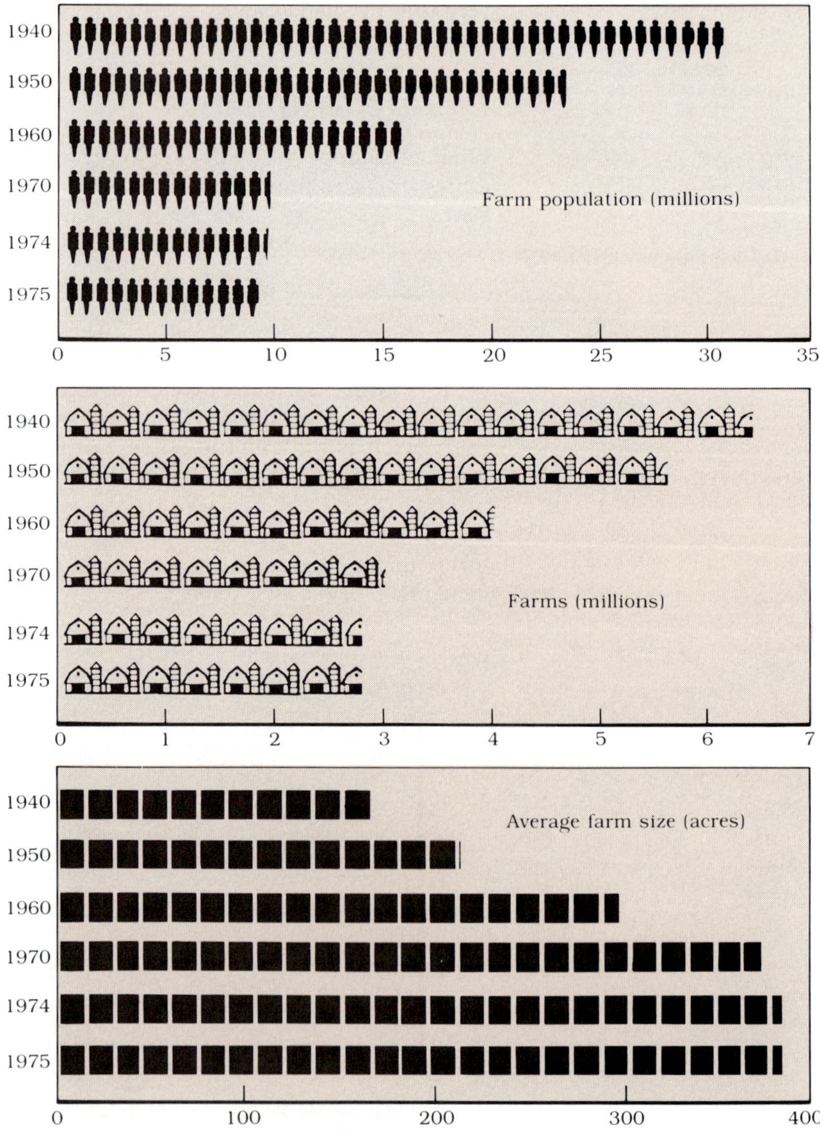

EXERCISES/Appendix A

1. The farm population as a percent of total U.S. population is shown by decade for the period 1880–1970. Construct a line graph for this information.

Year	Percent	Year	Percent
1880	44	1930	25
1890	42	1940	23
1900	42	1950	15
1910	35	1960	9
1920	30	1970	5

2. Prepare a pie chart showing federal government revenue, in billions of dollars, by tax source. Total revenue = $140 billion.

Source	Amount
Individual income tax	90
Corporate income tax	30
Excise and customs duty	15
Other (death, gift, etc.)	5

3. Show the number of immigrants to the United States from various parts of the world in 1970 on a bar graph.

Origin	Number
Europe	110,000
Asia	90,000
Western Hemisphere	160,000
Africa	7,000
Australasia	4,000

4. Make a line graph to display population trend for people 65 years of age or older.

Year	Population
1900	3,080,000
1930	6,634,000
1970	19,972,000
1975	22,400,000
2000	30,600,000

5. Make a bar graph showing the recoverable petroleum reserves, in billions of barrels, in the world.

Area	Reserves
North America	58.3
South America	17.5
Europe	23.0
Asia	108.4
Africa	34.3
Middle East	403.9

6. Construct a comparative bar graph for the information on the popular vote cast, in millions, for president in recent elections.

Year	Democrat	Republican
1944	25.6	22.0
1948	24.2	22.0
1952	27.3	34.0
1956	26.0	35.6
1960	34.2	34.1
1964	43.1	27.2
1968	31.3	31.8
1972	29.2	47.2
1976	40.3	38.5

7. Construct a vertical bar graph to show recoverable natural gas reserves (in trillions of cubic feet).

Area	Reserves
North America	304
South America	48
Europe	170
Asia	812
Africa	229
Middle East	670

8. The percent of the U.S. population represented by couples over age 65 with incomes in various categories is shown on the next page. Display this information on a line graph.

Income	Percent
Under $2000	2.4
$2000–$3999	16.7
$4000–$5999	24.3
$6000–$7999	18.3
$8000–$9999	11.8
Over $10,000	26.5

9. Draw a comparative line graph for the following information on average annual fuel consumption (in gallons).

Year	Autos	Buses	Trucks
1950	600	3750	1250
1955	650	3000	1300
1960	675	3050	1325
1965	650	2800	1350
1970	725	2500	1375
1975	675	1900	1275

10. Make a picture graph using drops of water to display the total amount of land area irrigated in the United States. Areas are in millions of acres.

Year	Area
1920	200
1930	250
1940	300
1950	300
1960	250
1970	210

11. Draw a picture graph of the following information on average farm size.

Year	Acreage
1930	150
1940	170
1950	210
1960	300
1970	370

12. Display the following information on U.S. transportation energy consumption, by type of vehicle, using a circle graph. Consumption is shown in millions of BTUs per capita.

Type	Consumption
Automobile	47.3
Truck	18.0
Aircraft	6.4
Train	2.8
Bus	0.4
Other	10.6

13. Display the annual budget, in dollars, of a family of four as a pie chart.

Item	Cost
Food	4450
Housing	4900
Transportation	1500
Medical Care	800
Clothing, personal items	2000
Taxes	4500
Other	2400

APPENDIX B
Summation notation—Σ

The Greek letter Σ (read "sigma") is the capital Greek S and is used to indicate a series of additions or sums. It is merely a shorthand way of saying: Add everything that comes after. Let us illustrate its use with the following set of data.

$$X\text{: } 1, 2, 7, 9, 3$$

The notation ΣX means: Add all the values of X.

$$\Sigma X = 1 + 2 + 7 + 9 + 3 = 22.$$

The notation $\Sigma 2X$ means: (1) Multiply each value of X by 2, and then (2) add these results.

$$\Sigma 2X = 2 \cdot 1 + 2 \cdot 2 + 2 \cdot 7 + 2 \cdot 9 + 2 \cdot 3$$

$$= 2 + 4 + 14 + 18 + 6 = 44.$$

Note that

$$\Sigma 2X = 2\Sigma X.$$

$$44 = 2 \cdot 22 = 44.$$

Thus, in general,

$$\boxed{\Sigma cX = c\Sigma X, \quad \text{where } c \text{ is a constant.}}$$

An important concept in working with summation notation is the order in which the operations are performed.

$\Sigma(X - 1)$ means: (1) Subtract 1 from each value of X, and then (2) add the results.

$$\Sigma(X - 1) = (1 - 1) + (2 - 1) + (7 - 1) + (9 - 1) + (3 - 1)$$

$$= 0 + 1 + 6 + 8 + 2 = 17.$$

We generally start in the middle and work our way out: We subtracted first, then added. Now consider $\Sigma X - 1$. Here, the summation symbol is connected only with the X, or $(\Sigma X) - 1$, so we get $22 - 1 = 21$; subtraction followed addition. The order of the operations performed is crucial.

The same requirement applies to operations involving exponents.

ΣX^2 means: (1) Square each value of X; that is, multiply it by itself, and then (2) add the results.

$$\Sigma X^2 = 1^2 + 2^2 + 7^2 + 9^2 + 3^2$$

$$= 1 + 4 + 49 + 81 + 9 = 144.$$

Now consider the same symbols rearranged.

$(\Sigma X)^2$ means: (1) Add the values of X, and then (2) square that result.

$$(\Sigma X)^2 = 22^2 = 484.$$

Note that

$$\Sigma X^2 \neq (\Sigma X)^2$$

$$144 \neq 484.$$

Now let us pair some values for another variable, Y, with the values for X. This occurs when we collect information on two characteristics from each member of a sample.

X:	1	2	7	9	3
Y:	0	4	6	2	1

This table tells us that the first member of the sample has an X value of 1 and a Y value of 0. In more realistic terms, this is similar to saying John Smith weighs 175 lb and is 5′10″ tall. Find ΣXY.

ΣXY means: (1) Multiply each value of X by the associated value of Y, and then

(2) add the results.

X:	1	2	7	9	3
Y:	0	4	6	2	1
XY:	0	8	42	18	4

$$\Sigma XY = 0 + 8 + 42 + 18 + 4 = 72.$$

Let us run through each of the operations discussed using another set of data.

S:	11	10	2	6	15
T:	100	101	99	80	70

$$\Sigma S = 11 + 10 + 2 + 6 + 15 = 44.$$

$$\Sigma(T - 90) = (100 - 90) + (101 - 90) + (99 - 90) + (80 - 90) + (70 - 90)$$
$$= 10 + 11 + 9 + (-10) + (-20)$$
$$= 0$$

$$\Sigma S^2 = 11^2 + 10^2 + 2^2 + 6^2 + 15^2$$
$$= 121 + 100 + 4 + 36 + 225 = 486.$$

$$\Sigma(T - 90)^2 = 10^2 + 11^2 + 9^2 + (-10)^2 + (-20)^2$$
$$= 100 + 121 + 81 + 100 + 400 = 802.$$

$$(\Sigma T)^2 = (100 + 101 + 99 + 80 + 70)^2 = (450)^2 = 202{,}500.$$

$$\Sigma ST = (11 \cdot 100) + (10 \cdot 101) + (2 \cdot 99) + (6 \cdot 80) + (15 \cdot 70)$$
$$= 1100 + 1010 + 198 + 480 + 1050$$
$$= 3838.$$

Considerably more algebraic manipulation can be done to derive rules for working with summation notation. We have demonstrated only the applications required in this text. If you will do the following exercises, you should have no difficulty in using the equations required for this study.

EXERCISES/Appendix B

Use the following data and calculate as indicated.

1. X: 10, 5, 12, 0, 2
   a) ΣX **b)** $\Sigma(X - 6)$ **c)** ΣX^2 **d)** $(\Sigma X)^2$

2. X: 10, 5, 12, 0, 2
   Y: 44, 4, 9, 3, 6
   a) ΣY **b)** $\Sigma(Y - 30)$ **c)** ΣY^2 **d)** $(\Sigma Y)^2$
   e)ΣXY

3. X: 50, 100, 150, 200, 250
   a) ΣX **b)** $\Sigma(X - 175)$ **c)** ΣX^2 **d)** $(\Sigma X)^2$

4. X: 50, 100, 150, 200, 250
   Y: 1390, 1680, 1700, 1840, 1890
   a) ΣY **b)** $\Sigma(Y - 1700)$ **c)** ΣY^2 **d)** $(\Sigma Y)^2$
   e) ΣXY

5. X: 0.24, 0.15, 0.89, 0.17, 0.66, 0.53
   Y: 10, 20, 8, 15, 30, 35
   a) ΣX **b)** ΣY **c)** ΣX^2 **d)** ΣY^2 **e)** ΣXY

APPENDIX C
Factorial notation—!

The symbol (!) after a whole number is used to indicate a series of multiplications and is read "factorial"; 6! is read "six factorial." It requires us to multiply all whole numbers in a series, starting with the number given through 1. Thus $5! = 5 \cdot 4 \cdot 3 \cdot 2 \cdot 1$. In general,

$$\boxed{n! = n \cdot (n-1) \cdot (n-2) \cdot \cdots \cdot 1.}$$

Frequently, this notation is used as part of further multiplication and division; in those cases, it may not be necessary to calculate the complete product each time. For example, further division may have the effect of canceling factors and reducing the work required to get a final answer.

EXAMPLE C.1 Find 6!/3!.

Solution

$$6! = 6 \cdot 5 \cdot 4 \cdot 3 \cdot 2 \cdot 1$$

$$3! = 3 \cdot 2 \cdot 1$$

$$\frac{6!}{3!} = \frac{6 \cdot 5 \cdot 4 \cdot \cancel{3} \cdot \cancel{2} \cdot \cancel{1}}{\cancel{3} \cdot \cancel{2} \cdot \cancel{1}}$$

$$= 6 \cdot 5 \cdot 4 = 120.$$

This is certainly easier than calculating $6! = 720$, $3! = 7$, and then dividing. □

EXAMPLE C.2 Find 10!/8!.

Solution

$$\frac{10!}{8!} = \frac{10 \cdot 9 \cdot \cancel{8} \cdot \cancel{7} \cdot \cancel{6} \cdot \cancel{5} \cdot \cancel{4} \cdot \cancel{3} \cdot \cancel{2} \cdot \cancel{1}}{\cancel{8} \cdot \cancel{7} \cdot \cancel{6} \cdot \cancel{5} \cdot \cancel{4} \cdot \cancel{3} \cdot \cancel{2} \cdot \cancel{1}}$$

$$= 10 \cdot 9 = 90. \quad □$$

Even this amount of work is unecessary. Note that $10! = 10 \cdot 9 \cdot 8!$, and thus

$$\frac{10!}{8!} = \frac{10 \cdot 9 \cdot \cancel{8!}}{\cancel{8!}} = 90.$$

In general,

$$\boxed{n! = n \cdot (n-1) \cdot (n-2) \cdot \cdots \cdot (n-r)!,}$$

where r is a whole number less than n.

EXAMPLE C.3 Find 8!/6!2!.

Solution

$$8! = 8 \cdot 7 \cdot 6!$$

Thus

$$\frac{8!}{6!2!} = \frac{8 \cdot 7 \cdot \cancel{6!}}{\cancel{6!} \cdot 2 \cdot 1} = \frac{8 \cdot 7}{2}$$

$$= 28. \quad \square$$

In Example C.3, the sum of the numbers in the denominator equals the number in the numerator. This type of calculation is required in Chapter 6. Let us examine some further examples of this nature.

EXAMPLE C.4 Find 12!/8!4!.

Solution

$$\frac{12!}{8!4!} = \frac{12 \cdot 11 \cdot 10 \cdot 9 \cdot \cancel{8!}}{\cancel{8!} \cdot 4 \cdot 3 \cdot 2 \cdot 1} = \frac{\cancel{12} \cdot 11 \cdot \overset{5}{\cancel{10}} \cdot 9}{\cancel{4} \cdot \cancel{3} \cdot \cancel{2}}$$

$$= 495. \quad \square$$

Note that all the results in these examples are whole numbers. So long as the number in the numerator is equal to or larger than the sum of the numbers in the denominator, the answer will be a whole number.

EXAMPLE C.5 Find 7!/5!(7 − 5)!.

Solution

$$\frac{7!}{5!(7-5)!} = \frac{7 \cdot 6 \cdot \cancel{5!}}{\cancel{5!} \cdot 2!} = \frac{7 \cdot 6}{2 \cdot 1}$$

$$= 21. \quad \square$$

EXAMPLE C.6 Find 8!/8!(8 − 8)!.

Solution

$$\frac{8!}{8!(8-8)!} = \frac{8!}{8! \cdot 0!} = \frac{1}{0!}.$$

Since the answer is to be a whole number, 0! is defined as equal to 1 for consistency with all the uses to which factorial notation is put.

$$\boxed{0! = 1.}$$

Thus

$$\frac{8!}{8!(8-8)!} = \frac{1}{0!} = 1. \quad \square$$

EXERCISES/Appendix C

Calculate the following.

1. $6!$
2. $4!$
3. $2!3!$
4. $\frac{4!}{3!}$
5. $\frac{12!}{9!}$
6. $\frac{15!}{13!}$
7. $\frac{11!}{7!}$
8. $\frac{8!}{6!}$
9. $\frac{7!}{4!}$
10. $\frac{9!}{7!}$
11. $\frac{11!}{7!4!}$
12. $\frac{10!}{6!4!}$
13. $\frac{7!}{3!4!}$
14. $\frac{9!}{9!0!}$
15. $\frac{12!}{10!2!}$

APPENDIX D
Tables

TABLE D.1 Binomial probabilities

		p										
n	*X*	.1	.2	.25	.3	.4	.5	.6	.7	.75	.8	.9
1	0	.9000	.8000	.7500	.7000	.6000	.5000	.4000	.3000	.2500	.2000	.1000
	1	.1000	.2000	.2500	.3000	.4000	.5000	.6000	.7000	.7500	.8000	.9000
2	0	.8100	.6400	.5625	.4900	.3600	.2500	.1600	.0900	.0625	.0400	.0100
	1	.1800	.3200	.3750	.4200	.4800	.5000	.4800	.4200	.3750	.3200	.1800
	2	.0100	.0400	.0625	.0900	.1600	.2500	.3600	.4900	.5625	.6400	.8100
3	0	.7290	.5120	.4219	.3430	.2160	.1250	.0640	.0270	.0156	.0080	.0010
	1	.2430	.3840	.4219	.4410	.4320	.3750	.2880	.1890	.1406	.0960	.0270
	2	.0270	.0960	.1406	.1890	.2880	.3750	.4320	.4410	.4219	.3840	.2430
	3	.0010	.0080	.0156	.0270	.0640	.1250	.2160	.3430	.4219	.5120	.7290
4	0	.6561	.4096	.3164	.2401	.1296	.0625	.0256	.0081	.0039	.0016	.0001
	1	.2916	.4096	.4219	.4116	.3456	.2500	.1536	.0756	.0469	.0256	.0036
	2	.0486	.1536	.2109	.2646	.3456	.3750	.3456	.2646	.2109	.1536	.0486
	3	.0036	.0256	.0469	.0756	.1536	.2500	.3456	.4116	.4219	.4096	.2916
	4	.0001	.0016	.0039	.0081	.0256	.0625	.1296	.2401	.3164	.4096	.6561
5	0	.5905	.3277	.2373	.1681	.0778	.0313	.0102	.0024	.0010	.0003	.0000
	1	.3281	.4096	.3955	.3602	.2592	.1563	.0768	.0284	.0146	.0064	.0005
	2	.0729	.2048	.2637	.3037	.3456	.3125	.2304	.1323	.0879	.0512	.0081
	3	.0081	.0512	.0879	.1323	.2304	.3125	.3456	.3087	.2637	.2048	.0729
	4	.0004	.0064	.0146	.0283	.0768	.1563	.2592	.3602	.3955	.4096	.3281
	5	.0000	.0003	.0010	.0024	.0102	.0313	.0778	.1681	.2373	.3277	.5905
6	0	.5314	.2621	.1780	.1176	.0467	.0156	.0041	.0007	.0002	.0001	.0000
	1	.3543	.3932	.3560	.3025	.1866	.0938	.0369	.0102	.0044	.0015	.0001
	2	.0984	.2458	.2966	.3241	.3110	.2344	.1382	.0595	.0330	.0154	.0012
	3	.0146	.0819	.1318	.1852	.2765	.3125	.2765	.1852	.1318	.0819	.0146
	4	.0012	.0154	.0330	.0595	.1382	.2344	.3110	.3241	.2966	.2458	.0984
	5	.0001	.0015	.0044	.0102	.0369	.0938	.1866	.3025	.3560	.3932	.3543
	6	.0000	.0001	.0002	.0007	.0041	.0156	.0467	.1176	.1780	.2621	.5314
7	0	.4783	.2097	.1335	.0824	.0280	.0078	.0016	.0002	.0001	.0000	.0000
	1	.3720	.3670	.3115	.2471	.1306	.0547	.0172	.0036	.0013	.0004	.0000
	2	.1240	.2753	.3115	.3177	.2613	.1641	.0774	.0250	.0115	.0043	.0002
	3	.0230	.1147	.1730	.2269	.2903	.2734	.1935	.0972	.0577	.0287	.0026
	4	.0026	.0287	.0577	.0972	.1935	.2734	.2903	.2269	.1730	.1147	.0230
	5	.0002	.0043	.0115	.0250	.0774	.1641	.2613	.3177	.3115	.2753	.1240
	6	.0000	.0004	.0013	.0036	.0172	.0547	.1306	.2471	.3115	.3670	.3720
	7	.0000	.0000	.0001	.0002	.0016	.0078	.0280	.0824	.1335	.2097	.4783

		p										
n	X	.1	.2	.25	.3	.4	.5	.6	.7	.75	.8	.9
8	0	.4305	.1678	.1001	.0576	.0168	.0039	.0007	.0001	.0000	.0000	.0000
	1	.3826	.3355	.2670	.1977	.0896	.0312	.0079	.0012	.0004	.0001	.0000
	2	.1488	.2936	.3115	.2965	.2090	.1094	.0413	.0100	.0038	.0011	.0000
	3	.0331	.1468	.2076	.2541	.2787	.2188	.1239	.0467	.0231	.0092	.0004
	4	.0046	.0459	.0865	.1361	.2322	.2734	.2322	.1361	.0865	.0459	.0046
	5	.0004	.0092	.0231	.0467	.1239	.2188	.2787	.2541	.2076	.1468	.0331
	6	.0000	.0011	.0038	.0100	.0413	.1094	.2090	.2965	.3115	.2936	.1488
	7	.0000	.0001	.0004	.0012	.0079	.0312	.0896	.1977	.2670	.3355	.3826
	8	.0000	.0000	.0000	.0001	.0007	.0039	.0168	.0576	.1001	.1678	.4305
9	0	.3874	.1342	.0751	.0404	.0101	.0020	.0003	.0000	.0000	.0000	.0000
	1	.3874	.3020	.2253	.1556	.0605	.0176	.0035	.0004	.0001	.0000	.0000
	2	.1722	.3020	.3003	.2668	.1612	.0703	.0212	.0039	.0012	.0003	.0000
	3	.0446	.1762	.2336	.2668	.2508	.1641	.0743	.0210	.0087	.0028	.0001
	4	.0074	.0661	.1168	.1715	.2508	.2461	.1672	.0735	.0389	.0165	.0008
	5	.0008	.0165	.0389	.0735	.1672	.2461	.2508	.1715	.1168	.0661	.0074
	6	.0001	.0028	.0087	.0210	.0743	.1641	.2508	.2668	.2336	.1762	.0446
	7	.0000	.0003	.0012	.0039	.0212	.0703	.1612	.2668	.3003	.3020	.1722
	8	.0000	.0000	.0001	.0004	.0035	.0176	.0605	.1556	.2253	.3020	.3874
	9	.0000	.0000	.0000	.0000	.0003	.0020	.0101	.0404	.0751	.1342	.3874
10	0	.3487	.1074	.0563	.0282	.0060	.0010	.0001	.0000	.0000	.0000	.0000
	1	.3874	.2684	.1877	.1211	.0403	.0098	.0016	.0001	.0000	.0000	.0000
	2	.1937	.3020	.2816	.2335	.1209	.0439	.0106	.0014	.0004	.0001	.0000
	3	.0574	.2013	.2503	.2668	.2150	.1172	.0425	.0090	.0031	.0008	.0000
	4	.0112	.0881	.1460	.2001	.2508	.2051	.1115	.0368	.0162	.0055	.0001
	5	.0015	.0264	.0584	.1029	.2007	.2461	.2007	.1029	.0584	.0264	.0015
	6	.0001	.0055	.0162	.0368	.1115	.2051	.2508	.2001	.1460	.0881	.0112
	7	.0000	.0008	.0031	.0090	.0425	.1172	.2150	.2668	.2503	.2013	.0574
	8	.0000	.0001	.0004	.0014	.0106	.0439	.1209	.2335	.2816	.3020	.1937
	9	.0000	.0000	.0000	.0001	.0016	.0098	.0403	.1211	.1877	.2684	.3874
	10	.0000	.0000	.0000	.0000	.0001	.0010	.0060	.0282	.0563	.1074	.3487
11	0	.3138	.0859	.0422	.0198	.0036	.0005	.0000	.0000	.0000	.0000	.0000
	1	.3835	.2362	.1549	.0932	.0266	.0054	.0007	.0000	.0000	.0000	.0000
	2	.2131	.2953	.2581	.1998	.0887	.0269	.0052	.0005	.0001	.0000	.0000
	3	.0710	.2215	.2581	.2568	.1774	.0806	.0234	.0037	.0011	.0002	.0000
	4	.0158	.1107	.1721	.2201	.2365	.1611	.0701	.0173	.0064	.0017	.0000
	5	.0025	.0388	.0803	.1321	.2207	.2256	.1471	.0566	.0268	.0097	.0003
	6	.0003	.0097	.0268	.0566	.1471	.2256	.2207	.1321	.0803	.0388	.0025
	7	.0000	.0017	.0064	.0173	.0701	.1611	.2365	.2201	.1721	.1107	.0158
	8	.0000	.0002	.0011	.0037	.0234	.0806	.1774	.2568	.2581	.2215	.0710
	9	.0000	.0000	.0001	.0005	.0052	.0269	.0887	.1998	.2581	.2953	.2131
	10	.0000	.0000	.0000	.0000	.0007	.0054	.0266	.0932	.1549	.2362	.3835
	11	.0000	.0000	.0000	.0000	.0000	.0005	.0036	.0198	.0422	.0859	.3138

TABLE D.1 (Cont.)

		p										
n	*X*	.1	.2	.25	.3	.4	.5	.6	.7	.75	.8	.9
12	0	.2824	.0687	.0317	.0138	.0022	.0002	.0000	.0000	.0000	.0000	.0000
	1	.3766	.2062	.1267	.0712	.0174	.0029	.0003	.0000	.0000	.0000	.0000
	2	.2301	.2835	.2323	.1678	.0639	.0161	.0025	.0002	.0000	.0000	.0000
	3	.0852	.2362	.2581	.2397	.1419	.0537	.0125	.0015	.0004	.0001	.0000
	4	.0213	.1329	.1936	.2311	.2128	.1208	.0420	.0078	.0024	.0005	.0000
	5	.0038	.0532	.1032	.1585	.2270	.1934	.1009	.0291	.0115	.0033	.0000
	6	.0005	.0155	.0401	.0792	.1766	.2256	.1766	.0792	.0401	.0155	.0005
	7	.0000	.0033	.0115	.0291	.1009	.1934	.2270	.1585	.1032	.0532	.0038
	8	.0000	.0005	.0024	.0078	.0420	.1208	.2128	.2311	.1936	.1329	.0213
	9	.0000	.0001	.0004	.0015	.0125	.0537	.1419	.2397	.2581	.2362	.0852
	10	.0000	.0000	.0000	.0002	.0025	.0161	.0639	.1678	.2323	.2835	.2301
	11	.0000	.0000	.0000	.0000	.0003	.0029	.0174	.0712	.1267	.2062	.3766
	12	.0000	.0000	.0000	.0000	.0000	.0002	.0022	.0138	.0317	.0687	.2824
13	0	.2542	.0550	.0238	.0097	.0013	.0001	.0000	.0000	.0000	.0000	.0000
	1	.3672	.1787	.1029	.0540	.0113	.0016	.0001	.0000	.0000	.0000	.0000
	2	.2448	.2680	.2059	.1388	.0453	.0095	.0012	.0001	.0000	.0000	.0000
	3	.0997	.2457	.2517	.2181	.1107	.0349	.0065	.0006	.0001	.0000	.0000
	4	.0277	.1535	.2097	.2337	.1845	.0873	.0243	.0034	.0009	.0001	.0000
	5	.0055	.0691	.1258	.1803	.2214	.1571	.0656	.0142	.0047	.0011	.0000
	6	.0008	.0230	.0559	.1030	.1968	.2095	.1312	.0442	.0186	.0058	.0001
	7	.0001	.0058	.0186	.0442	.1312	.2095	.1968	.1030	.0559	.0230	.0008
	8	.0000	.0011	.0047	.0142	.0656	.1571	.2214	.1803	.1258	.0691	.0055
	9	.0000	.0001	.0009	.0034	.0243	.0873	.1845	.2337	.2097	.1535	.0277
	10	.0000	.0000	.0001	.0006	.0065	.0349	.1107	.2181	.2517	.2457	.0997
	11	.0000	.0000	.0000	.0001	.0012	.0095	.0453	.1388	.2059	.2680	.2448
	12	.0000	.0000	.0000	.0000	.0001	.0016	.0113	.0540	.1029	.1787	.3672
	13	.0000	.0000	.0000	.0000	.0000	.0001	.0013	.0097	.0238	.0550	.2542
14	0	.2288	.0440	.0178	.0068	.0008	.0001	.0000	.0000	.0000	.0000	.0000
	1	.3559	.1539	.0832	.0407	.0073	.0009	.0001	.0000	.0000	.0000	.0000
	2	.2570	.2501	.1802	.1134	.0317	.0056	.0005	.0000	.0000	.0000	.0000
	3	.1142	.2501	.2402	.1943	.0845	.0222	.0033	.0002	.0000	.0000	.0000
	4	.0349	.1720	.2202	.2290	.1549	.0611	.0136	.0014	.0003	.0000	.0000
	5	.0078	.0860	.1468	.1963	.2066	.1222	.0408	.0066	.0018	.0003	.0000
	6	.0013	.0322	.0734	.1262	.2066	.1833	.0918	.0232	.0082	.0020	.0000
	7	.0002	.0092	.0280	.0618	.1574	.2095	.1574	.0618	.0280	.0092	.0002
	8	.0000	.0020	.0082	.0232	.0918	.1833	.2066	.1262	.0734	.0322	.0013
	9	.0000	.0003	.0018	.0066	.0408	.1222	.2066	.1963	.1468	.0860	.0078
	10	.0000	.0000	.0003	.0014	.0136	.0611	.1549	.2290	.2202	.1720	.0349
	11	.0000	.0000	.0000	.0002	.0033	.0222	.0845	.1943	.2402	.2501	.1142
	12	.0000	.0000	.0000	.0000	.0005	.0056	.0317	.1134	.1802	.2501	.2570
	13	.0000	.0000	.0000	.0000	.0001	.0009	.0073	.0407	.0832	.1539	.3559
	14	.0000	.0000	.0000	.0000	.0000	.0001	.0008	.0068	.0178	.0440	.2288

		p										
n	*X*	.1	.2	.25	.3	.4	.5	.6	.7	.75	.8	.9
15	0	.2059	.0352	.0134	.0047	.0005	.0000	.0000	.0000	.0000	.0000	.0000
	1	.3432	.1319	.0668	.0305	.0047	.0005	.0000	.0000	.0000	.0000	.0000
	2	.2669	.2309	.1559	.0916	.0219	.0032	.0003	.0000	.0000	.0000	.0000
	3	.1285	.2501	.2252	.1700	.0634	.0139	.0016	.0001	.0000	.0000	.0000
	4	.0428	.1876	.2252	.2186	.1268	.0417	.0074	.0006	.0001	.0000	.0000
	5	.0105	.1032	.1651	.2061	.1859	.0916	.0245	.0030	.0007	.0001	.0000
	6	.0019	.0430	.0917	.1472	.2066	.1527	.0612	.0116	.0034	.0007	.0000
	7	.0003	.0138	.0393	.0811	.1771	.1964	.1181	.0348	.0131	.0035	.0000
	8	.0000	.0035	.0131	.0348	.1181	.1964	.1771	.0811	.0393	.0138	.0003
	9	.0000	.0007	.0034	.0116	.0612	.1527	.2066	.1472	.0917	.0430	.0019
	10	.0000	.0001	.0007	.0030	.0245	.0916	.1859	.2061	.1651	.1032	.0105
	11	.0000	.0000	.0001	.0006	.0074	.0417	.1268	.2186	.2252	.1876	.0428
	12	.0000	.0000	.0000	.0001	.0016	.0139	.0634	.1700	.2252	.2501	.1285
	13	.0000	.0000	.0000	.0000	.0003	.0032	.0219	.0916	.1559	.2309	.2669
	14	.0000	.0000	.0000	.0000	.0000	.0005	.0047	.0305	.0668	.1319	.3432
	15	.0000	.0000	.0000	.0000	.0000	.0000	.0005	.0047	.0134	.0352	.2059
20	0	.1216	.0115	.0032	.0008	.0000	.0000	.0000	.0000	.0000	.0000	.0000
	1	.2702	.0576	.0211	.0068	.0005	.0000	.0000	.0000	.0000	.0000	.0000
	2	.2852	.1369	.0669	.0278	.0031	.0002	.0000	.0000	.0000	.0000	.0000
	3	.1901	.2054	.1339	.0716	.0123	.0011	.0000	.0000	.0000	.0000	.0000
	4	.0898	.2182	.1897	.1304	.0350	.0046	.0003	.0000	.0000	.0000	.0000
	5	.0319	.1746	.2023	.1789	.0746	.0148	.0013	.0000	.0000	.0000	.0000
	6	.0089	.1091	.1686	.1916	.1244	.0370	.0049	.0002	.0000	.0000	.0000
	7	.0020	.0545	.1124	.1643	.1659	.0739	.0146	.0010	.0002	.0000	.0000
	8	.0004	.0222	.0609	.1144	.1797	.1201	.0355	.0039	.0008	.0001	.0000
	9	.0001	.0074	.0271	.0654	.1597	.1602	.0710	.0120	.0030	.0005	.0000
	10	.0000	.0020	.0099	.0308	.1171	.1762	.1171	.0308	.0099	.0020	.0000
	11	.0000	.0005	.0030	.0120	.0710	.1602	.1597	.0654	.0271	.0074	.0001
	12	.0000	.0001	.0008	.0039	.0355	.1201	.1797	.1144	.0609	.0222	.0004
	13	.0000	.0000	.0002	.0010	.0146	.0739	.1659	.1643	.1124	.0545	.0020
	14	.0000	.0000	.0000	.0002	.0049	.0370	.1244	.1916	.1686	.1091	.0089
	15	.0000	.0000	.0000	.0000	.0013	.0148	.0746	.1789	.2023	.1746	.0319
	16	.0000	.0000	.0000	.0000	.0003	.0046	.0350	.1304	.1897	.2182	.0898
	17	.0000	.0000	.0000	.0000	.0000	.0011	.0123	.0716	.1339	.2054	.1901
	18	.0000	.0000	.0000	.0000	.0000	.0002	.0031	.0278	.0669	.1369	.2852
	19	.0000	.0000	.0000	.0000	.0000	.0000	.0005	.0068	.0211	.0576	.2702
	20	.0000	.0000	.0000	.0000	.0000	.0000	.0000	.0008	.0032	.0115	.1216

TABLE D.1 (Cont.)

		p										
n	X	.1	.2	.25	.3	.4	.5	.6	.7	.75	.8	.9
25	0	.0718	.0038	.0008	.0001	.0000	.0000	.0000	.0000	.0000	.0000	.0000
	1	.1994	.0236	.0063	.0014	.0000	.0000	.0000	.0000	.0000	.0000	.0000
	2	.2659	.0708	.0251	.0074	.0004	.0000	.0000	.0000	.0000	.0000	.0000
	3	.2265	.1358	.0641	.0243	.0019	.0001	.0000	.0000	.0000	.0000	.0000
	4	.1384	.1867	.1175	.0572	.0071	.0004	.0000	.0000	.0000	.0000	.0000
	5	.0646	.1960	.1645	.1030	.0199	.0016	.0000	.0000	.0000	.0000	.0000
	6	.0239	.1633	.1828	.1472	.0442	.0053	.0002	.0000	.0000	.0000	.0000
	7	.0072	.1108	.1654	.1712	.0800	.0143	.0009	.0000	.0000	.0000	.0000
	8	.0018	.0623	.1241	.1651	.1200	.0322	.0031	.0001	.0000	.0000	.0000
	9	.0004	.0294	.0781	.1336	.1511	.0609	.0088	.0004	.0000	.0000	.0000
	10	.0001	.0118	.0417	.0916	.1612	.0974	.0212	.0013	.0002	.0000	.0000
	11	.0000	.0040	.0189	.0536	.1465	.1328	.0434	.0042	.0007	.0001	.0000
	12	.0000	.0012	.0074	.0268	.1140	.1550	.0760	.0115	.0025	.0003	.0000
	13	.0000	.0003	.0025	.0115	.0760	.1550	.1140	.0268	.0074	.0012	.0000
	14	.0000	.0001	.0007	.0042	.0434	.1328	.1465	.0536	.0189	.0040	.0000
	15	.0000	.0000	.0002	.0013	.0212	.0974	.1612	.0916	.0417	.0118	.0001
	16	.0000	.0000	.0000	.0004	.0088	.0609	.1511	.1336	.0781	.0294	.0004
	17	.0000	.0000	.0000	.0001	.0031	.0322	.1200	.1651	.1241	.0623	.0018
	18	.0000	.0000	.0000	.0000	.0009	.0143	.0800	.1712	.1654	.1108	.0072
	19	.0000	.0000	.0000	.0000	.0002	.0053	.0442	.1472	.1828	.1633	.0239
	20	.0000	.0000	.0000	.0000	.0000	.0016	.0199	.1030	.1645	.1960	.0646
	21	.0000	.0000	.0000	.0000	.0000	.0004	.0071	.0572	.1175	.1867	.1384
	22	.0000	.0000	.0000	.0000	.0000	.0001	.0019	.0243	.0641	.1358	.2265
	23	.0000	.0000	.0000	.0000	.0000	.0000	.0004	.0074	.0251	.0708	.2659
	24	.0000	.0000	.0000	.0000	.0000	.0000	.0000	.0014	.0063	.0236	.1994
	25	.0000	.0000	.0000	.0000	.0000	.0000	.0000	.0001	.0008	.0038	.0718

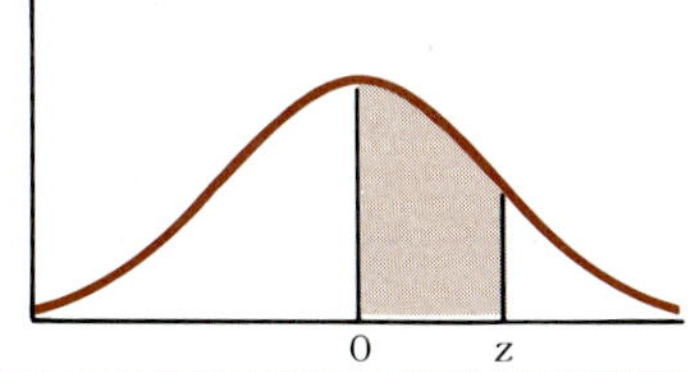

TABLE D.2 The standard normal (z) distribution

z	.00	.01	.02	.03	.04	.05	.06	.07	.08	.09
0.0	.0000	.0040	.0080	.0120	.0160	.0199	.0239	.0279	.0319	.0359
0.1	.0398	.0438	.0478	.0517	.0557	.0596	.0636	.0675	.0714	.0753
0.2	.0793	.0832	.0871	.0910	.0948	.0987	.1026	.1064	.1103	.1141
0.3	.1179	.1217	.1255	.1293	.1331	.1368	.1406	.1443	.1480	.1517
0.4	.1554	.1591	.1628	.1664	.1700	.1736	.1772	.1808	.1844	.1879
0.5	.1915	.1950	.1985	.2019	.2054	.2088	.2123	.2157	.2190	.2224
0.6	.2257	.2291	.2324	.2357	.2389	.2422	.2454	.2486	.2517	.2549
0.7	.2580	.2611	.2642	.2673	.2704	.2734	.2764	.2794	.2823	.2852
0.8	.2881	.2910	.2939	.2967	.2995	.3023	.3051	.3078	.3106	.3133
0.9	.3159	.3186	.3212	.3238	.3264	.3289	.3315	.3340	.3365	.3389
1.0	.3413	.3438	.3461	.3485	.3508	.3531	.3554	.3577	.3599	.3621
1.1	.3643	.3665	.3686	.3708	.3729	.3749	.3770	.3790	.3810	.3830
1.2	.3849	.3869	.3888	.3907	.3925	.3944	.3962	.3980	.3997	.4015
1.3	.4032	.4049	.4066	.4082	.4099	.4115	.4131	.4147	.4162	.4177
1.4	.4192	.4207	.4222	.4236	.4251	.4265	.4279	.4292	.4306	.4319
1.5	.4332	.4345	.4357	.4370	.4382	.4394	.4406	.4418	.4429	.4441
1.6	.4452	.4463	.4474	.4484	.4495	.4505	.4515	.4525	.4535	.4545
1.7	.4554	.4564	.4573	.4582	.4591	.4599	.4608	.4616	.4625	.4633
1.8	.4641	.4649	.4656	.4664	.4671	.4678	.4686	.4693	.4699	.4706
1.9	.4713	.4719	.4726	.4732	.4738	.4744	.4750	.4756	.4761	.4767
2.0	.4772	.4778	.4783	.4788	.4793	.4798	.4803	.4808	.4812	.4817
2.1	.4821	.4826	.4830	.4834	.4838	.4842	.4846	.4850	.4854	.4857
2.2	.4861	.4864	.4868	.4871	.4875	.4878	.4881	.4884	.4887	.4890
2.3	.4893	.4896	.4898	.4901	.4904	.4906	.4909	.4911	.4913	.4916
2.4	.4918	.4920	.4922	.4925	.4927	.4929	.4931	.4932	.4934	.4936
2.5	.4938	.4940	.4941	.4943	.4945	.4946	.4948	.4949	.4951	.4952
2.6	.4953	.4955	.4956	.4957	.4959	.4960	.4961	.4962	.4963	.4964
2.7	.4965	.4966	.4967	.4968	.4969	.4970	.4971	.4972	.4973	.4974
2.8	.4974	.4975	.4976	.4977	.4977	.4978	.4979	.4979	.4980	.4981
2.9	.4981	.4982	.4982	.4983	.4984	.4984	.4985	.4985	.4986	.4986
3.0	.4987	.4987	.4987	.4988	.4988	.4989	.4989	.4989	.4990	.4990

Reprinted by permission from Mario F. Triola, *Elementary Statistics.* Menlo Park, Calif.: Benjamin/Cummings, 1983.

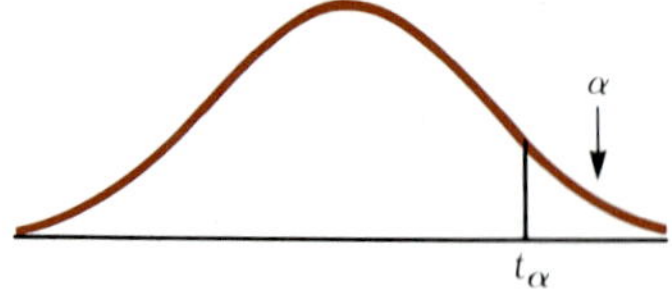

TABLE D.3 Student's t distribution

df	$t_{.10}$	$t_{.05}$	$t_{.025}$	$t_{.01}$	$t_{.005}$
1	3.08	6.31	12.71	31.82	63.66
2	1.89	2.92	4.30	6.96	9.92
3	1.64	2.35	3.18	4.54	5.84
4	1.53	2.13	2.78	3.75	4.60
5	1.48	2.02	2.57	3.36	4.03
6	1.44	1.94	2.45	3.14	3.71
7	1.42	1.89	2.36	3.00	3.50
8	1.40	1.86	2.31	2.90	3.36
9	1.38	1.83	2.26	2.82	3.25
10	1.37	1.81	2.23	2.76	3.17
11	1.36	1.80	2.20	2.72	3.11
12	1.36	1.78	2.18	2.68	3.05
13	1.35	1.77	2.16	2.65	3.01
14	1.35	1.76	2.14	2.62	2.98
15	1.34	1.75	2.13	2.60	2.95
16	1.34	1.75	2.12	2.58	2.92
17	1.33	1.74	2.11	2.57	2.90
18	1.33	1.73	2.10	2.55	2.88
19	1.33	1.73	2.09	2.54	2.86
20	1.33	1.72	2.09	2.53	2.85
21	1.32	1.72	2.08	2.52	2.83
22	1.32	1.72	2.07	2.51	2.82
23	1.32	1.71	2.07	2.50	2.81
24	1.32	1.71	2.06	2.49	2.80
25	1.32	1.71	2.06	2.49	2.79
26	1.32	1.71	2.06	2.48	2.78
27	1.31	1.70	2.05	2.47	2.77
28	1.31	1.70	2.05	2.47	2.76
29	1.31	1.70	2.05	2.46	2.76
z	1.28	1.64	1.96	2.33	2.58

Adapted from Donald B. Owen, *Handbook of Statistical Tables.* Copyright © 1962, U.S. Department of Energy; published by Addison-Wesley, Reading, Mass. Adapted from Neil Weiss and Matthew Hassett, *Introductory Statistics.* Reading, Mass.: Addison-Wesley, 1982. Reprinted with permission.

TABLE D.4 The chi-square (χ^2) distribution

	α									
df	**0.995**	**0.99**	**0.975**	**0.95**	**0.90**	**0.10**	**0.05**	**0.025**	**0.01**	**0.005**
1	—	—	0.001	0.004	0.016	2.706	3.841	5.024	6.635	7.879
2	0.010	0.020	0.051	0.103	0.211	4.605	5.991	7.378	9.210	10.597
3	0.072	0.115	0.216	0.352	0.584	6.251	7.815	9.348	11.345	12.838
4	0.207	0.297	0.484	0.711	1.064	7.779	9.488	11.143	13.277	14.860
5	0.412	0.554	0.831	1.145	1.610	9.236	11.071	12.833	15.086	16.750
6	0.676	0.872	1.237	1.635	2.204	10.645	12.592	14.449	16.812	18.548
7	0.989	1.239	1.690	2.167	2.833	12.017	14.067	16.013	18.475	20.278
8	1.344	1.646	2.180	2.733	3.490	13.362	15.507	17.535	20.090	21.955
9	1.735	2.088	2.700	3.325	4.168	14.684	16.919	19.023	21.666	23.589
10	2.156	2.558	3.247	3.940	4.865	15.987	18.307	20.483	23.209	25.188
11	2.603	3.053	3.816	4.575	5.578	17.275	19.675	21.920	24.725	26.757
12	3.074	3.571	4.404	5.226	6.304	18.549	21.026	23.337	26.217	28.299
13	3.565	4.107	5.009	5.892	7.042	19.812	22.362	24.736	27.688	29.819
14	4.075	4.660	5.629	6.571	7.790	21.064	23.685	26.119	29.141	31.319
15	4.601	5.229	6.262	7.261	8.547	22.307	24.996	27.488	30.578	32.801
16	5.142	5.812	6.908	7.962	9.312	23.542	26.296	28.845	32.000	34.267
17	5.697	6.408	7.564	8.672	10.085	24.769	27.587	30.191	33.409	35.718
18	6.265	7.015	8.231	9.390	10.865	25.989	28.869	31.526	34.805	37.156
19	6.844	7.633	8.907	10.117	11.651	27.204	30.144	32.852	36.191	38.582
20	7.434	8.260	9.591	10.851	12.443	28.412	31.410	34.170	37.566	39.997
21	8.034	8.897	10.283	11.591	13.240	29.615	32.671	35.479	38.932	41.401
22	8.643	9.542	10.982	12.338	14.042	30.813	33.924	36.781	40.289	42.796
23	9.260	10.196	11.689	13.091	14.848	32.007	35.172	38.076	41.638	44.181
24	9.886	10.856	12.401	13.848	15.659	33.196	36.415	39.364	42.980	45.559
25	10.520	11.524	13.120	14.611	16.473	34.382	37.652	40.646	44.314	46.928
26	11.160	12.198	13.844	15.379	17.292	35.563	38.885	41.923	45.642	48.290
27	11.808	12.879	14.573	16.151	18.114	36.741	40.113	43.194	46.963	49.645
28	12.461	13.565	15.308	16.928	18.939	37.916	41.337	44.461	48.278	50.993
29	13.121	14.257	16.047	17.708	19.768	39.087	42.557	45.722	49.588	52.336
30	13.787	14.954	16.791	18.493	20.599	40.256	43.773	46.979	50.892	53.672
40	20.707	22.164	24.433	26.509	29.051	51.805	55.758	59.342	63.691	66.766
50	27.991	29.707	32.357	34.764	37.689	63.167	67.505	71.420	76.154	79.490
60	35.534	37.485	40.482	43.188	46.459	74.397	79.082	83.298	88.379	91.952
70	43.275	45.442	48.758	51.739	55.329	85.527	90.531	95.023	100.425	104.215
80	51.172	53.540	57.153	60.391	64.278	96.578	101.879	106.629	112.329	116.321
90	59.196	61.754	65.647	69.126	73.291	107.565	113.145	118.136	124.116	128.299
100	67.328	70.065	74.222	77.929	82.358	118.498	124.342	129.561	135.807	140.169

Reprinted by permission from Mario F. Triola, *Elementary Statistics*. Menlo Park, Calif.: Benjamin/Cummings, 1983.

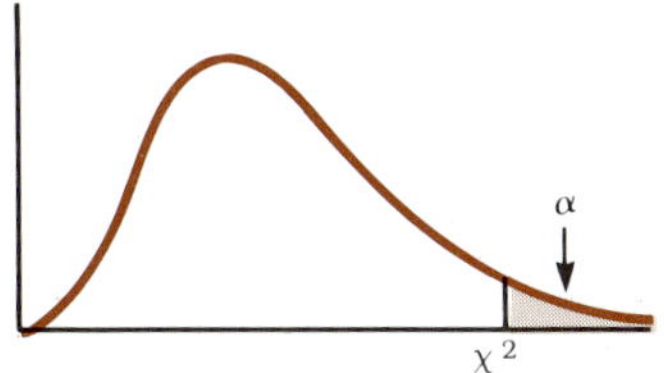

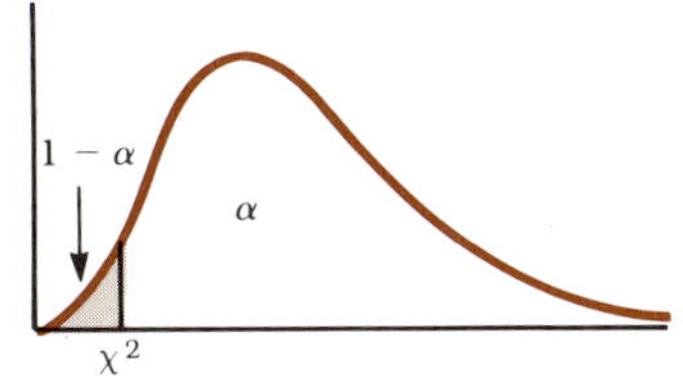

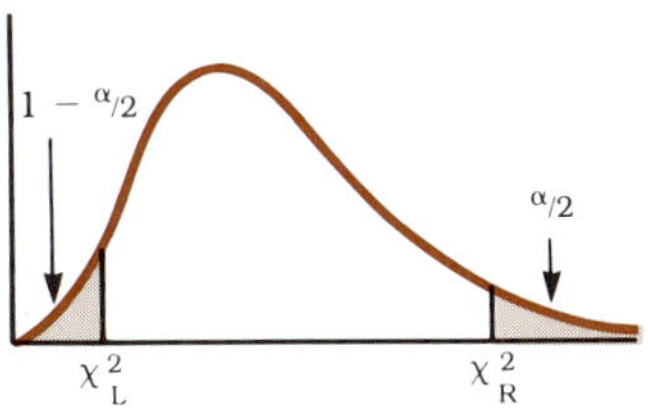

TABLE D.5 *F* Distribution

Values of $F_{0.01}$

df for Denominator	df for Numerator 1	2	3	4	5	6	7	8	9
1	4052	4999.5	5403	5625	5764	5859	5928	5981	6022
2	98.50	99.00	99.17	99.25	99.30	99.33	99.36	99.37	99.39
3	34.12	30.82	29.46	28.71	28.24	27.91	27.67	27.49	27.35
4	21.20	18.00	16.69	15.98	15.52	15.21	14.98	14.80	14.66
5	16.26	13.27	12.06	11.39	10.97	10.67	10.46	10.29	10.16
6	13.75	10.92	9.78	9.15	8.75	8.47	8.26	8.10	7.98
7	12.25	9.55	8.45	7.85	7.46	7.19	6.99	6.84	6.72
8	11.26	8.65	7.59	7.01	6.63	6.37	6.18	6.03	5.91
9	10.56	8.02	6.99	6.42	6.06	5.80	5.61	5.47	5.35
10	10.04	7.56	6.55	5.99	5.64	5.39	5.20	5.06	4.94
11	9.65	7.21	6.22	5.67	5.32	5.07	4.89	4.74	4.63
12	9.33	6.93	5.95	5.41	5.06	4.82	4.64	4.50	4.39
13	9.07	6.70	5.74	5.21	4.86	4.62	4.44	4.30	4.19
14	8.86	6.51	5.56	5.04	4.69	4.46	4.28	4.14	4.03
15	8.68	6.36	5.42	4.89	4.56	4.32	4.14	4.00	3.89
16	8.53	6.23	5.29	4.77	4.44	4.20	4.03	3.89	3.78
17	8.40	6.11	5.18	4.67	4.34	4.10	3.93	3.79	3.68
18	8.29	6.01	5.09	4.58	4.25	4.01	3.84	3.71	3.60
19	8.18	5.93	5.01	4.50	4.17	3.94	3.77	3.63	3.52
20	8.10	5.85	4.94	4.43	4.10	3.87	3.70	3.56	3.46
21	8.02	5.78	4.87	4.37	4.04	3.81	3.64	3.51	3.40
22	7.95	5.72	4.82	4.31	3.99	3.76	3.59	3.45	3.35
23	7.88	5.66	4.76	4.26	3.94	3.71	3.54	3.41	3.30
24	7.82	5.61	4.72	4.22	3.90	3.67	3.50	3.36	3.26
25	7.77	5.57	4.68	4.18	3.85	3.63	3.46	3.32	3.22
26	7.72	5.53	4.64	4.14	3.82	3.59	3.42	3.29	3.18
27	7.68	5.49	4.60	4.11	3.78	3.56	3.39	3.26	3.15
28	7.64	5.45	4.57	4.07	3.75	3.53	3.36	3.23	3.12
29	7.60	5.42	4.54	4.04	3.73	3.50	3.33	3.20	3.09
30	7.56	5.39	4.51	4.02	3.70	3.47	3.30	3.17	3.07
40	7.31	5.18	4.31	3.83	3.51	3.29	3.12	2.99	2.89
60	7.08	4.98	4.13	3.65	3.34	3.12	2.95	2.82	2.72
120	6.85	4.79	3.95	3.48	3.17	2.96	2.79	2.66	2.56
∞	6.63	4.61	3.78	3.32	3.02	2.80	2.64	2.51	2.41

df for Numerator

10	12	15	20	24	30	40	60	120	∞
6056	6106	6157	6209	6235	6261	6287	6313	6339	6366
99.40	99.42	99.43	99.45	99.46	99.47	99.47	99.48	99.49	99.50
27.23	27.05	26.87	26.69	26.60	26.50	26.41	26.32	26.22	26.13
14.55	14.37	14.20	14.02	13.93	13.84	13.75	13.65	13.56	13.46
10.05	9.89	9.72	9.55	9.47	9.38	9.29	9.20	9.11	9.02
7.87	7.72	7.56	7.40	7.31	7.23	7.14	7.06	6.97	6.88
6.62	6.47	6.31	6.16	6.07	5.99	5.91	5.82	5.74	5.65
5.81	5.67	5.52	5.36	5.28	5.20	5.12	5.03	4.95	4.86
5.26	5.11	4.96	4.81	4.73	4.65	4.57	4.48	4.40	4.31
4.85	4.71	4.56	4.41	4.33	4.25	4.17	4.08	4.00	3.91
4.54	4.40	4.25	4.10	4.02	3.94	3.86	3.78	3.69	3.60
4.30	4.16	4.01	3.86	3.78	3.70	3.62	3.54	3.45	3.36
4.10	3.96	3.82	3.66	3.59	3.51	3.43	3.34	3.25	3.17
3.94	3.80	3.66	3.51	3.43	3.35	3.27	3.18	3.09	3.00
3.80	3.67	3.52	3.37	3.29	3.21	3.13	3.05	2.96	2.87
3.69	3.55	3.41	3.26	3.18	3.10	3.02	2.93	2.84	2.75
3.59	3.46	3.31	3.16	3.08	3.00	2.92	2.83	2.75	2.65
3.51	3.37	3.23	3.08	3.00	2.92	2.84	2.75	2.66	2.57
3.43	3.30	3.15	3.00	2.92	2.84	2.76	2.67	2.58	2.49
3.37	3.23	3.09	2.94	2.86	2.78	2.69	2.61	2.52	2.42
3.31	3.17	3.03	2.88	2.80	2.72	2.64	2.55	2.46	2.36
3.26	3.12	2.98	2.83	2.75	2.67	2.58	2.50	2.40	2.31
3.21	3.07	2.93	2.78	2.70	2.62	2.54	2.45	2.35	2.26
3.17	3.03	2.89	2.74	2.66	2.58	2.49	2.40	2.31	2.21
3.13	2.99	2.85	2.70	2.62	2.54	2.45	2.36	2.27	2.17
3.09	2.96	2.81	2.66	2.58	2.50	2.42	2.33	2.23	2.13
3.06	2.93	2.78	2.63	2.55	2.47	2.38	2.29	2.20	2.10
3.03	2.90	2.75	2.60	2.52	2.44	2.35	2.26	2.17	2.06
3.00	2.87	2.73	2.57	2.49	2.41	2.33	2.23	2.14	2.03
2.98	2.84	2.70	2.55	2.47	2.39	2.30	2.21	2.11	2.01
2.80	2.66	2.52	2.37	2.29	2.20	2.11	2.02	1.92	1.80
2.63	2.50	2.35	2.20	2.12	2.03	1.94	1.84	1.73	1.60
2.47	2.34	2.19	2.03	1.95	1.86	1.76	1.66	1.53	1.38
2.32	2.18	2.04	1.88	1.79	1.70	1.59	1.47	1.32	1.00

TABLE D.5 (Cont.)

Values of $F_{0.025}$

df for Denominator	df for Numerator: 1	2	3	4	5	6	7	8	9
1	647.79	799.50	864.16	899.58	921.85	937.11	948.22	956.66	963.28
2	38.51	39.00	39.17	39.25	39.30	39.33	39.36	39.37	39.39
3	17.44	16.04	15.44	15.10	14.89	14.74	14.62	14.54	14.47
4	12.22	10.65	9.98	9.60	9.36	9.20	9.07	8.98	8.90
5	10.00	8.43	7.76	7.39	7.15	6.98	6.85	6.76	6.68
6	8.81	7.26	6.60	6.23	5.99	5.82	5.70	5.60	5.52
7	8.07	6.54	5.89	5.52	5.29	5.12	4.99	4.90	4.82
8	7.57	6.06	5.42	5.05	4.82	4.65	4.53	4.43	4.36
9	7.21	5.71	5.08	4.72	4.48	4.32	4.20	4.10	4.03
10	6.94	5.46	4.83	4.47	4.24	4.07	3.95	3.85	3.78
11	6.72	5.26	4.63	4.28	4.04	3.88	3.76	3.66	3.59
12	6.55	5.10	4.47	4.12	3.89	3.73	3.61	3.51	3.44
13	6.41	4.97	4.35	4.00	3.77	3.60	3.48	3.39	3.31
14	6.30	4.86	4.24	3.89	3.66	3.50	3.38	3.29	3.21
15	6.20	4.77	4.15	3.80	3.58	3.41	3.29	3.20	3.12
16	6.12	4.69	4.08	3.73	3.50	3.34	3.22	3.12	3.05
17	6.04	4.62	4.01	3.66	3.44	3.28	3.16	3.06	2.98
18	5.98	4.56	3.95	3.61	3.38	3.22	3.10	3.01	2.93
19	5.92	4.51	3.90	3.56	3.33	3.17	3.05	2.96	2.88
20	5.87	4.46	3.86	3.51	3.29	3.13	3.01	2.91	2.84
21	5.83	4.42	3.82	3.48	3.25	3.09	2.97	2.87	2.80
22	5.79	4.38	3.78	3.44	3.22	3.05	2.93	2.84	2.76
23	5.75	4.35	3.75	3.41	3.18	3.02	2.90	2.81	2.73
24	5.72	4.32	3.72	3.38	3.15	2.99	2.87	2.78	2.70
25	5.69	4.29	3.69	3.35	3.13	2.97	2.85	2.75	2.68
26	5.66	4.27	3.67	3.33	3.10	2.94	2.82	2.73	2.65
27	5.63	4.24	3.65	3.31	3.08	2.92	2.80	2.71	2.63
28	5.61	4.22	3.63	3.29	3.06	2.90	2.78	2.69	2.61
29	5.59	4.20	3.61	3.27	3.04	2.88	2.76	2.67	2.59
30	5.57	4.18	3.59	3.25	3.03	2.87	2.75	2.65	2.57
40	5.42	4.05	3.46	3.13	2.90	2.74	2.62	2.53	2.45
60	5.29	3.93	3.34	3.01	2.79	2.63	2.51	2.41	2.33
120	5.15	3.80	3.23	2.89	2.67	2.52	2.39	2.30	2.22
∞	5.02	3.69	3.12	2.79	2.57	2.41	2.29	2.19	2.11

df for Numerator

10	12	15	20	24	30	40	60	120	∞
968.63	976.71	984.87	993.10	997.25	1001.4	1005.6	1009.8	1014.0	1018.3
39.40	39.42	39.43	39.45	39.46	39.47	39.47	39.48	39.49	39.50
14.42	14.34	14.25	14.17	14.12	14.08	14.04	13.99	13.95	13.90
8.84	8.75	8.66	8.56	8.51	8.46	8.41	8.36	8.31	8.26
6.62	6.52	6.43	6.33	6.28	6.23	6.18	6.12	6.07	6.02
5.46	5.37	5.27	5.17	5.12	5.07	5.01	4.96	4.90	4.85
4.76	4.67	4.57	4.47	4.42	4.36	4.31	4.25	4.20	4.14
4.30	4.20	4.10	4.00	3.95	3.89	3.84	3.78	3.73	3.67
3.96	3.87	3.77	3.67	3.61	3.56	3.51	3.45	3.39	3.33
3.72	3.62	3.52	3.42	3.37	3.31	3.26	3.20	3.14	3.08
3.53	3.43	3.33	3.23	3.17	3.12	3.06	3.00	2.94	2.88
3.37	3.28	3.18	3.07	3.02	2.96	2.91	2.85	2.79	2.72
3.25	3.15	3.05	2.95	2.89	2.84	2.78	2.72	2.66	2.60
3.15	3.05	2.95	2.84	2.79	2.73	2.67	2.61	2.55	2.49
3.06	2.96	2.86	2.76	2.70	2.64	2.59	2.52	2.46	2.40
2.99	2.89	2.79	2.68	2.63	2.57	2.51	2.45	2.38	2.32
2.92	2.82	2.72	2.62	2.56	2.50	2.44	2.38	2.32	2.25
2.87	2.77	2.67	2.56	2.50	2.44	2.38	2.32	2.26	2.19
2.82	2.72	2.62	2.51	2.45	2.39	2.33	2.27	2.20	2.13
2.77	2.68	2.57	2.46	2.41	2.35	2.29	2.22	2.16	2.09
2.73	2.64	2.53	2.42	2.37	2.31	2.25	2.18	2.11	2.04
2.70	2.60	2.50	2.39	2.33	2.27	2.21	2.14	2.08	2.00
2.67	2.57	2.47	2.36	2.30	2.24	2.18	2.11	2.04	1.97
2.64	2.54	2.44	2.33	2.27	2.21	2.15	2.08	2.01	1.94
2.61	2.51	2.41	2.30	2.24	2.18	2.12	2.05	1.98	1.91
2.59	2.49	2.39	2.28	2.22	2.16	2.09	2.03	1.95	1.88
2.57	2.47	2.36	2.25	2.19	2.13	2.07	2.00	1.93	1.85
2.55	2.45	2.34	2.23	2.17	2.11	2.05	1.98	1.91	1.83
2.53	2.43	2.32	2.21	2.15	2.09	2.03	1.96	1.89	1.81
2.51	2.41	2.31	2.20	2.14	2.07	2.01	1.94	1.87	1.79
2.39	2.29	2.18	2.07	2.01	1.94	1.88	1.80	1.72	1.64
2.27	2.17	2.06	1.94	1.88	1.82	1.74	1.67	1.58	1.48
2.16	2.05	1.95	1.82	1.76	1.69	1.61	1.53	1.43	1.31
2.05	1.94	1.83	1.71	1.64	1.57	1.48	1.39	1.27	1.00

TABLE D.5 (Cont.)

Values of $F_{0.05}$

df for Denominator	df for Numerator: 1	2	3	4	5	6	7	8	9
1	161.4	199.5	215.7	224.6	230.2	234.0	236.8	238.9	240.5
2	18.51	19.00	19.16	19.25	19.30	19.33	19.35	19.37	19.38
3	10.13	9.55	9.28	9.12	9.01	8.94	8.89	8.85	8.81
4	7.71	6.94	6.59	6.39	6.26	6.16	6.09	6.04	6.00
5	6.61	5.79	5.41	5.19	5.05	4.95	4.88	4.82	4.77
6	5.99	5.14	4.76	4.53	4.39	4.28	4.21	4.15	4.10
7	5.59	4.74	4.35	4.12	3.97	3.87	3.79	3.73	3.68
8	5.32	4.46	4.07	3.84	3.69	3.58	3.50	3.44	3.39
9	5.12	4.26	3.86	3.63	3.48	3.37	3.29	3.23	3.18
10	4.96	4.10	3.71	3.48	3.33	3.22	3.14	3.07	3.02
11	4.84	3.98	3.59	3.36	3.20	3.09	3.01	2.95	2.90
12	4.75	3.89	3.49	3.26	3.11	3.00	2.91	2.85	2.80
13	4.67	3.81	3.41	3.18	3.03	2.92	2.83	2.77	2.71
14	4.60	3.74	3.34	3.11	2.96	2.85	2.76	2.70	2.65
15	4.54	3.68	3.29	3.06	2.90	2.79	2.71	2.64	2.59
16	4.49	3.63	3.24	3.01	2.85	2.74	2.66	2.59	2.54
17	4.45	3.59	3.20	2.96	2.81	2.70	2.61	2.55	2.49
18	4.41	3.55	3.16	2.93	2.77	2.66	2.58	2.51	2.46
19	4.38	3.52	3.13	2.90	2.74	2.63	2.54	2.48	2.42
20	4.35	3.49	3.10	2.87	2.71	2.60	2.51	2.45	2.39
21	4.32	3.47	3.07	2.84	2.68	2.57	2.49	2.42	2.37
22	4.30	3.44	3.05	2.82	2.66	2.55	2.46	2.40	2.34
23	4.28	3.42	3.03	2.80	2.64	2.53	2.44	2.37	2.32
24	4.26	3.40	3.01	2.78	2.62	2.51	2.42	2.36	2.30
25	4.24	3.39	2.99	2.76	2.60	2.49	2.40	2.34	2.28
26	4.23	3.37	2.98	2.74	2.59	2.47	2.39	2.32	2.27
27	4.21	3.35	2.96	2.73	2.57	2.46	2.37	2.31	2.25
28	4.20	3.34	2.95	2.71	2.56	2.45	2.36	2.29	2.24
29	4.18	3.33	2.93	2.70	2.55	2.43	2.35	2.28	2.22
30	4.17	3.32	2.92	2.69	2.53	2.42	2.33	2.27	2.21
40	4.08	3.23	2.84	2.61	2.45	2.34	2.25	2.18	2.12
60	4.00	3.15	2.76	2.53	2.37	2.25	2.17	2.10	2.04
120	3.92	3.07	2.68	2.45	2.29	2.17	2.09	2.02	1.96
∞	3.84	3.00	2.60	2.37	2.21	2.10	2.01	1.94	1.88

df for Numerator

10	12	15	20	24	30	40	60	120	∞
241.9	243.9	245.9	248.0	249.1	250.1	251.1	252.2	253.3	254.3
19.40	19.41	19.43	19.45	19.45	19.46	19.47	19.48	19.49	19.50
8.79	8.74	8.70	8.66	8.64	8.62	8.59	8.57	8.55	8.53
5.96	5.91	5.86	5.80	5.77	5.75	5.72	5.69	5.66	5.63
4.74	4.68	4.62	4.56	4.53	4.50	4.46	4.43	4.40	4.36
4.06	4.00	3.94	3.87	3.84	3.81	3.77	3.74	3.70	3.67
3.64	3.57	3.51	3.41	3.41	3.38	3.34	3.30	3.27	3.23
3.35	3.28	3.22	3.15	3.12	3.08	3.04	3.01	2.97	2.93
3.14	3.07	3.01	2.94	2.90	2.86	2.83	2.79	2.75	2.71
2.98	2.91	2.85	2.77	2.74	2.70	2.66	2.62	2.58	2.54
2.85	2.79	2.72	2.65	2.61	2.57	2.53	2.49	2.45	2.40
2.75	2.69	2.62	2.54	2.51	2.47	2.43	2.38	2.34	2.30
2.67	2.60	2.53	2.46	2.42	2.38	2.34	2.30	2.25	2.21
2.60	2.53	2.46	2.39	2.35	2.31	2.27	2.22	2.18	2.13
2.54	2.48	2.40	2.33	2.29	2.25	2.20	2.16	2.11	2.07
2.49	2.42	2.35	2.28	2.24	2.19	2.15	2.11	2.06	2.01
2.45	2.38	2.31	2.23	2.19	2.15	2.10	2.06	2.01	1.96
2.41	2.34	2.27	2.19	2.15	2.11	2.06	2.02	1.97	1.92
2.38	2.31	2.23	2.16	2.11	2.07	2.03	1.98	1.93	1.88
2.35	2.28	2.20	2.12	2.08	2.04	1.99	1.95	1.90	1.84
2.32	2.25	2.18	2.10	2.05	2.01	1.96	1.92	1.87	1.81
2.30	2.23	2.15	2.07	2.03	1.98	1.94	1.89	1.84	1.78
2.27	2.20	2.13	2.05	2.01	1.96	1.91	1.86	1.81	1.76
2.25	2.18	2.11	2.03	1.98	1.94	1.89	1.84	1.79	1.73
2.24	2.16	2.09	2.01	1.96	1.92	1.87	1.82	1.77	1.71
2.22	2.15	2.07	1.99	1.95	1.90	1.85	1.80	1.75	1.69
2.20	2.13	2.06	1.97	1.93	1.88	1.84	1.79	1.73	1.67
2.19	2.12	2.04	1.96	1.91	1.87	1.82	1.77	1.71	1.65
2.18	2.10	2.03	1.94	1.90	1.85	1.81	1.75	1.70	1.64
2.16	2.09	2.01	1.93	1.89	1.84	1.79	1.74	1.68	1.62
2.08	2.00	1.92	1.84	1.79	1.74	1.69	1.64	1.58	1.51
1.99	1.92	1.84	1.75	1.70	1.65	1.59	1.53	1.47	1.39
1.91	1.83	1.75	1.66	1.61	1.55	1.50	1.43	1.35	1.25
1.83	1.75	1.67	1.57	1.52	1.46	1.39	1.32	1.22	1.00

TABLE D.6 Critical values of the sample correlation coefficient r

df \ α	.10	.05	.02	.01
1	.988	.997	.9995	.9999
2	.900	.950	.980	.990
3	.805	.878	.934	.959
4	.729	.811	.882	.917
5	.669	.754	.833	.874
6	.622	.707	.789	.834
7	.582	.666	.750	.798
8	.549	.632	.716	.765
9	.521	.602	.685	.735
10	.497	.576	.658	.708
11	.476	.553	.634	.684
12	.458	.532	.612	.661
13	.441	.514	.592	.641
14	.426	.497	.574	.623
15	.412	.482	.558	.606
16	.400	.468	.543	.590
17	.389	.456	.528	.575
18	.378	.444	.516	.561
19	.369	.433	.503	.549
20	.360	.423	.492	.537
21	.352	.413	.482	.526
22	.344	.404	.472	.515
23	.337	.396	.462	.505
24	.330	.388	.453	.496
25	.323	.381	.445	.487
26	.317	.374	.437	.479
27	.311	.367	.430	.471
28	.306	.361	.423	.463
29	.301	.355	.416	.456
30	.296	.349	.409	.449
40	.257	.304	.358	.393
50	.231	.273	.322	.354
60	.211	.250	.295	.325
70	.195	.232	.274	.302
80	.183	.217	.257	.283
90	.173	.205	.242	.267
100	.164	.195	.230	.254

Other entries derived using the relation $t_{df} = r\sqrt{df/(1 - r^2)}$.

Note: A value given in the table is the *right-hand* critical value for a *two-tail* test at the significance level indicated. The left-hand critical value is just the negative of the right-hand critical value.

Adapted from Donald B. Owen, *Handbook of Statistical Tables.* Copyright © 1962, U.S. Department of Energy; published by Addison-Wesley, Reading, Mass. Adapted from Neil Weiss and Matthew Hassett, *Introductory Statistics.* Reading, Mass.: Addison-Wesley, 1982. Reprinted with permission.

APPENDIX E
Answers to Odd-Numbered Exercises

EXERCISES/Chapter 2

9. The data are not given any context. There are many more civilians than military personnel. Thus the 120,000 civilians may be a very small percent of that group, whereas 18,000 military personnel may be a much larger percent of that group.

EXERCISES/Section 3.1

3.

X	f
0	6
1	5
2	9
3	2
4	3
5	4
6	1

5. a)

Class	f	X
6– 8	1	7
9–11	9	10
12–14	7	13
15–17	4	16
18–20	5	19
21–23	4	22
24–26	2	25
27–29	1	28
30–32	2	31

b) 5.5–8.5 c) 3 d) 7

7. a)

Class	f	X
100–124	2	112
125–149	6	137
150–174	5	162
175–199	6	187
200–224	4	212
225–249	3	237
250–274	1	262
275–299	2	287
300–324	1	312

b) 25 c) 124.5–149.5 d) 112

9.

Class	f	X
6.0– 7.9	3	6.95
8.0– 9.9	9	8.95
10.0–11.9	16	10.95
12.0–13.9	11	12.95
14.0–15.9	7	14.95
16.0–17.9	4	16.95

EXERCISES/Section 3.2

1.

		f
0	5 2 2 6 3 1 5 1 8	9
1	9 2 2 6 6 9 6	7
2	1 9 9 2 2	5
3	1 6 6	3
4	3 6 0 7 1 5	6
5	2 1 9 4 4 1	6
6	8 2	2
7	1 0	2

3.

		f
18	0 5	2
19	6 4	2
20	2 7 0 9	4
21	2 7 5 8 4 0 6 5	8
22	0 3	2
23	1 8	2
24	0 1 8	3
25	8 0 4 1 3 2 5	7
26	0 3 6 7 7 8 7 1	8
27	7 9 6 1 0 0 3 2 0 0 4	11
28	9 0 2 1	4
29	0 5 9 9	4
30		0
31	1	1
32	5 1	2

5.

		f
0	1 0 0 2 6 2 3 2 0 9 7 0 2 3 4 8 0 0 3 1 4 3 2 6	24
1	8 7 8	3
2	1 1	2
3	7 5	2
4	8 2 6 5	4
5	9 7	2
6	0	1
7	5 5	2

EXERCISES/Section 3.3

1. **3.**

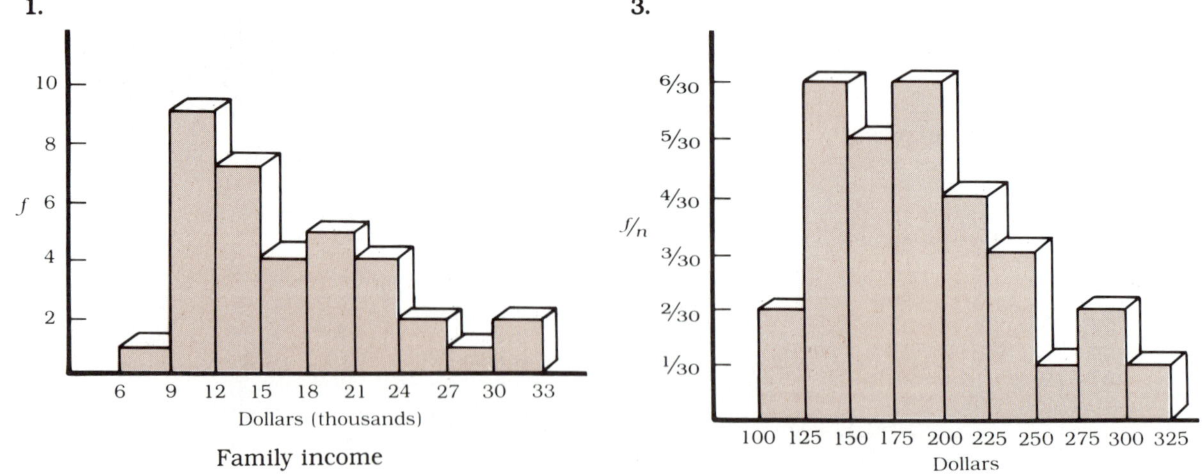

Family income

Starting weekly salaries in data processing

5.

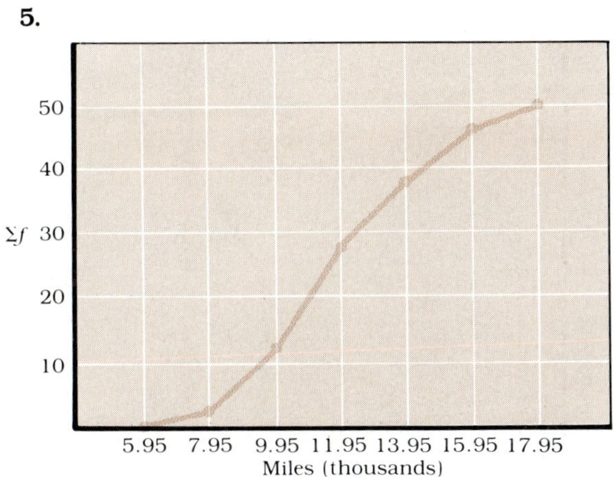

Tire life

7.

10
8
f 6
4
2
0 10 20 30 40 50 60 70 80
Inches

Lengths of legs of prehistoric animals

9.

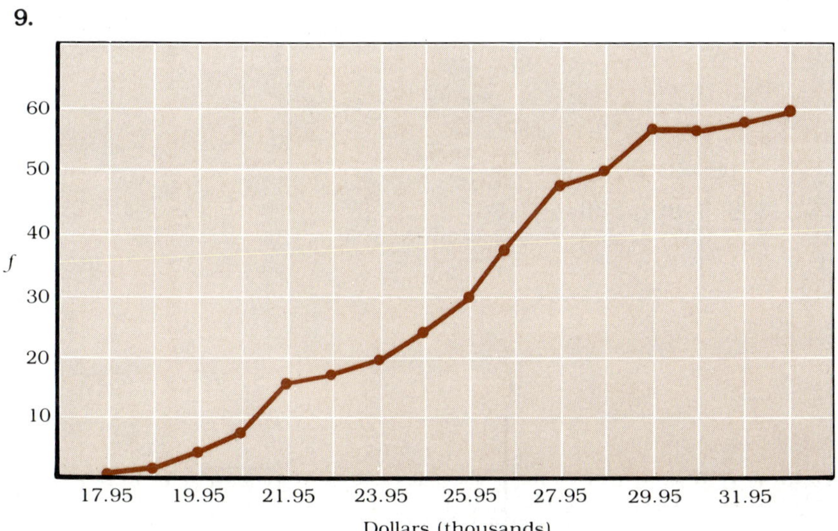

Annual salaries in engineering fields

REVIEW PROBLEMS/Chapter 3

1.

Class	f	Σf	Class	f	Σf
30–39	3	3	70–79	3	13
40–49	1	4	80–89	8	21
50–59	1	5	90–99	4	25
60–69	5	10			

3. 10

5.

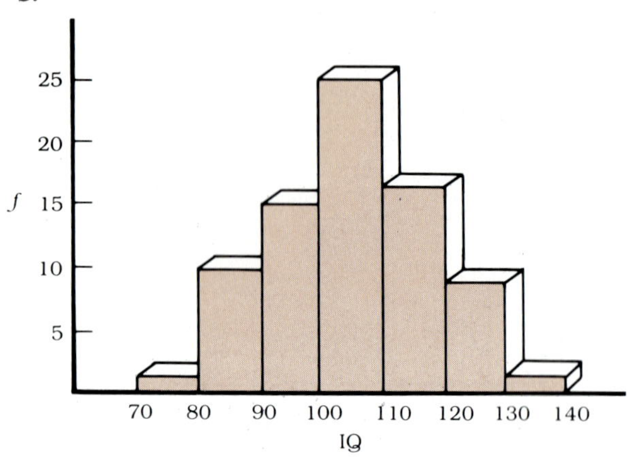

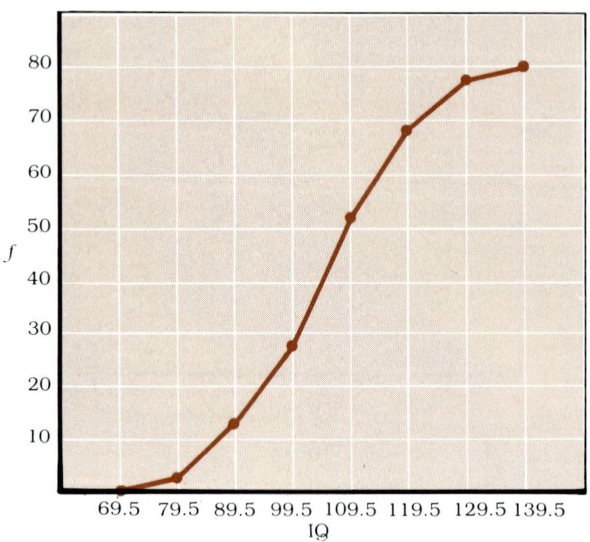

7. a)

		f
1	9	1
2	2 9 4 2 8 6 4 7 9 5 6 2 8 4 3 8 4 7 8	19
3	1 4 6 4 7 7 2 0 5 7 4 0 7 3 2 4 2 2 1 2 4 4 3 4	24
4	8 7 2 6 0 1 2 1 2 0 1 3 1 3 1	15
5	1	1

b)

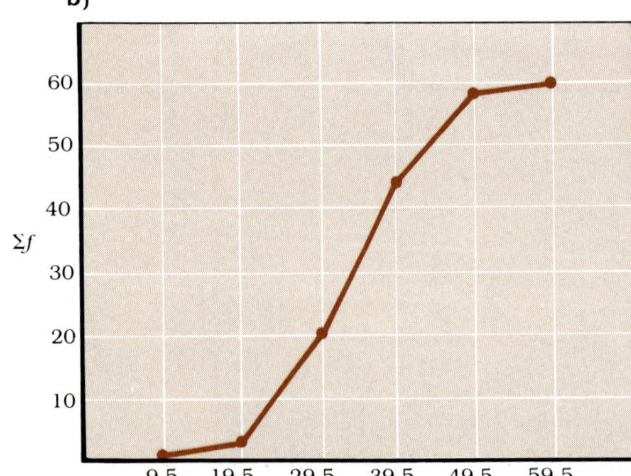

Sales records for cat food

c)

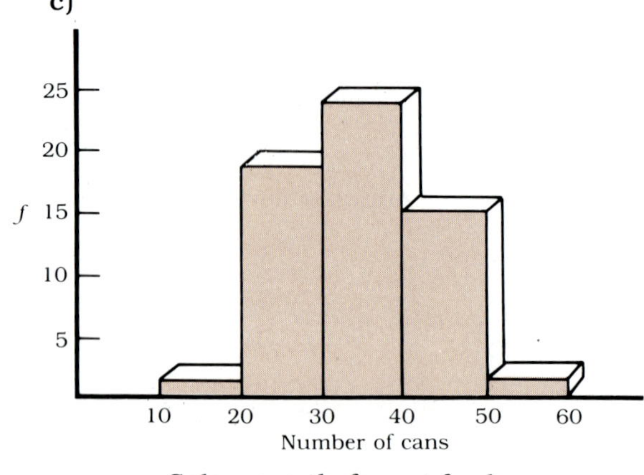

Sales records for cat food

9. **a)**

Class	f
0– 29	11
30– 59	10
60– 89	4
90–119	0
120–149	0
150–179	0
180–209	1
210–239	0
240–269	4

b)

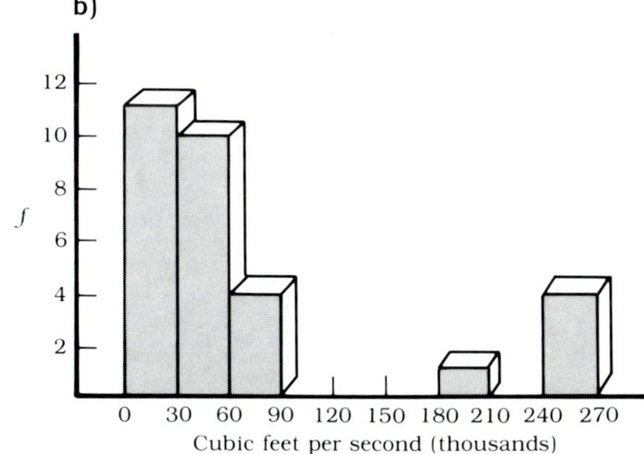

Average discharge area of major rivers

c)

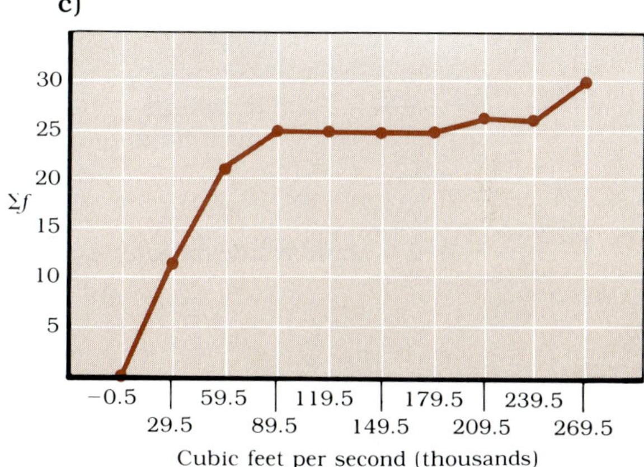

Average discharge area of major rivers

EXERCISES/Section 4.1

1. Qualitative
3. Quantitative, discrete
5. Quantitative, discrete
7. Quantitative, continuous
9. Quantitative, continuous
11. Quantitative, continuous
13. Qualitative
15. Quantitative, continuous
17. Quantitative, continuous
19. Quantitative, discrete

EXERCISES/Section 4.2

1. Mean 5.3
Median 6.5
Mode 7

3. **a)** $44,333.33 **b)** $20,000 **c)** $20,000

5. **a)** 69.2 **b)** 71.5 **c)** 63

7. **a)** 12.0 **b)** 11.5 **c)** 9

9. **a)** 939.5 **b)** 928.1 **c)** 924.5

11. **a)** 84.5 **b)** 86.2 **c)** 89.5

13.

0	0 7 8 6 0 9 7
1	9 7 8 7 1 6 7 4 4
2	3 4 7 1 4 2 8
3	2 2 5
4	1 0 1

Median = 17.

15. 25.8

EXERCISES/Section 4.3

1. a) 8
   b) $s^2 = 8.97.$
      $s = 3.00.$
3. a) $240,000
   b) $s^2 = 3{,}328{,}411{,}800.$
      $s = 57{,}692.4.$
5. a) 11.34 b) 11.34
7. a) 3.13 b) 3.13
9. a) 3512.6 b) 59.27
11. a) $s^2 = 510.17.$ b) 22.59
13. a) $s_X = s_Y.$ **b)** $s_X < s_Y.$ **c)** $s_X > s_Y.$

EXERCISES/Section 4.4

1. $\frac{7.5 - 15}{3} = \frac{-7.5}{3} = -2.5$ standard deviations

$\frac{22.5 - 15}{3} = 2.5$ standard deviations.

Chebyshev's Theorem says that

$$1 - \frac{1}{2.5^2} = 0.84, \text{ or}$$

at least 84% of the distribution lies between 7.5 and 22.5.

EXERCISES/Section 4.5

1. a) 18.0 b) 23.1 c) 18.1, 22.5, 24.6
   d) 21.35 e) 6.5
3. a) 27 b) 18 c) 15, 24, 34 d) 24.5
   e) 19
5. a) 82.4 b) 62.4 c) 67.8, 73.9, 81.2
   d) 74.5 e) 13.
7. a) 12. b) 11. c) 13. d) 13.
9. a) 22. b) 27. c) 21. d) 27.
11. a) 2.33, −3.67, −1.33, 3.66
    b) 28.50, 27.39, 28.70, 27.58
13. a) 1.6, −1.5, 0.4, −0.2
    b) 0.0435, 0.075, −0.0423, −0.0609
15. a) 1.38, 3.10, −2.76, −1.03
    b) 29.09, 31.99, 18.65, 21.55

REVIEW PROBLEMS/Chapter 4

	Mean	Median	Mode	Range	Variance	Standard Deviation
1.	12.	12	12, 15	6	4.19	2.05
3.	48.	45	40, 42	28	84.25	9.18
5.	123.	122.9	130.9	22	70.24	8.38
7.	2.	2	2	4	3.47	1.86
9.	79.	83.7	84.5		219.51	14.82
11.	7356.6	6888.39	6499.5		2,997,256.9	1731.26

13. a) 1. b) −1 c) 0 d) −1.8 e) 2.8
15. a) 128 b) 119 c) 80 d) 107 e) 110
17. $17,500, $500
19. a) 51, 74, 87 b) 28, 81, 96 c) 69, 36

EXERCISES/Section 5.2

1. **a)** 3/36 **b)** 4/36 **c)** 6/36 **d)** 3/36 **e)** 18/36 **f)** 18/36 **g)** 6/36 **h)** 30/36
i) 20/36 **j)** 9/36 **k)** 6/36 **l)** 11/36 **m)** 11/36

3. **a)** 12/66 **b)** 24/66 **c)** 18/66 **d)** 12/66 **e)** 36/66

5. **a)** 1/24 **b)** 1/24 **c)** 1/24 **d)** 3/24
e) 3/24

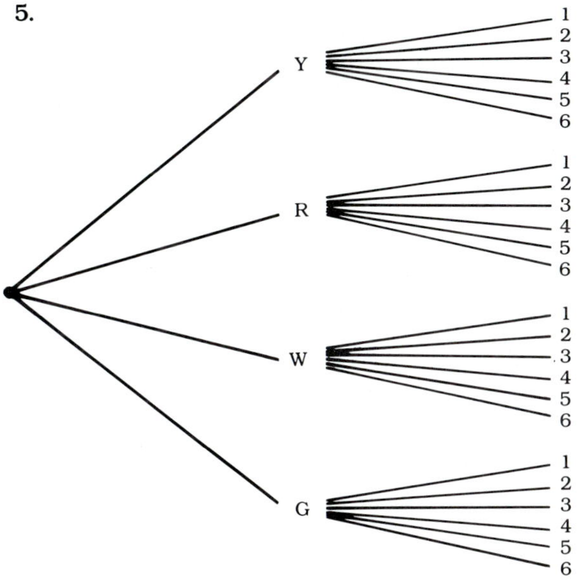

7. **a)** 18/37 **b)** 18/37 **c)** 18/37
d) 18/37 **e)** 1/37 **f)** 1/37

9. **a)** 29/81 **b)** 20/81

11. **a)** 4/32 **b)** 20/32 **c)** 22/32 **d)** 9/32
e) 28/32 **f)** 12/32 **g)** 10/32
h) 23/32 **i)** 10/32 **j)** 22/32

EXERCISES/Section 5.3

1. 35/36

3. **a)** No; 14/36 **b)** No; 16/52 **c)** Yes; 0 **d)** No; 1 **e)** No; 12/36

5. **a)** 16/52 **b)** 22/52 **c)** 28/52 **d)** 40/52 **e)** 40/52

EXERCISES/Section 5.4

1. **a)** Yes **b)** Yes **c)** No **d)** No **e)** Yes

3. **a)** 16/2704 **b)** 169/2704 **c)** 52/2704 **d)** 52/2704

5. 240/2704, 240/2652

7. **a)** 1/1024 **b)** 1/1024 **c)** 1/1024 **d)** 1/1024 **e)** 1/1024

9. **a)** 1 **b)** 1/2 **c)** 4/40 **d)** 4/20 **e)** 1

11. **a)** 50/90 **b)** 55/100 **c)** 0.05 **d)** 0.05 **e)** Yes

13. **a)** 20/25 **b)** 18/25 **c)** 22/25 **d)** 12/25 **e)** 8/25 **f)** 10/25 **g)** 13/25
h) 17/25 **i)** 15/25

15. 336/720

REVIEW PROBLEMS/Chapter 5

1. **a)** 300/960 **b)** 75/960 **c)** 125/960
   d) 210/960 **e)** 250/960

3. **a)** 4/52 **b)** 24/52 **c)** 24/52
   d) 12/52 **e)** 13/52

5.

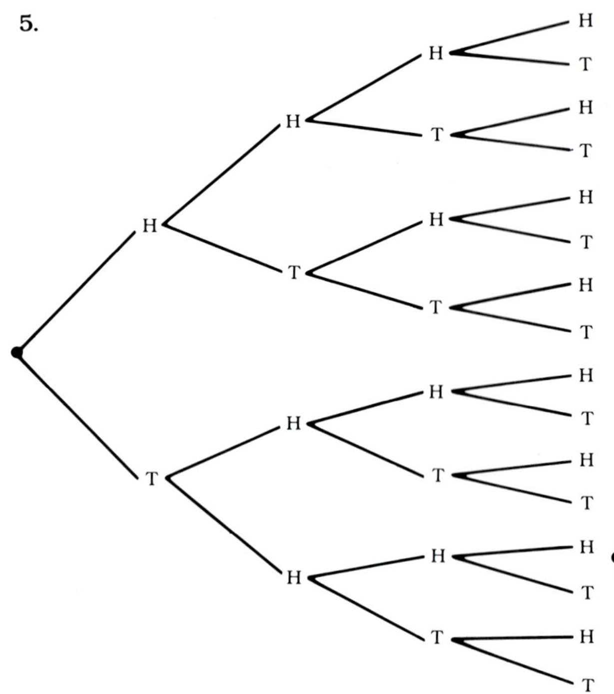

   a) 1/16 **b)** 4/16 **c)** 6/16 **d)** 4/16
   e) 1/16

7. **a)** Tree diagram ▶
   1) 27/64 **2)** 9/64 **3)** 1/64
   4) 9/64 **5)** 1/64 **6)** 1/64 **7)** 1/64
   8) 1/64
   b) See tree diagram on page opposite.
   1) 18/24 **2)** 0 **3)** 0 **4)** 0 **5)** 0
   6) 0 **7)** 0 **8)** 1/24

9. **a)** 12/2682 **b)** 16/2682

11. **a)** 15/20 **b)** 5/20 **c)** 4/20 **d)** 6/20
    e) 5/20 **f)** 12/20

13. **a)** 26/36 **b)** 28/36 **c)** 6/36 **d)** 10/36
    e) 0 **f)** 1 **g)** 22/36 **h)** 22/36

15. **a)** 100/500 **e)** 200/500 **i)** 140/250
    b) 250/500 **f)** 65/300 **j)** 95/300
    c) 150/500 **g)** 65/100 **k)** 95/150
    d) 300/500 **h)** 140/300

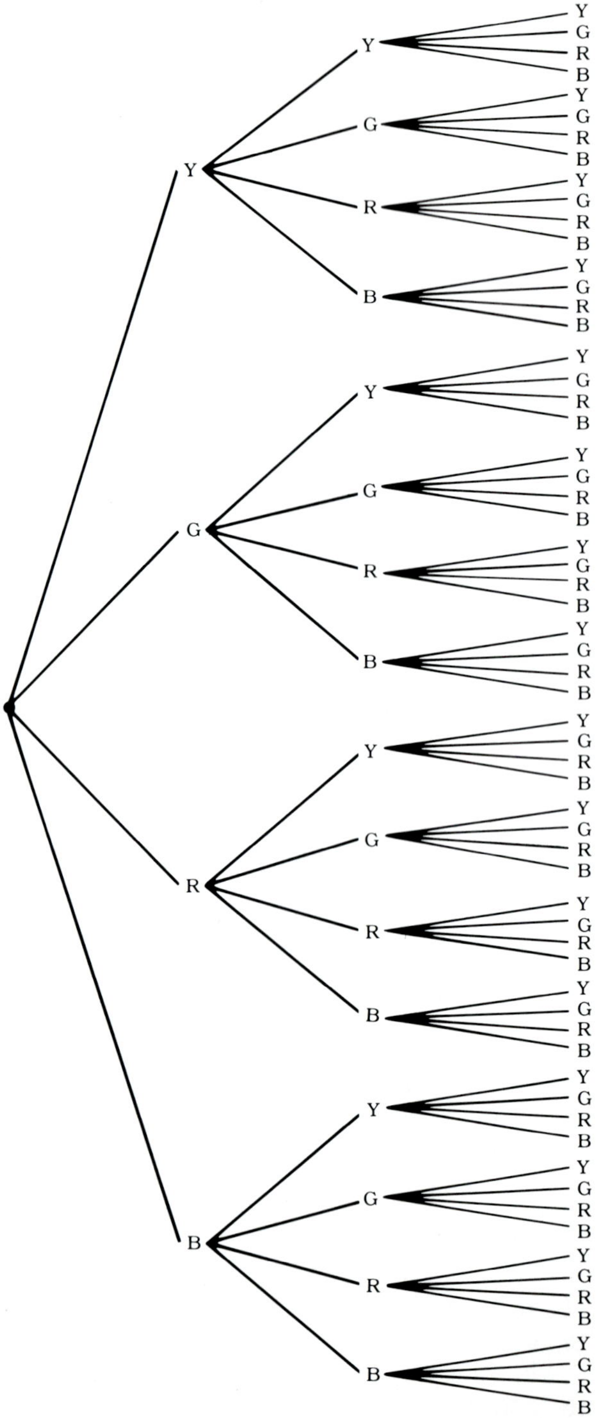

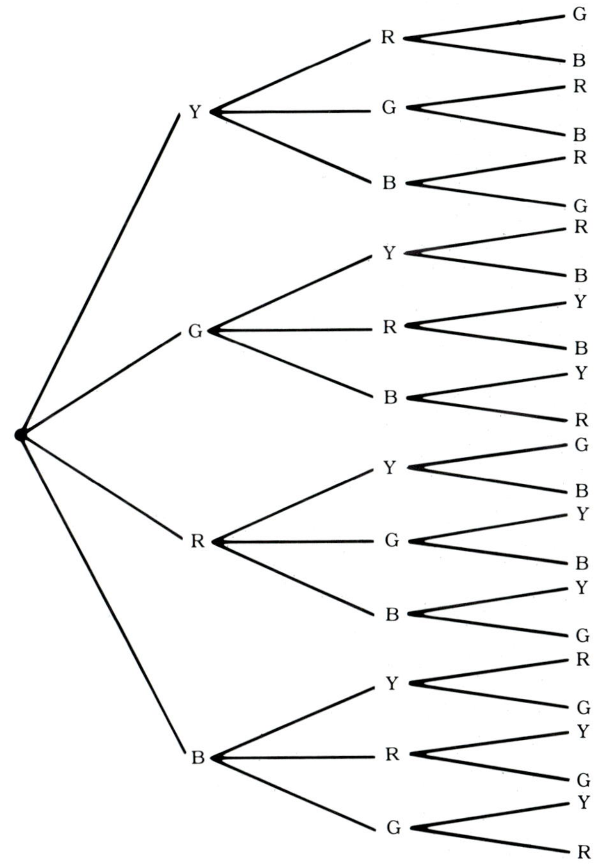

17. a) 6/10 b) 4/10 c) 1/6 d) 1 e) 2/10 f) 2/10 g) 2/10 h) 2/10 i) 1/10 j) 1/10 k) 1/4 l) 1/2 m) 0 n) 0

19. a) 0.92 b) 0.0000002 c) 0.9999998 d) 0.0000314

21. a) 0.00000016 b) 0.0196

EXERCISES/Section 6.1

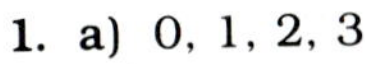

1. a) 0, 1, 2, 3

b)

X	$P(X)$	X	$P(X)$
0	1/8	2	3/8
1	3/8	3	1/8

c)

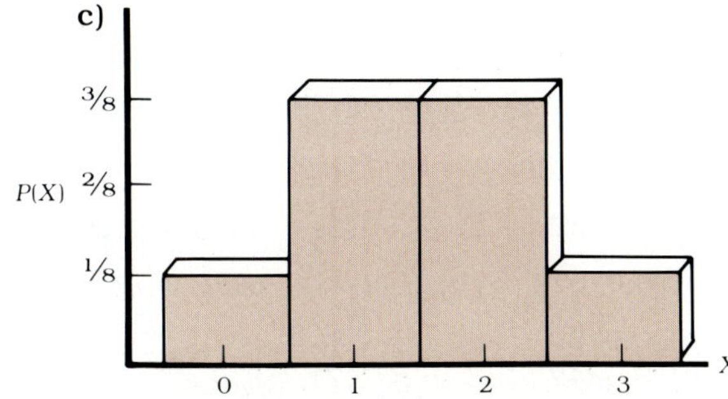

d) $\mu = 1.5.$
$\sigma = 0.87.$

3. a) 1, 5, 10, 20

b)

X	$P(X)$
1	5/9
5	2/9
10	1/9
20	1/9

c) $\mu = 5.$
$\sigma = 6.15.$

5. a) $\mu = 2.8.$
$\sigma = 1.17.$

b)

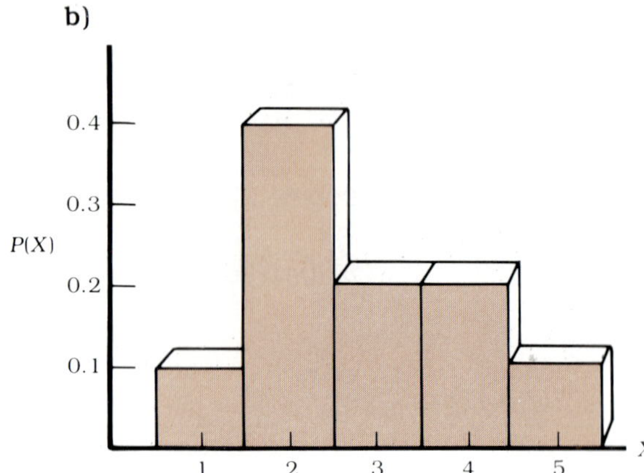

7. a)

X	$P(X)$
0	6/18
1	5/18
2	4/18
3	3/18

b)

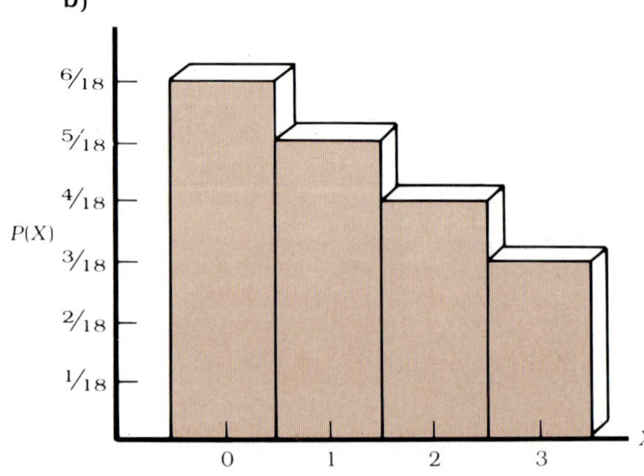

c) $\mu = 22/18$.
$\sigma = 1.25$.

9. a) $n = 6$.

X	$P(X)$
2	0
3	1/6
4	2/6
5	3/6

b)

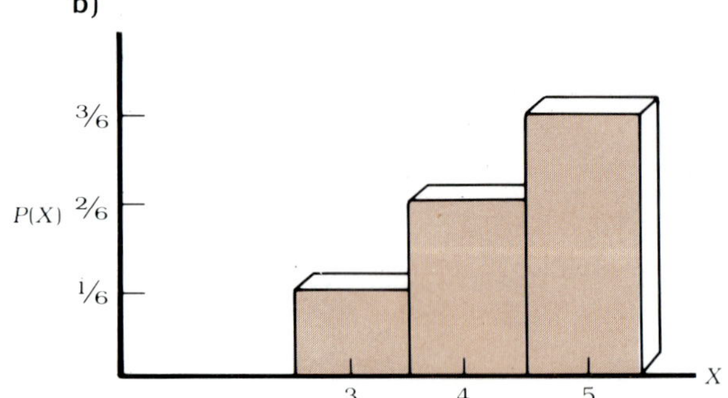

c) $\mu = 26/6$.
$\sigma = 0.75$.

EXERCISES/Section 6.2

1. a) 0.0469

b)

X	$P(X)$
0	0.3164
1	0.4219
2	0.2109
3	0.0469
4	0.0039

c) See bar graph at top left, page opposite.
d) $\mu = 1$.
$\sigma = 0.866$.

3. a) 0.3432 b) $\mu = 1.5$. c) $\sigma = 1.16$.

5. a) 0.01074 b) $(0.2)^{10}$ c) 2 d) 2

7. a) 0.0232 c) $\mu = 9.8$.
b) 0.5842 $\sigma = 1.71$.

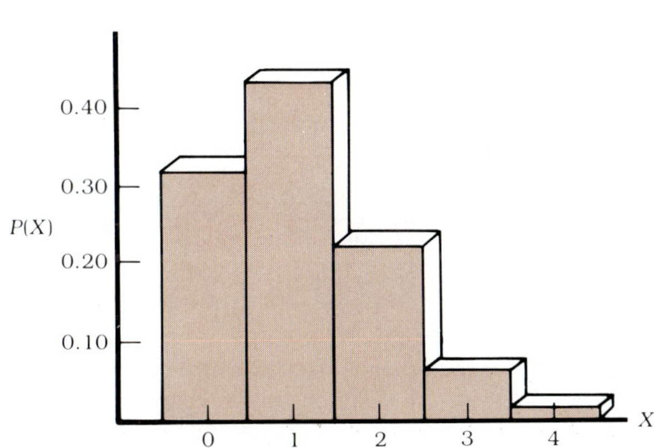

c)

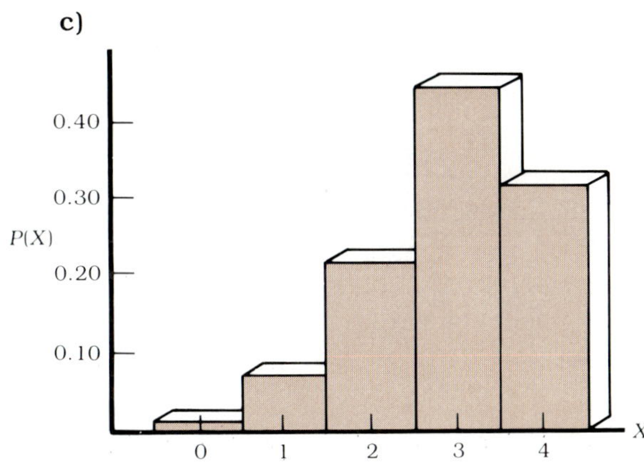

d) $\mu = 3$.
$\sigma = 0.866$.

9. a) 0.3164

b)

X	$P(X)$
0	0.0039
1	0.0469
2	0.2109
3	0.4219
4	0.3164

EXERCISES/Section 6.3

1. a) 0.4599 b) 0.4927 c) 0.2967 d) 0.2673 e) 0.4761
3. a) 0.6347 b) 0.9060 c) 0.9485 d) 0.3721 e) 0.9500
5. a) 0.1501 b) 0.1195 c) 0.1853 d) 0.0872 e) 0.2546
7. a) 0.5 b) 0.4772 c) 0.0049 d) 0.0166 e) 0.9332 f) 0.0013
9. a) 0.8790 b) 0.7938 c) 0.0228 d) 0.0918 e) 0.1934 f) 0.1193
11. a) 0.8413 b) 0.3085 c) 0.0062 d) 0.4000 e) 648 f) 767
13. a) 129–150 b) 120–128 c) 107–119 d) 95–106
15. a) 0.0853 b) 0.7854 c) 0.9750 d) 25.6 e) 22.1

EXERCISES/Section 6.4

1. 0.0102 3. a) 0.1071 b) 0.1020 c) 0.8788 5. 0.352
7. 0.8207 9. 0.0023 11. 0.0606 13. 0.0029 15. 0.0104

REVIEW PROBLEMS/Chapter 6

1. **a)** 1, 5, 10, 25

b)

X	P(X)
1	8/20
5	6/20
10	3/20
25	3/20

c) $\mu = 7.15$.
$\sigma = 8.09$.

3. **a)** 2–10

b)

X	P(X)	X	P(X)
2	1/25	7	4/25
3	2/25	8	3/25
4	3/25	9	2/25
5	4/25	10	1/25
6	5/25		

c) $\mu = 6$.
$\sigma = 2$.

5. $\mu = 3.8$.
$\sigma = 2.6$.

7. **a)** 0.209 **b)** 0.0373 **c)** 0.4757
d) 0.1954

9. **a)** 0.2013 **b)** 0.0328 **c)** 0.6778
d) 0.8925

11. **a)** 0.4192 **b)** 0.3849 **c)** 0.2195
d) 0.3806 **e)** 0.5671 **f)** 0.2734
g) 0.1736 **h)** 0.0081 **i)** 0.1104
j) 0.6772

13. **a)** 27.43% **b)** 18.41% **c)** 35.96%
d) 104.8

15. **a)** 0.0192 **b)** 0.4247 **c)** 0.9545
d) 0.8577 **e)** 0.5729

17. **a)** 0.0823 **b)** 0.3745 **c)** 0.6911

EXERCISES/Section 7.1

1. **a)** 0.5 **b)** 0.9313

3. **a)** 0.1271 **b)** 3.5–1.98 **c)** 0.0021

5. **a)** 0.2847 **b)** 0.9751

7. 0.9924 **9.** 0.294

EXERCISES/Section 7.2

1. **a)** −1.73 **b)** −2.10, +2.10
c) 2.60 **d)** −2.47 **e)** −3.25, +3.25

3. 5900 **5.** 166.59

REVIEW PROBLEMS/Chapter 7

1. **a)**

0, 0	1, 0	2, 0	3, 0
0, 1	1, 1	2, 1	3, 1
0, 2	1, 2	2, 2	3, 2
0, 3	1, 3	2, 3	3, 3

b)

0	0.5	1	1.5
0.5	1	1.5	2
1	1.5	2	2.5
1.5	2	2.5	3

c)

$\bar{X}$	$P(\bar{X})$
0	1/16
0.5	2/16
1	3.16
1.5	4/16
2	3/16
2.5	2/16
3	1/16

d)

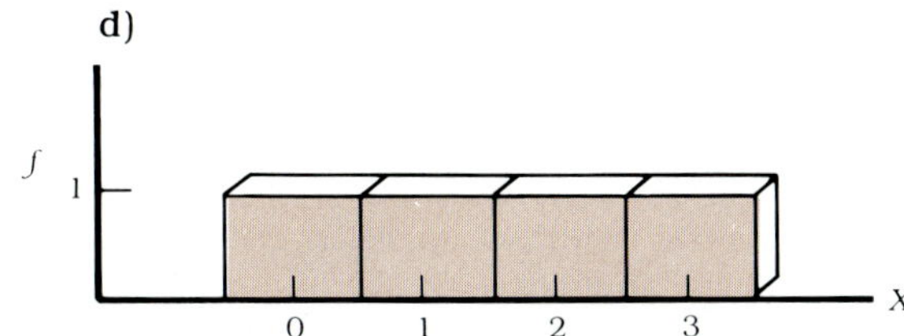

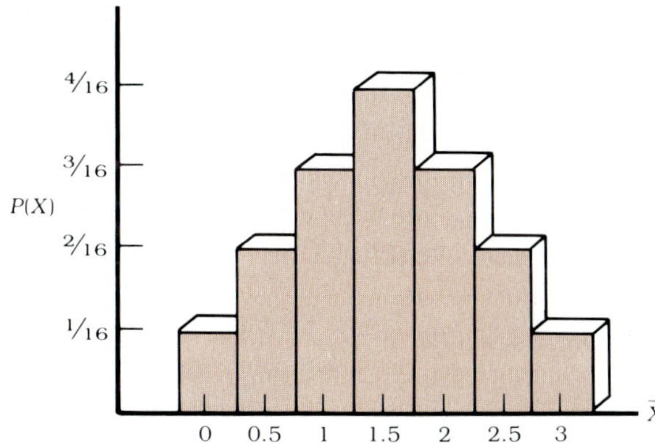

e) $\mu_{\bar{X}} = 1.5$
$\sigma_{\bar{X}} = 0.79.$

3. a) 0.0228 **b)** 0.1587 **c)** 0.8185

5. a) 0.9499 **b)** 0.8133 **c)** 0.8096

7. 0.023 **9.** $11.14

EXERCISES/Section 8.2

1. a) Rejecting
b) A guilty verdict when, in fact, the person is innocent.
c) A verdict of innocence when, in fact, the person is guilty.

3. a) Correct decisions: The food is wholesome and you decide to eat it. The food is not wholesome and you do not eat it.
Type I error: You decide not to eat wholesome food.
Type II error: You eat unwholesome food.
b) You decide not to eat food, which in reality is wholesome.
c) You eat unhealthy food.
d) Type II
e) 1

EXERCISES/Section 8.3

1. a) 2.33 **b)** −1.65 **c)** −2.58, 2.58
d) −1.96, 1.96 **e)** 1.65 **f)** −2.33

3. a) −2.58, 2.58 **b)** 2.33 **c)** −2.33
d) −1.65 **e)** −1.96, 1.96 **f)** 1.65

5. a) H_0: $\mu = 60.$ H_A: $\mu \neq 60.$
b) H_0: $\mu = 9.$ H_A: $\mu < 9.$
c) H_0: $\mu = 55.3.$ H_A: $\mu \neq 55.3.$
d) H_0: $\mu = 20.$ H_A: $\mu > 20.$
e) H_0: $\mu = 15.78.$ H_A: $\mu \neq 15.78.$
f) H_0: $\mu = 0.8.$ H_A: $\mu > 0.8.$
g) H_0: $\mu = 2.$ H_A: $\mu > 2.$
h) H_0: $\mu = 200.$ H_A: $\mu < 200.$
i) H_0: $\mu = 12.$ H_A: $\mu < 12.$
j) H_0: $\mu = 10.$ H_A: $\mu \neq 10.$

EXERCISES/Section 8.4

1. H_0: $\mu = 22.7$.
   H_A: $\mu > 22.7$.
   Test region: Reject H_0 when $z^* > 1.65$.
   Test statistic: $z^* = 1.79$.
   Conclusion: This winter had more severe snowfall.

3. H_0: $\mu = 2050$.
   H_A: $\mu \neq 2050$.
   Test region: Reject H_0 when $z^* < -1.96$ or $z^* > 1.96$.
   Test statistic: $z^* = 1.89$.
   Conclusion: We cannot reject H_0.

5. H_0: $\mu = 6$.
   H_A: $\mu < 6$.
   Test region: Reject H_0 when $z^* < -1.65$.
   Test statistic: $z^* = -3.07$.
   Conclusion: Reject H_0; yes, the chief made his case.

7. H_0: $\mu = 2.54$.
   H_A: $\mu > 2.54$.
   Test region: Reject H_0 when $z^* > 1.65$.
   Test statistic: $z^* = 1.48$.
   Conclusion: We cannot reject H_0.

9. H_0: $\mu = 0.9$.
   H_A: $\mu > 0.9$.
   Test region: Reject H_0 when $t^* > 1.74$.
   Test statistic: $t^* = 3.26$.
   Conclusion: Reject H_0.

EXERCISES/Section 8.5

1. 45.2 ± 0.83
   $44.37 < \mu < 46.03$.

3. 106.7 ± 5.77
   $100.93 < \mu < 112.47$.

5. 14.2 ± 0.15
   $14.05 < \mu < 14.35$.

7. 73.5 ± 5.11
   $68.39 < \mu < 78.61$.

9. a) 5.2
   b) 5.2 ± 0.44
      $4.76 < \mu < 5.64$.
   c) 5.2 ± 0.58
      $4.62 < \mu < 5.78$.

11. a) 30.2
    b) 30.2 ± 0.99
       $29.21 < \mu < 31.19$.
    c) 30.2 ± 1.34
       $28.86 < \mu < 31.54$.

13. 10 ± 2.21
    $7.79 < \mu < 12.21$.

15. 77 ± 2.31
    $74.69 < \mu < 79.31$.

17. 25.4 ± 3.36
    $72.04 < \mu < 78.76$.

19. 40

21. 42

REVIEW PROBLEMS/Chapter 8

1. H_0: $\mu = 72$.
   H_A: $\mu > 72$.
   Test region: Reject H_0 when $z^* > 1.65$.
   Test statistic: $z^* = 2.11$.
   Conclusion: Reject H_0; yes, this group is superior.

3. H_0: $\mu = 68$.
   H_A: $\mu > 68$.
   Test region: Reject H_0 when $t^* > 2.09$.
   Test statistic: $t^* = 2.61$.
   Conclusion: Reject H_0; she can reject the seedman's claim.

5. H_0: $\mu = 75$
H_A: $\mu > 75$
Test region: Reject H_0 when $z^* > 1.65$.
Test statistic: $z^* = 2.36$.
Conclusion: Reject H_0.

7. H_0: $\mu = 16$.
H_A: $\mu \neq 16$.
Test region: Reject H_0 when $t^* > 3.25$ or $t^* < -3.25$.
Test statistic: $t^* = 0.84$.
Conclusion: We cannot reject H_0.

9. 98.50 ± 3.3
$95.20 < \mu < 101.80$.

11. 5.6 ± 0.77
$4.83 < \mu < 6.37$.

13. 71 ± 7.9
$63.1 < \mu < 78.9$.

15. 39

17. 97

19. 27.75 ± 1.41
$26.34 < \mu < 29.16$.

21. 9750 ± 1001.26
$\$8748.74 < \mu < \$10{,}751.26$.

EXERCISES/Section 9.1

1. H_0: $\mu_{GB} - \mu_{US} = 0$.
H_A: $\mu_{GB} - \mu_{US} \neq 0$.
Test region: Reject H_0 when $z^* < -1.96$ or $z^* > 1.96$.
Test statistic: $z^* = 2.84$.
Conclusion: Reject H_0; there is a difference.

3. H_0: $\mu_M - \mu_W = 0$.
H_A: $\mu_M - \mu_W > 0$.
Test region: Reject H_0 when $z^* > 2.58$.
Test statistic: $z^* = 2.85$.
Conclusion: Reject H_0; there is a difference.

5. H_0: $\mu_W - \mu_E = 0$.
H_A: $\mu_W - \mu_E \neq 0$.
Test region: Reject H_0 when $z^* > 1.96$ or $z^* < -1.96$.
Test statistic: $z^* = -2.34$.
Conclusion: Reject H_0; there is a difference.

7. 0.31 ± 0.22
$0.09 < \mu_B - \mu_A < 0.53$.

9. 3.1 ± 1.63
$1.47 < \mu_P - \mu_N < 4.73$.

EXERCISES/Section 9.2

1. $s_p = 3.36$.
H_0: $\mu_1 - \mu_2 = 0$.
H_A: $\mu_1 - \mu_2 > 0$.
Test region: Reject H_0 when $t^* > 1.70$.
Test statistic: $t^* = 1.96$.
Conclusion: Reject H_0; there was a decline.

3. $s_p = 5.57$.
H_0: $\mu_A - \mu_B = 0$.
H_A: $\mu_A - \mu_B > 0$.
Test region: Reject H_0 when $t^* > 2.50$.
Test statistic: $t^* = 0.70$.
Conclusion: We cannot reject H_0.

5. $s_p = 4.34$.
H_0: $\mu_S - \mu_N = 0$.
H_A: $\mu_S - \mu_N > 0$.
Test region: Reject H_0 when $t^* > 2.76$.
Test statistic: $t^* = 3.19$.
Conclusion: Reject H_0; Storebrand sales exceed.

7. $s_p = 7.97$.
23.1 ± 9.29
$13.81 < \mu_A - \mu_B < 32.39$.

9. $s_p = 8.36$.
74.5 ± 10.76
$63.74 < \mu_D - \mu_I < 85.26$.

EXERCISES/Section 9.3

1. H_0: $\mu_d = 0$.
H_A: $\mu_d > 0$.
Test region: Reject H_0 when $t^* > 1.74$.
Test statistic: $t^* = 3.95$.
Conclusion: Reject H_0; the drug produced a decrease.

3. $s_d = 7.35$.
H_0: $\mu_d = 0$.
H_A: $\mu_d > 0$.
Test region: Reject H_0 when $t^* > 2.72$.
Test statistic: $t^* = 1.33$.
Conclusion: We cannot reject H_0.

5. $s_d = 8.61$.
H_0: $\mu_d = 0$.
H_A: $\mu_d \neq 0$.
Test region: Reject H_0 when $t^* > 2.78$ or $t^* < -2.78$.
Test statistic: $t^* = 1.71$.
Conclusion: We cannot reject H_0.

7. $s_d = 0.88$
0.76 ± 0.63
$0.13 < \mu_d < 1.39$.

9. $s_d = 323.95$.
310.16 ± 205.74
$104.42 < \mu_d < 515.90$.

REVIEW PROBLEMS/Chapter 9

1. 2.21 ± 0.96
$1.25 < \mu_M - \mu_W < 3.17$.

3. H_0: $\mu_A - \mu_B = 0$.
H_A: $\mu_A - \mu_B \neq 0$.
Test region: Reject H_0 when $z^* > 1.96$ or $z^* < -1.96$.
Test statistic: $z^* = 3.99$.
Conclusion: Reject H_0; there is a difference.

5. H_0: $\mu_M - \mu_N = 0$.
H_A: $\mu_M - \mu_N \neq 0$.
Test region: Reject H_0 when $z^* < -1.65$ or $z^* > 1.65$.
Test statistic: $z^* = -1.92$.
Conclusion: Reject H_0; there is a difference.

7. 0.75 ± 0.71
$0.04 < \mu_V - \mu_B < 1.46$.

9. $s_p = 0.91$.
H_0: $\mu_A - \mu_B = 0$.
H_A: $\mu_A - \mu_B \neq 0$.
Test region: Reject H_0 when $t^* > 2.10$ or $t^* < -2.10$.
Test statistic: $t^* = 0.61$.
Conclusion: We cannot reject H_0.

11. $s_p = 6.19$.
H_0: $\mu_X - \mu_Y = 0$.
H_A: $\mu_X - \mu_Y > 0$.
Test region: Reject H_0 when $t^* > 2.33$.
Test statistic: $t^* = 0.95$.
Conclusion: We cannot reject H_0.

13. $s_p = 1.38$.
1.3 ± 1.22
$0.08 < \mu_D - \mu_F < 2.52$.

15. $s_d = 2.36$.
H_0: $\mu_d = 0$.
H_A: $\mu_d > 0$.
Test region: Reject H_0 when $t^* > 1.83$.
Test statistic: $t^* = 2.68$.
Conclusion: Reject H_0; music makes a difference.

17. 2 ± 1.69.
$0.31 < \mu_d < 3.69$.

EXERCISES/Section 10.1

1. H_0: $p = 0.5$.
H_A: $p < 0.5$.
Test region: Reject H_0 when $z^* < -2.33$.
Test statistics: $z^* = -1.39$.
Conclusion: We cannot reject H_0.

3. H_0: $p = 0.25$.
H_A: $p \neq 0.25$.
Test region: Reject H_0 when $z^* > 2.58$ or $z^* < -2.58$.
Test statistic: $z^* = -1.22$.
Conclusion: We cannot reject H_0.

5. H_0: $p = 0.45$.
H_A: $p < 0.45$.
Test region: Reject H_0 when $z^* < -1.65$.
Test statistic: $z^* = -2.17$.
Conclusion: Reject H_0; the claim was not justified.

7. 0.76 ± 0.022
$0.738 < p < 0.782$.

9. 0.28 ± 0.029
$0.251 < p < 0.309$.

11. **a)** $n = 97$. **b)** $n = 79$.

13. **a)** $n = 66{,}564$. **b)** $n = 20{,}043$.

15. $n = 10{,}651$.

EXERCISES/Section 10.2

1. H_0: $p_U - p_S = 0$.
H_A: $p_U - p_S \neq 0$.
Test region: Reject H_0 when $z^* > 2.58$ or $z^* < -2.58$.
Test statistic: $z^* = 0.83$.
Conclusion: We cannot reject H_0.

3. H_0: $p_M - p_W = 0$.
H_A: $p_M - p_W \neq 0$.
Test region: Reject H_0 when $z^* > 2.58$ or $z^* < -2.58$.
Test statistic: $z^* = -0.42$.
Conclusion: We cannot reject H_0.

5. H_0: $p_W - p_M = 0$.
H_A: $p_W - p_M > 0$.
Test region: Reject H_0 when $z^* > 2.33$.
Test statistic: $z^* = 4.33$.
Conclusion: Reject H_0; more women agree.

7. 0.05 ± 0.036
$0.014 < p_W - p_M < 0.086$.

9. 0.24 ± 0.05
$0.19 < p_B - p_W < 0.29$.

EXERCISES/Section 10.3

1. **a)** 16.812 **b)** 5.229 **c)** 7.434, 39.997 **d)** 118.498 **e)** 24.433

3. H_0: $\sigma = 12$.
H_A: $\sigma < 12$.
Test region: Reject H_0 when $\chi^{2*} < 18.493$.
Test statistic: $\chi^{2*} = 20.833$.
Conclusion: We cannot reject H_0.

5. H_0: $\sigma = 2$.
H_A: $\sigma > 2$.
Test region: Reject H_0 when $\chi^{2*} > 67.505$.
Test statistic: $\chi^{2*} = 76.56$.
Conclusion: Reject H_0.

7. H_0: $\sigma = 200$.

H_A: $\sigma > 200$.

Test region: Reject H_0 when $\chi^{2*} > 16.919$.

Test statistic: $\chi^{2*} = 9.92$.

Conclusion: We cannot reject H_0.

9. $0.0069 < \sigma < 0.0183$.

11. $3343.0 < \sigma < 9267.7$.

EXERCISES/Section 10.4

1. a) 7.19 b) 2.84 c) 2.01 d) 2.14 e) 1.84

3. H_0: $\sigma_F^2 = \sigma_S^2$

H_A: $\sigma_F^2 > \sigma_S^2$

Test region: Reject H_0 when $F^* > 1.90$.

Test statistic: $F^* = 1.78$.

Conclusion: We cannot reject H_0.

5. H_0: $\sigma_O^2 = \sigma_N^2$

H_A: $\sigma_O^2 < \sigma_N^2$

Test region: Reject H_0 when $F^* > 1.69$.

Test statistic: $F^* = 3.06$.

Conclusion: Reject H_0; variability has increased.

7. H_0: $s_{SAT}^2 = \sigma_{SUN}^2$

H_A: $\sigma_{SAT}^2 > \sigma_{SUN}^2$

Test region: Reject H_0 when $F^* > 4.90$.

Test statistic: $F^* = 5.57$.

Conclusion: Reject H_0; variation is greater on Saturdays.

REVIEW PROBLEMS/Chapter 10

1. 0.233 ± 0.127
$0.106 < p < 0.360$.

3. 0.24 ± 0.156
$0.084 < p < 0.396$.

5. H_0: $p = 0.30$.

H_A: $p < 0.30$.

Test region: Reject H_0 when $z^* < -2.58$.

Test statistic: $z^* = -1.85$.

Conclusion: We cannot reject H_0.

7. H_0: $p_W - p_N = 0$.

H_A: $p_W - p_N > 0$.

Test region: Reject H_0 when $z^* > 1.65$.

Test statistic: $z^* = 0.41$.

Conclusion: We cannot reject H_0.

9. H_0: $p_A - p_E = 0$.

H_A: $p_A - p_E \neq 0$.

Test region: Reject H_0 when $z^* < -1.65$ or $z^* > 1.65$.

Test statistic: $z^* = 0.42$.

Conclusion: We cannot reject H_0.

11. 0.44 ± 0.244
$0.196 < p_O - p_U < 0.684$.

13. H_0: $\sigma = 5$.

H_A: $\sigma > 5$.

Test region: Reject H_0 when $\chi^{2*} > 19.675$.

Test statistic: $\chi^{2*} = 19.166$.

Conclusion: We cannot reject H_0.

15. H_0: $\sigma = 5$.

H_A: $\sigma > 5$.

Test region: Reject H_0 when $\chi^{2*} > 45.559$.

Test statistic: $\chi^{2*} = 77.76$.

Conclusion: Reject H_0.

17. $2.55 < \sigma < 5.76$.

19. $27.79 < \sigma < 47.96$.

21. H_0: $\sigma_A^2 = \sigma_B^2$

H_A: $\sigma_A^2 \neq \sigma_B^2$

Test region: Reject H_0 when $F^* > 1.80$.

Test statistic: $F^* = 2.78$.

Conclusion: Reject H_0; there is a difference.

EXERCISES/Section 11.1

1. $s_{\bar{X}}^2 = 0.0771;$ $s_p^2 = 0.1022.$

H_0: The means are all equal.

H_A: The means are not all equal.

Test region: Reject H_0 when $F^* > 3.89$.

Test statistic: $F^* = 3.77$.

Conclusion: We cannot reject H_0.

3. $s_{\bar{X}}^2 = 35.08;$ $s_p^2 = 25.22.$

H_0: The means are all equal.

H_A: The means are not all equal.

Test region: Reject H_0 when $F^* > 4.26$.

Test statistic: $F^* = 5.56$.

Conclusion: Reject H_0; days do not all have equal production rates.

EXERCISES/Section 11.2

1. ANOVA

Source	df	SS	MS
Between	2	22.06	11.03
Within	9	5.38	0.598
Total	11		

H_0: The means are all equal.

H_A: The means are not all equal.

Test region: Reject H_0 when $F^* > 8.02$.

Test statistic: $F^* = 18.44$.

Conclusion: Reject H_0; there is a difference between types of ad campaigns.

3. ANOVA

Source	df	SS	MS
Between	3	1969.932	656.644
Within	16	863.716	53.982
Total	19		

H_0: The means are all equal.

H_A: The means are not all equal.

Test region: Reject H_0 when $F^* > 3.24$.

Test statistic: $F^* = 12.16$.

Conclusion: Reject H_0; there is a difference between mean battery lives.

REVIEW PROBLEMS/Chapter 11

1. ANOVA

Source	df	SS	MS
Between	3	1113.59	371.20
Within	8	231.33	28.92
Total	11		

H_0: The means are all equal.

H_A: The means are not all equal.

Test region: Reject H_0 when $F^* > 4.07$.

Test statistic: $F^* = 12.84$.

Conclusion: Reject H_0; the mean weight gains are not all the same.

3. ANOVA

Source	df	SS	MS
Between	3	29.94	9.98
Within	12	70.96	5.91
Total	15		

H_0: The means are all equal.

H_A: The means are not all equal.

Test region: Reject H_0 when $F^* > 3.49$.

Test statistic: $F^* = 1.69$.

Conclusion: We cannot reject H_0.

EXERCISES/Section 12.1

1. a)

Y

X

b) 0.970

c) $H_0: \rho = 0.$

$H_A: \rho \neq 0.$

Test region: Reject H_0 when $r^* > 0.874$ or $r^* < -0.874$.

Test statistic: $r^* = 0.970$.

Conclusion: Reject H_0; X and Y are related.

3. a)

Number of theatres

Average tax rate (% of personal income)

Tax rate versus number of theatres

b) -0.739

c) $H_0: \rho = 0.$

$H_A: \rho \neq 0.$

Test region: Reject H_0 when $r^* > 0.632$ or $r^* < -0.632$.

Test statistic: $r^* = -0.739$.

Conclusion: Reject H_0; acreage tax rate on personal income and the number of movie theatres in a country are related.

5. a)

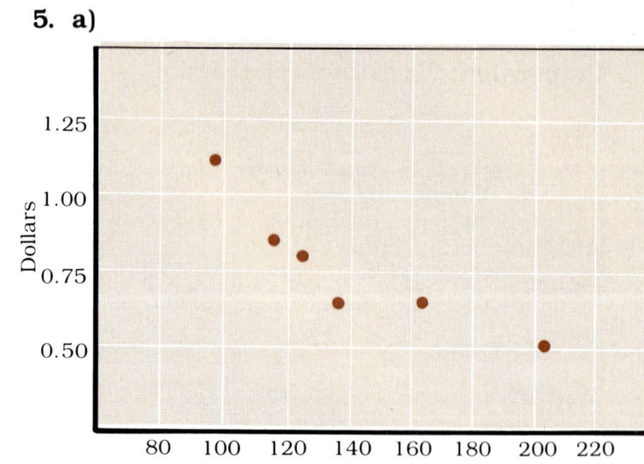

Annual rainfall versus price of a loaf of white bread

b) -0.876

c) $H_0: \rho = 0.$

$H_A: \rho \neq 0.$

Test region: Reject H_0 when $r^* > 0.811$ or $r^* < -0.811$.

Test statistic: $r^* = -0.876$.

Conclusion: Reject H_0; amount of rainfall and price of bread are related.

7. a)

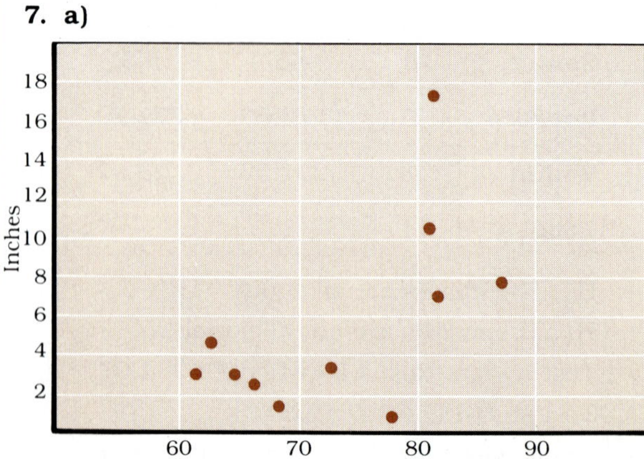

July temperatures versus precipitation

b) 0.598

c) $H_0: \rho = 0$.

$H_A: \rho \neq 0$.

Test region: Reject H_0 when $r^* > 0.685$ or $r^* < -0.685$.

Test statistic: $r^* = 0.598$.

Conclusion: We cannot reject H_0.

9. a)

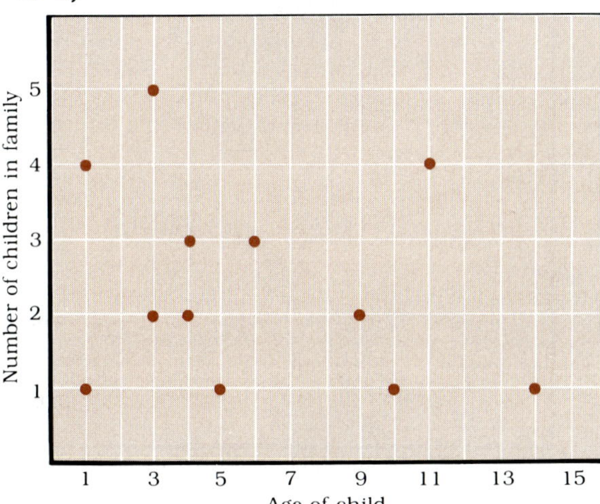

Incidence of child abuse

b) −0.262

c) $H_0: \rho = 0$.

$H_A: \rho \neq 0$.

Test region: Reject H_0 when $r^* > 0.576$ or $r^* < -0.576$.

Test statistic: $r^* = -0.262$.

Conclusion: We cannot reject H_0.

11. a) Graph at top of right-hand column.

b) 0.340

c) $H_0: \rho = 0$.

$H_A: \rho \neq 0$.

Test region: Reject H_0 when $r^* > 0.666$ or $r^* < -0.666$.

Test statistic: $r^* = 0.340$.

Conclusion: We cannot reject H_0.

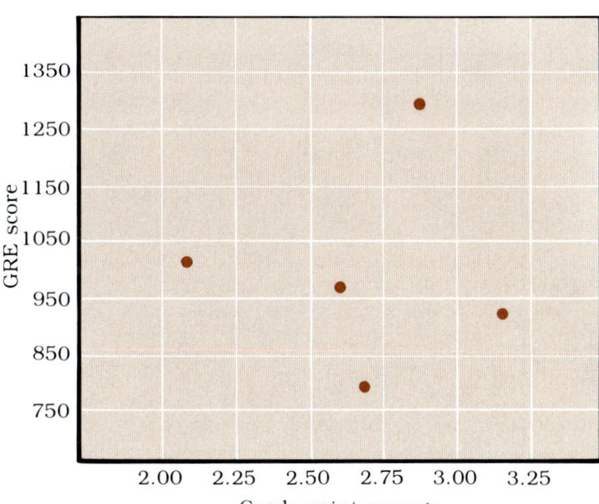

Grade point average versus GRE score

13. a)

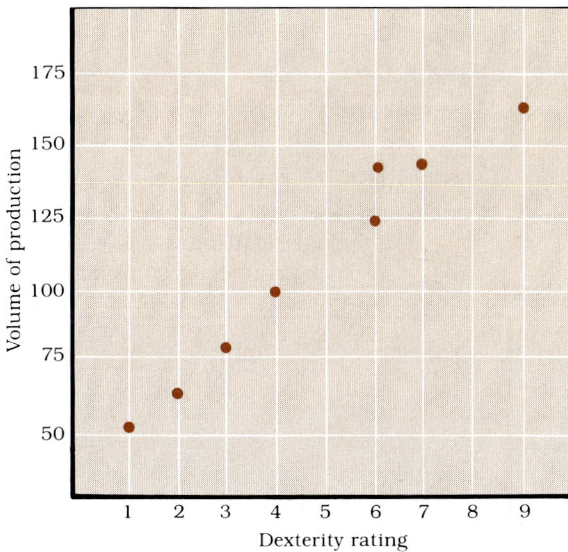

New employees' ratings by supervisors

b) 0.986

c) $H_0: \rho = 0$.

$H_A: \rho \neq 0$.

Test region: Reject H_0 when $r^* > 0.707$ or $r^* < -0.707$.

Test statistic: $r^* = 0.986$.

Conclusion: Reject H_0; dexterity and production level are related.

15. a)

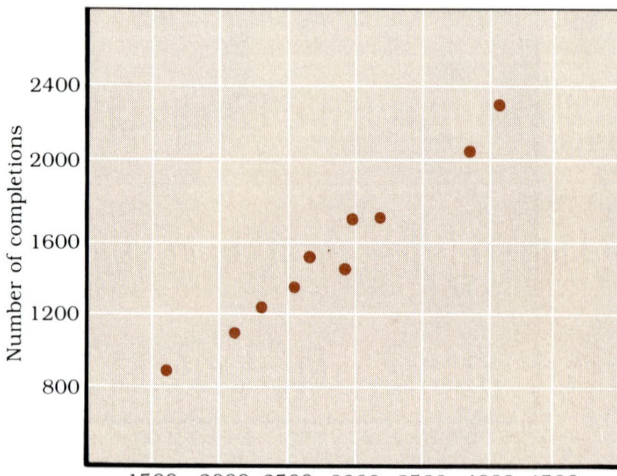

NFL statistics for leading passers

b) 0.993

c) $H_0: \rho = 0.$

$H_A: \rho \neq 0.$

Test region: Reject H_0 when $r^* > 0.765$ or $r^* < -0.765$.

Test statistic: $r^* = 0.993$.

Conclusion: Reject H_0; the number of passing attempts is related to the number of completions.

17. a)

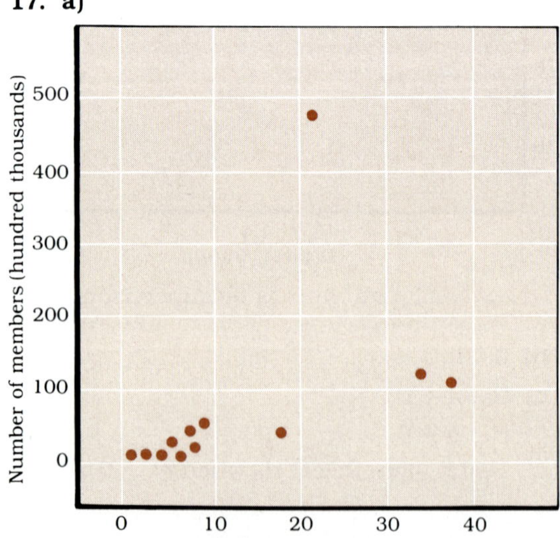

Church membership

b) 0.517

c) $H_0: \rho = 0.$

$H_A: \rho \neq 0.$

Test region: Reject H_0 when $r^* > 0.576$ or $r^* < -0.576$.

Test statistic: $r^* = 0.517$.

Conclusion: We cannot reject H_0.

19. a)

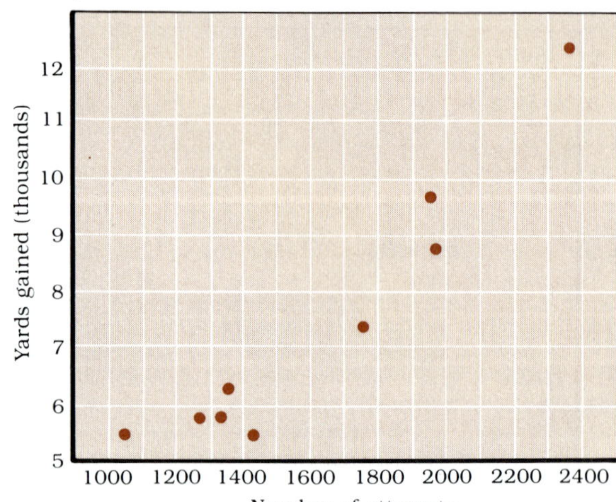

NFL statistics for rushers

b) 0.964

c) $H_0: \rho = 0.$

$H_A: \rho \neq 0.$

Test region: Reject H_0 when $r^* > 0.798$ or $r^* < -0.798$.

Test statistic: $r^* = 0.964$.

Conclusion: Reject H_0; number of attempts and yards gained are related.

EXERCISES/Section 12.2

1. 4.5 **3.** 2.2 **5.** 0.78 **7.** — **15.** 1530.1 **17.** — **19.** 9.6
9. — **11.** — **13.** 110.9

REVIEW PROBLEMS/Chapter 12

1. **a)** −0.976 **b)** $\hat{Y} = 3.73 - 0.42X$. **c)** 2.68

3. 0.956

5. **a)**

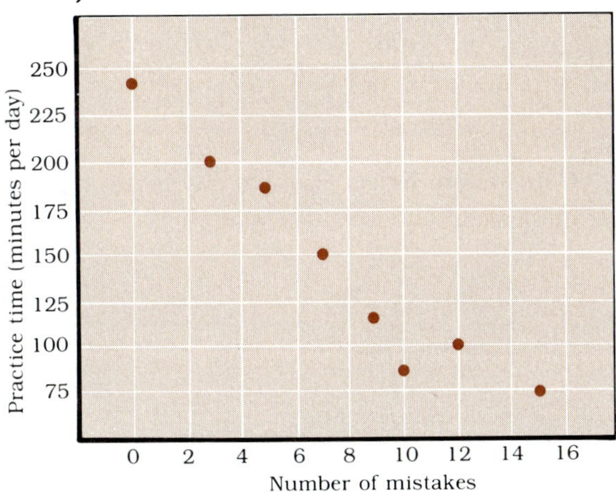

Piano practicing

b) −0.574

c) $H_0: \rho = 0$.

$H_A: \rho \neq 0$.

Test region: Reject H_0 when $r^* > 0.707$ or $r^* < -0.707$.

Test statistic: $r^* = -0.574$.

Conclusion: We cannot reject H_0.

d) $\hat{Y} = 233.06 - 11.63X$.
221.43

EXERCISES/Section 13.1

1. H_0: All days are equally likely to have the same number of tardy workers.

H_A: All days are not equally likely.

Test region: Reject H_0 when $\chi^{2*} > 9.488$.

Test statistic:

$$\chi^{2*} = \frac{(62-52)^2}{52} + \frac{(37-52)^2}{52} + \frac{(52-52)^2}{52} + \frac{(49-52)^2}{52} + \frac{(60-52)^2}{52} = 7.654.$$

Conclusion: We cannot reject H_0.

3. H_0: All brands are equally popular.

H_A: All brands are not equally popular.

Test region: Reject H_0 when $\chi^{2*} < 5.991$.

Test statistic:

$$\chi^{2*} = \frac{(150-133.33)^2}{133.33} + \frac{(130-133.33)^2}{133.33} + \frac{(120-133.33)^2}{133.33} = 3.50.$$

Conclusion: We cannot reject H_0.

5. H_0: Production day makes no difference in quality.

H_A: Production day makes a difference in quality.

Test region: Reject H_0 when $\chi^{2*} > 13.277$.

Test statistic:

$$\chi^{2*} = \frac{(73-63)^2}{63} + \frac{(52-63)^2}{63} + \frac{(63-63)^2}{63} + \frac{(61-63)^2}{63} + \frac{(66-63)^2}{63} = 3.714.$$

Conclusion: We cannot reject H_0.

7. H_0: The distribution of frequency is binomial with $p = 0.4$.

   H_A: The distribution of frequency is not binomial.

   Test region: Reject H_0 when $\chi^{2*} > 13.277$.

   Test statistic:

$$\chi^{2*} = \frac{(30-25.92)^2}{25.92} + \frac{(75-69.12)^2}{69.12} + \frac{(60-69.12)^2}{69.12} + \frac{(25-30.72)^2}{30.72} + \frac{(10-5.12)^2}{5.12} = 8.06.$$

   Conclusion: We cannot reject H_0.

9. H_0: This city is typical of the country as a whole.

   H_A: This city is not typical of the country as a whole.

   Test region: Reject H_0 when $\chi^{2*} > 9.488$.

   Test statistic:

$$\chi^{2*} = \frac{(50-85.4)^2}{85.4} + \frac{(150-166.6)^2}{166.6} + \frac{(230-218.4)^2}{218.4} + \frac{(150-136.5)^2}{136.5} + \frac{(120-93.1)^2}{93.1}$$

$$= 26.05.$$

   Conclusion: Reject H_0; this city is not typical.

EXERCISES/Section 13.2

1.

	Democrat	Republican	Other	
Blue Collar Worker	180 (143.3)	120 (151.3)	45 (50.4)	345
White Collar Worker	90 (126.7)	165 (133.7)	50 (44.6)	305
	270	285	95	650

H_0: Political affiliation and occupation are independent.

H_A: Political affiliation and occupation are not independent.

Test region: Reject H_0 when $\chi^{2*} > 5.991$.

Test statistic: $\chi^{2*} = 35.06$.

Conclusion: Reject H_0; there is a relationship.

3.

	A, B	C	D, F	
Engineering	41 (33.3)	9 (12.6)	4 (8.1)	54
Liberal Arts	286 (264.0)	92 (99.5)	50 (64.5)	428
Business	130 (160.4)	70 (60.5)	60 (39.1)	260
Science	55 (54.3)	22 (20.4)	11 (13.1)	88
				830

H_0: Major and math grade are independent.
H_A: Major and math grade are not independent.
Test region: Reject H_0 when $\chi^{2*} > 12.592$.
Test statistic: $\chi^{2*} = 28.0$.
Conclusion: Reject H_0; major and math grade are related.

5.

	A	B	C	D	
1	52 (27.3)	32 (27.8)	19 (28.6)	6 (25.3)	109
2	34 (30.7)	49 (31.4)	25 (32.3)	15 (28.6)	123
3	14 (42.0)	21 (42.8)	61 (44.1)	72 (39.1)	168
	100	102	105	93	400

H_0: Test results and job performance are independent.
H_A: Test results and job performance are not independent.
Test region: Reject H_0 when $\chi^{2*} > 12.592$.
Test statistic: $\chi^{2*} = 123.2$.
Conclusion: Reject H_0; test results and job performance are related.

7.

	18 – 29	30 – 59	over 60	
Passbook	9 (23.5)	20 (30.6)	60 (34.8)	89
Statement	41 (26.5)	45 (34.4)	14 (39.2)	100
	50	65	74	189

H_0: Method and age are independent.

H_A: Method and age are not independent.

Test region: Reject H_0 when $\chi^{2*} > 5.991$.

Test statistic: $\chi^{2*} = 58.3$.

Conclusion: Reject H_0; age and method are related.

REVIEW PROBLEMS/Chapter 13

1. H_0: Incidence of airport terrorism is uniformly distributed.

H_A: Incidence of airport terrorism is not uniformly distributed.

Test region: Reject H_0 when $\chi^{2*} > 16.812$.

Test statistic: $\chi^{2*} = 51.371$.

Conclusion: Reject H_0; air transport terrorism has not been a uniform occurrence.

3.

	Opinion			
	Agree	**Disagree**	**Don't Know**	
Men	61 (62.5)	31 (29)	8 (8.5)	100
Women	64 (62.5)	27 (29)	9 (8.5)	100
	125	58	17	200

H_0: Opinion is independent of sex.

H_A: Opinion is not independent of sex.

Test region: Reject H_0 when $\chi^{2*} > 5.991$.

Test statistic: $\chi^{2*} = 0.407$.

Conclusion: We cannot reject H_0.

5.

Age	Sex: Men	Sex: Women	
0–24	1 (0.99)	1 (1.01)	2
25–44	3 (1.98)	1 (2.02)	4
45–54	8 (5.94)	4 (6.06)	12
55–64	18 (13.86)	10 (14.14)	28
65 and Over	69 (76.23)	85 (77.77)	154
	99	101	200

This problem should combine categories in such a way that E for each cell is greater than S.

H_0: Age at death is independent of sex.

H_A: Age at death is not independent of sex.

Test region: Reject H_0 when $\chi^{2*} > 13.277$.

Test statistic: $\chi^{2*} = 6.262$.

Conclusion: We cannot reject H_0.

7.

Age	Observed	Expected
14–24	20	12.3
25–34	40	31.5
35–44	30	25.1
45–54	25	27.3
55–64	20	23.8
65 and over	15	30.1

H_0: The community's age distribution reflects the national distribution.

H_A: The community's age distribution does not reflect the national distribution.

Test region: Reject H_0 when $\chi^{2*} > 11.071$.

Test statistic: $\chi^{2*} = 16.446$.

Conclusion: Reject H_0; the community's age distribution does not reflect that of the nation.

EXERCISES/Section 14.1

1. H_0: Med = 65.

H_A: Med > 65.

Test region: Reject H_0 if there are 10 or more plus signs.

Test statistic: There are 9 plus signs.

Conclusion: We cannot reject H_0.

3. H_0: Med = 2.5.

H_A: Med ≠ 2.5.

Test region: Reject H_0 when there are 0 minus signs or 7 minus signs.

Test statistic: There are 6 minus signs.

Conclusion: We cannot reject H_0.

5. H_0: Med = 30 mph.

H_A: Med > 30 mph.

Test region: Reject H_0 when $z^* > 1.65$.

Test statistic:

$$z^* = \frac{667 - 760/2}{\sqrt{760}/2} = 20.82.$$

Conclusion: Reject H_0.

7. H_0: Med = 130.

H_A: Med > 130.

Test region: Reject H_0 when $z^* > 1.65$.

Test statistic:

$$z^* = \frac{75 - 85/2}{\sqrt{85}/2} = 7.05.$$

Conclusion: Reject H_0; Med > 130.

9. H_0: Med = 0.

H_A: Med < 0.

Test region: Reject H_0 if there are 11 or 12 minuses.

Test statistic: There are 9 minuses.

Conclusion: We cannot reject H_0.

11. H_0: Med = 0.

H_A: Med > 0.

Test region: Reject H_0 if there are 5 plus signs.

Test statistic: There are 4 plus signs.

Conclusion: We cannot reject H_0.

EXERCISES/Section 14.2

1. H_0: $\mu_V - \mu_T = 0$.

H_A: $\mu_V - \mu_T \neq 0$.

Test region: Reject H_0 when $z^* < -1.96$ or $z^* > 1.96$.

Test statistic: $R_V = 96.5$; $R_T = 179.5$;

$$z^* = \frac{96.5 - 132}{16.25} = -2.18.$$

Conclusion: Reject H_0; there is a difference in mean grade-point averages.

3. H_0: $\mu_I - \mu_F = 0$.

H_A: $\mu_I - \mu_F \neq 0$.

Test region: Reject H_0 when $z^* < -1.96$ or $z^* > 1.96$.

Test statistic: $R_I = 246$; $R_F = 180$.

$$z^* = \frac{246 - 188.5}{21.71} = 2.65.$$

Conclusion: Reject H_0; age affects mean diastolic blood pressure.

5. H_0: $\mu_X - \mu_Y = 0$.

H_A: $\mu_X - \mu_Y \neq 0$.

Test region: Reject H_0 when $z^* > 1.96$ or $z^* < -1.96$.

Test statistic: $R_X = 160.5$; $R_Y = 190.5$.

$$z^* = \frac{160.5 - 189}{19.44} = -1.47.$$

Conclusion: We cannot reject H_0.

REVIEW PROBLEMS/Chapter 14

1. H_0: Med $= 0.10$.
   H_A: Med > 0.10.
   Test region: Reject H_0 when there are more than 8 plus signs.
   Test statistic: There are 7 plus signs.
   Conclusion: We cannot reject H_0.
3. H_0: Med $= 0.270$.
   H_A: Med < 0.270.
   Test region: Reject H_0 when $z^* > 1.65$.
   Test statistic:
   $$z^* = \frac{21 - 25/2}{\sqrt{25}/2} = 3.4.$$
   Conclusion: Reject H_0; the team is batting significantly lower.
5. H_0: Med $= 0$.
   H_A: Med > 0.
   Test region: Reject H_0 when there are 8 or more plusses.
   Test statistic: There are 8 plusses.
   Conclusion: Reject H_0; music makes a difference.
7. H_0: $\mu_A - \mu_B = 0$.
   H_A: $\mu_A - \mu_B \neq 0$.
   Test region: Reject H_0 when $z^* > 1.96$ or $z^* < -1.96$.
   Test statistic: $R_A = 126.5$; $R_B = 173.5$.
   $$z^* = \frac{126.5 - 137.5}{17.26} = -0.64.$$
   Conclusion: We cannot reject H_0.
9. H_0: $\mu_1 - \mu_2 = 0$.
   H_A: $\mu_1 - \mu_2 \neq 0$.
   Test region: Reject H_0 when $z^* > 1.96$ or $z^* < -1.96$.
   Test statistic: $R_1 = 244$; $R_2 = 163$.
   $$z^* = \frac{244 - 217.5}{21.71} = 1.22.$$
   Conclusion: We cannot reject H_0.

EXERCISES/Appendix B

1. a) 29 b) −1 c) 273 d) 841
3. a) 750 b) −125 c) 137,500 d) 562,500
5. a) 2.64 b) 118 c) 1.6176 d) 2914

Index

D

E

F

G

H

I

L

M

T

U

V

W

Z

Student's *t* distribution

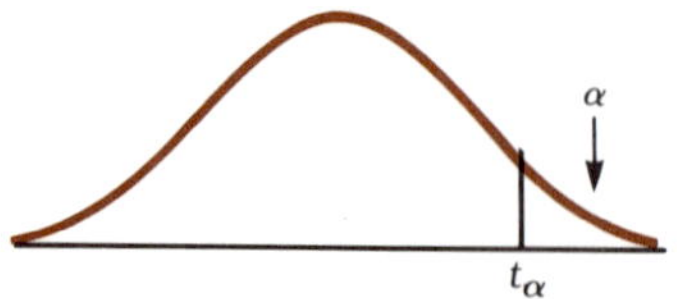